Advances in Global Change Research

Volume 55

AF832386

More information about this series at http://www.springer.com/series/5588

Stefan Brönnimann

Climatic Changes Since 1700

Springer

Stefan Brönnimann
Institute of Geography
University of Bern
Bern, Switzerland

ISSN 1574-0919 ISSN 2215-1621 (electronic)
Advances in Global Change Research
ISBN 978-3-319-34742-4 ISBN 978-3-319-19042-6 (eBook)
DOI 10.1007/978-3-319-19042-6

Springer Cham Heidelberg New York Dordrecht London

Softcover re-print of the Hardcover 1st edition 2015

Cover Illustration: © Morphart / fotolia

Printed on acid-free paper

Springer International Publishing AG Switzerland is part of Springer Science+Business Media (www.springer.com)

Preface

The Earth's climate is undergoing profound changes that will affect the future of our economies and societies. "The key to learning about the future lies in the past" is a common saying. Indeed, by studying past changes in climate, scientists can learn how the climate system works, test their models, and develop theories. At the same time, past climates exert a fascination for scientists as well as the public. They form a background for legends and are part of our culture.

When the existence of ice ages in ancient times was discovered in the nineteenth century, this stimulated both public imagination and scientific research:

> "The belief that it [the weather] has improved or worsened in the course of time quite persistently holds up. There is hardly any human being, even the most common, who has not tried, owing to the experiences of his life, to attribute the eternally changing, chaotic, unpredictable character of the weather with some law. [...] In the Ice Age, however, all miracles of this weather seem to unite. Something like an age-old fear of our ancestors seems to come to life again: of a global winter that destroyed everything. At the same time one believes that whoever solved the secret of the Ice Age would have to understand the magic of today's weather." (translated from p. 13 of Bölsche 1919)

A century later, this quote from a popular scientific booklet appears almost visionary as many climate theories, including the CO_2 greenhouse effect, are rooted in the ice age problem.

The debate on the ice age theory was particularly lively at the end of the nineteenth century. A leading figure in this debate was Eduard Brückner[1] (see Fig. 1), who co-published a standard textbook on ice ages ("Die Alpen im Eiszeitalter", 1909). Brückner also was a pioneer in the study of climatic changes on multidecadal time scales and their effects on society (von Storch and Stehr 2000). In 1890—at that time Brückner was a professor of Geography at the University of Bern—he published a book entitled "Klima-Schwankungen seit 1700" ("Climatic

[1]Eduard Brückner, 1862–1927, was a German geographer and climatologist. He was professor at the University of Bern from 1888 to 1904, then in Halle and from 1906 on in Vienna (see Stehr and von Storch 2000).

Fig. 1 Photo of Eduard Brückner (provided by the Institute of Geography, University of Bern)

Changes Since 1700") that was influential for generations of climatologists. In this landmark publication, several chapters of which were translated by Stehr and von Storch (2000), Brückner promoted the view of climatic fluctuations (against the conception of a stable climate or slow, progressive change) and proposed a 35-year cycle of global climate (see Sect. 4.1.2). He based his work on meteorological or hydrological observations, documentary evidence as well as indirect evidence or "proxy data", as we would say today. Brückner, likewise, considered studies of past climate to be a key to the future and, most importantly, saw their societal relevance:

> "There are numerous hypotheses and theories about climate change. Quite naturally they have caught the public attention, as any proof of past climatic change points to the possibility of future climate change, which inevitably will have significant implications for global economics." (translated from p. 3 in Brückner 1890)

Over a century later, the field of palaeoclimatology is undergoing profound changes. First, progress in analytical and sampling techniques as well as a better understanding of climate proxies has led to a wealth of information. Second, increased modelling capabilities and computing power has enabled us to address palaeoclimatological problems with full-fledged computer models. Third, powerful techniques such as data assimilation have been introduced to palaeoclimatology. Together with increasingly refined research questions in the light of global climate change, this technical progress has caused a constant reinvention and revaluation of historical data and approaches, accompanied by strong efforts in data recovery.

These developments are about to change our view of the past, specifically of the past ca. 300 years—the period studied by Brückner. This period is of particular interest because the radiative forcings changed considerably during this period.

The intention of this book, which is entitled "Climatic Changes Since 1700" in honour of Brückner's book, is to discuss mechanisms behind interannual-to-multidecadal climate variability and to analyse climatic changes since 1700 in light of new tools and data. The book shows *how* we can learn from climatic changes during the past 300 years and *what* we have learned from these changes. It does not replace a textbook, even though it is based on one of my courses and can be used in a lecture hall. There are excellent textbooks on climate and climatic changes, which are referenced at the appropriate places. It is also not a review of climate and climatic changes, which would be an impossible undertaking. Excellent reviews exist for individual topics, and for a comprehensive assessment of the literature, the reader is referred to the Fifth Assessment Report of the Intergovernmental Panel on Climate Change (IPCC 2013).

The book aims at readers interested in climatology and climatic changes, particularly graduate students in geography or climate sciences but also palaeo-climatologists, palaeoecologists, historians or other scientists interested in climatic changes. It starts with an introduction and a chapter on data and methods used in climatology and palaeoclimatology—an area in which our research group is strongly engaged. The chapter also introduces key data sets and some main concepts used in the later parts of the book. The third chapter introduces the most important mechanisms causing climatic changes on interannual-to-multidecadal time scales. These first chapters of the book will appeal to less-advanced readers as they start at a rather basic level. However, it is important, in my view, to discuss past climate changes by building upon a solid background in climatological methods and climate physics.

The fourth and main chapter of the book discusses major climatic variations since 1700, some of which are important in their own right, while others exemplify cases that are typical or characteristic for a certain period. Rather than to review the literature on these events, one main goal of this book is to reproduce and interpret these variations in the new data sets and the new approaches introduced in the second chapter. These analyses, though imperfect and preliminary, serve the purpose of illustrating what palaeoclimatology can do. The book concludes with a summary.

The choice of topics and of climatic events included in this book is not objective or comprehensive—this would be beyond my expertise. The selection of climatic events, data sets and models is therefore strongly biased by the work of our research group, and it has a focus on Europe and Switzerland. The book thus conveys a personal view of how and why the climate developed over the past 300 years; it does not do justice to all the excellent work that has been done by others in the field.

Bern, Switzerland
August 2015

Stefan Brönnimann

Acknowledgements

This book is the result of a number of projects in which I was fortunate to participate over the past decade. These projects were funded by the Swiss National Science Foundation (my SNF professorship, NCCR Climate project PALVAREXIII, research projects EVALUATE and TWIST and sinergia projects FUPSOL and FUPSOL 2), COST Action "Home", the EC FP7 projects ERA-CLIM and ERA-CLIM 2, ERC grant TITAN, MeteoSwiss, the cogito foundation, the Bretscher Stiftung, the Holderbank and Janggen-Poehn Foundations, and ETH Zurich. The CLIMANDES project made me draw simple schematic figures. Computing time was provided by the Swiss Supercomputer Centre CSCS. I also acknowledge the great support of the Oeschger Centre for Climate Change Research at the University of Bern and of the IT staff at our Institute: Basilio Ferrante and Andrey Martynov.

The book uses many data sets that are based on long-term efforts of numerous individuals and institutions such as NCEP/NCAR, the European Centre for Medium-Range Weather Forecast (ECMWF), NOAA/CIRES CDC, NASA, the Hadley Centre of the UK Met Office, the Climatic Research Unit of the University of East Anglia, Norwich, MeteoSwiss and many others.

The photos displayed in Figs. 2.2, 2.5, 2.6, 2.11, 2.27 and 5.3 are from Andrea Kaiser.

I would like to thank those who made the research possible, collaborated in the projects, hosted me during various stays or administered the projects. I thank Monique Stewart for language editing and Céline Dizerens, Matthias Röthlisberger and Yuri Brugnara for formatting. Finally, my greatest thanks goes to all those who performed the actual research on which this book is based, all of the Masters and Ph.D. students and the scientists whose results constitute the progress in our field and all colleagues and scientists with whom I have had the pleasure to work with: Rob Allan, Julien Anet, Florian Arfeuille, Renate Auchmann, Jürg Beer, Tom Bodenmann, Petra Breitenmoser, Philip Brohan, Yuri Brugnara, Marcel Bühler, Gilbert P. Compo, Silke Dierer, Tracy Ewen, Andreas Fischer, Doris Folini, David Frank, Jörg Franke, Andrea Grant, Thomas Griesser, Martin Grosjean,

Jenny Grütter, Gertrude Hirsch Hadorn, Ena Hirschi, Dogushan Kilic, Matthias Röthlisberger, Matthias Rutishauser, Stefan Hunziker, Martin Jacques-Coper, Sina Lenggenhager, Jürg Luterbacher, Brigitte Müller, Stefan Muthers, Raphael Neukom, Thomas Peter, Christoph Raible, Marco Rohrer, Olivia Romppainen, Eugene Rozanov, Margit Schwikowski, Alexander Stickler, Constantin Streit, Peter Stucki, Christian Vogler, Esther Volken, Heinz Wanner, Richard Wartenburger, Martin Wegmann, Helmut Weissert, Martin Wild, Sebastian Wolfinger, Heinz Zumbühl and many others.

Contents

1 Introduction 1

2 The Basis: Past Climate Observations and Methods 9

2.1 Observations of Weather and Climate 9

2.2 Historical Climate Observations 12

2.2.1 Documentary and Early Instrumental Data 12

2.2.2 From National Weather Services to a Global Observing System 15

2.3 Upper-Air and Satellite Observations 17

2.4 Data Dissemination in the Course of Time 19

2.5 Uncertainties in Climatic Data 23

2.5.1 Uncertainties and Errors in Measurement Series 23

2.5.2 Inhomogeneities and Homogenisation 26

2.5.3 The Chain of Uncertainties 29

2.6 Data Products and Dynamical-Statistical Methods 30

2.6.1 Spatial Information and Geostatistical Methods 30

2.6.2 Data Assimilation and Reanalyses 34

2.7 Climate Models 38

2.7.1 Characteristics of Climate Models 38

2.7.2 Types of Climate Models and Experiments 44

2.7.3 Downscaling, Nudging, and Other Techniques 47

2.8 Palaeoclimate Information and Techniques 48

2.8.1 Climate Proxies 48

2.8.2 Problems Arising When Working with Proxies 52

2.8.3 Calibration and Modelling of Proxies 54

2.8.4 Climate Reconstructions 56

2.8.5 Data Assimilation in Palaeoclimatology 59

2.9 Datasets Used in This Book 62

2.9.1 Overview 62

2.9.2 Observations and Reanalyses 62

2.9.3 Model Simulations 64

2.9.4 Reconstructions and Offline Assimilation 65
2.9.5 Conclusions 68

3 The Machinery: Mechanisms Behind Climatic Changes 71
3.1 Basic Climate Physics: The Mean State 72
3.1.1 Energy Budget and Heat Transport 72
3.1.2 The General Circulation of the Atmosphere 76
3.1.3 Stratospheric Circulation 85
3.1.4 The Ocean, Air–Sea and Land-Sea Interactions 93
3.2 Expressions and Mechanisms of Climate Variability 101
3.2.1 Statistical and Physical Perspectives of Climate Variability 101
3.2.2 Overview of Circulation Variability Modes 103
3.2.3 The North Atlantic Oscillation (NAO) 104
3.2.4 Variability Modes in the Pacific and the Southern Hemisphere 108
3.2.5 Extratropical Circulation Variability Modes and the Stratosphere 111
3.2.6 Tropical Variability Modes: El Niño–Southern Oscillation (ENSO) 113
3.2.7 Variability Modes in the Atlantic and Indian Ocean 119
3.2.8 Decadal Climate Variability Modes 120
3.3 Forced Climatic Changes 123
3.3.1 Volcanic Effects 123
3.3.2 Solar Influences 136
3.3.3 Well-Mixed Greenhouse Gases 145
3.3.4 Tropospheric Aerosols and Reactive Trace Gases 150
3.3.5 Land Surface 157
3.4 Coupling Between Systems and Feedback Mechanisms 161
3.4.1 Interaction Within the Machinery 161
3.4.2 Feedbacks Involving Water Vapour and Clouds 163
3.4.3 Feedbacks Involving Freezing Water: Arctic Amplification and the Cryosphere 164
3.4.4 Water as a Reactant: The Role of the Hydroxyl Radical 164
3.4.5 Feedbacks Involving Evaporating Water: Land Surface Feedbacks 165

4 Climatic Changes Since 1700 167
4.1 Climatic Changes of the Past Centuries: An Overview 168
4.1.1 The Last Millennium 168
4.1.2 An Overview of Climatic Changes Since 1700: Brükner's View 171
4.1.3 Climatic Changes Since 1700 in Climate Reconstructions 174
4.1.4 Climatic Changes in CCC400 and EKF400 185
4.1.5 Drivers of Global Climate Since 1600 193

4.2 The Period from 1700 to 1890: The Little Ice Age 197
4.2.1 Climate of the 18th and 19th Centuries 198
4.2.2 The Maunder Minimum: A Compelling Case for Solar Forcing? 201
4.2.3 The 1790s: Temporary Warmth and Strong Tropical Circulation 205
4.2.4 Tambora Eruption, Dalton Minimum, and the Year Without a Summer of 1816 210
4.2.5 The Maximum State of Alpine Glaciers in the 1850s 216
4.2.6 Global Droughts During 1876–1878 and El Niño 221
4.2.7 Frequent Flooding in the Central European Alps: 1830–1880s 226
4.2.8 The Broad Lines: The Little Ice Age 229
4.3 The Period from 1890 to 1945: Out of the Cold 230
4.3.1 Global Drivers and Global Changes 231
4.3.2 The Warming of the European Arctic from the 1910s to the 1940s 237
4.3.3 The "Dust Bowl" Droughts of the 1930s 243
4.3.4 Global Climate Anomalies During the 1939–1942 El Niño 250
4.3.5 The Broad Lines: Out of the Cold 254
4.4 The Period from 1945 to 1985: Delusive Stability 259
4.4.1 Global Drivers and Global Changes 260
4.4.2 Changes in Atmospheric Constituents 262
4.4.3 European Summers of 1945–1949 267
4.4.4 Sahel Pluvial and Drought 273
4.4.5 The Climate Shift of 1976/1977 276
4.4.6 The Broad Lines: Delusive Stability 279
4.5 The Period Since 1985: Accelerated and Slowed Warming 283
4.5.1 Global Drivers and Global Changes 284
4.5.2 The Ozone Hole 289
4.5.3 Eruption of Pinatubo 296
4.5.4 Warm European Winters, Increased Storminess and Positive NAO in the 1990s 300
4.5.5 The El Niño of 1997/1998, Forest Fires and the Atmospheric Brown Cloud 305
4.5.6 Droughts in Australia and the Northern Subtropics 308
4.5.7 Megaheatwaves 310
4.5.8 Recent Arctic Warming 312
4.5.9 The Global Warming Hiatus 316

5 Conclusions 323

References 327

List of Boxes

Box 2.1 Data rescue 14
Box 2.2 When meteorology was not considered science 22
Box 2.3 What is a good dataset? 32

Box 3.1 Statistical and physical interpretation of circulation variability modes 104
Box 3.2 Learning from volcanic eruptions 124
Box 3.3 Detection and attribution of climate change 149
Box 3.4 Negative feedbacks: The Daisy World, CLAW, and Gaia 161

Box 4.1 Francis Galton, Gilbert Walker, and Felix Exner: Statistics in climatology 224
Box 4.2 Early greenhouse effect theories: A bright future 256
Box 4.3 The Great Fear 280
Box 4.4 The Intergovernmental Panel on Climate Change (IPCC) 294
Box 4.5 Climate Engineering 299
Box 4.6 A historical perspective to assess storm risks 303

Chapter 1
Introduction

Climate is all about variability. Imagine a world in which temperature and other meteorological elements would not vary in time and space. There would be no need for the term "climate". In reality, it may be cool today but warm tomorrow (Fig. 1.1, left). Travelling a few hundred kilometres away, one may find a very different weather situation upon arrival (Fig. 1.1, right). These variations may not be predictable more than a few days ahead, but there are typical spatial patterns and temporal evolutions in this variability. The term climate describes these typical variations and aims at bringing an order to them. Climatic changes, then, are changes in these typical variations over time.

In the following, we will take a closer look at the time and space scales of climatic variations. Figure 1.2 shows annual-mean temperatures from the longest instrumental series in the world, the Central England temperature series (Manley 1974; Parker et al. 1992). The series reaches back to 1659 (Sect. 2.2 will introduce the history of this and other early instrumental series). Annual means of the Central England temperatures show frequent ups and downs appearing almost at random. However, there are also slower changes such as an increase in temperature during the first half of the 20th century and again during the past 50 years. The black time series in Fig. 1.2 is the global-mean temperature plotted on the same scale. It shows much-reduced variability as compared to the Central England temperature. Individual peaks in the two curves do not generally coincide, but some of the multidecadal changes do agree between the two curves, both in phase and amplitude. This may be due to common causes. In this case, the global curve averages out most of the local variations, emphasising the mechanisms acting on a global scale. On the other hand, global mechanisms are not the dominant factor explaining changes in Central England temperature. Due to local variations, we may find 30-year warming trends in Central England (e.g., in the early 18th century) that exceed recent warming trends. However, large-scale variations such as global warming during the past 50 years also show up in Central England.

S. Brönnimann, *Climatic Changes Since 1700*, Advances in Global Change Research 55, DOI 10.1007/978-3-319-19042-6_1

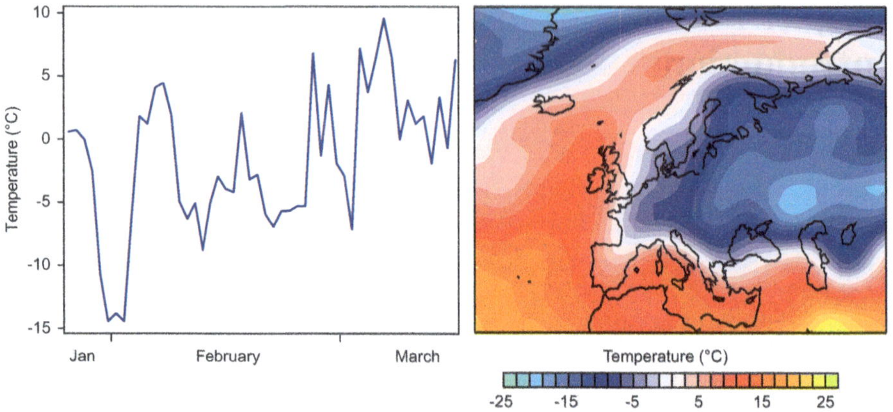

Fig. 1.1 Surface air temperature variations (*left*) over time at a given location (Geneva, Switzerland, temperature near sunrise in 1816, from Auchmann et al. 2012) and (*right*) in space on a specific day (NCEP/NCAR Global Reanalysis from Kalnay et al. 1996)

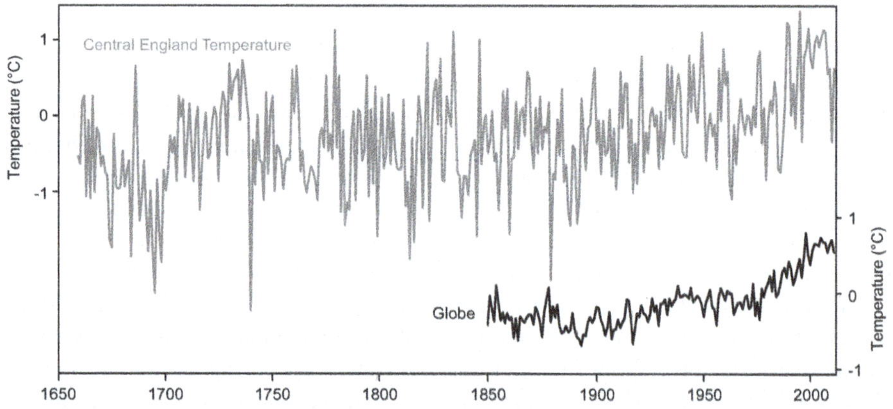

Fig. 1.2 Annual-mean temperature anomalies (with respect to 1961–1990) for (*grey*) Central England (Parker et al. 1992) and (*black*) the global average (HadCRUT4 data; Morice et al. 2012). Both datasets are from the Climatic Research Unit, University of East Anglia

How can we approach this problem scientifically? One viewpoint is statistical. For instance, we could ask: What is the relation between temperature trends and their space and time scales? A possible answer is given in Fig. 1.3. Trends[1] in annual-mean temperature from 1979 to 2011 as a function of the area considered are shown in the left panel of Fig. 1.3. To create this figure, the globe was divided into 1666 grid cells of (almost) equal area. By always adding the nearest neighbour cell, we searched for those areas of 1, 2, 3, . . ., 1665 grid cells that exhibit the largest (red) and smallest (blue) trends among all possible combinations.

[1] All trends in this book, unless otherwise noted, were obtained by linear least squares regression.

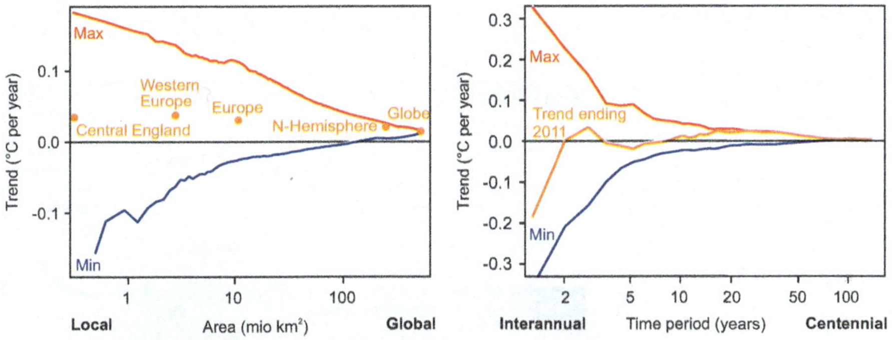

Fig. 1.3 (*left*) Dependence of annual-mean temperature trends from 1979 to 2011 on the spatial scale (*red* = global maximum, *blue* = global minimum) for close-to-circular areas (NCEP/NCAR Reanalysis data interpolated to equal-area grid). *Orange dots* show trends for selected areas. (*right*) Dependence of Northern Hemisphere annual-mean temperature trends within the 1881–2011 window for the chosen time period (HadCRUT4; Jones et al. 2012). The *orange curve* shows the trends for all periods ending in 2011. Note the logarithmic x axes on both plots (Based on work by Sebastian Wolfinger)

At regional scales (e.g., from 1×10^6 to 10×10^6 km^2), trends scatter considerably. There are many locations on the globe where temperature has not increased during the last 33 years. The trend for Central England (orange dot) is larger than that for the global mean. As the size of the area increases, trends converge. The largest area whose average temperature has a negative trend is around 100×10^6 km^2. Europe warms faster than the globe, but some regions (e.g., the Arctic) warm even faster. The Northern Hemisphere temperature trend is very close to the maximum trend that can be found for any hemisphere (the opposite must be true for the Southern Hemisphere, by implication).

Figure 1.3 (right) shows the time dependence of climate trends using the example of the Northern Hemisphere annual-mean temperature. For periods of ever-increasing length, the record of the past 131 years (1880–2011) was screened for time windows with the largest and smallest trends, respectively (see also Liebmann et al. 2010). By definition, the trends converge for a period of 131 years. The orange line shows the trend ending in 2011. For periods longer than 20 years, this is close to the maximum trend. However, short trends ending in 2011 are not necessarily high due to interannual-to-decadal variability. This statistical perspective helps to explain some of the differences between the Central England and global-mean temperature outlined in Fig. 1.2—and many other examples of space–time variability that will be presented in this book.

An alternative way to consider the differences would be to explore the processes causing temperature variations in Central England or at the global scale, and their expression at different time scales. Here, it is helpful to introduce a system view in which climate is seen as one global and bio–physico–chemical entity. The climate system is divided into spheres (Fig. 1.4) and climate processes are expressed through the storage and exchange of properties between these spheres (Fig. 1.5).

Fig. 1.4 The climate system and its components

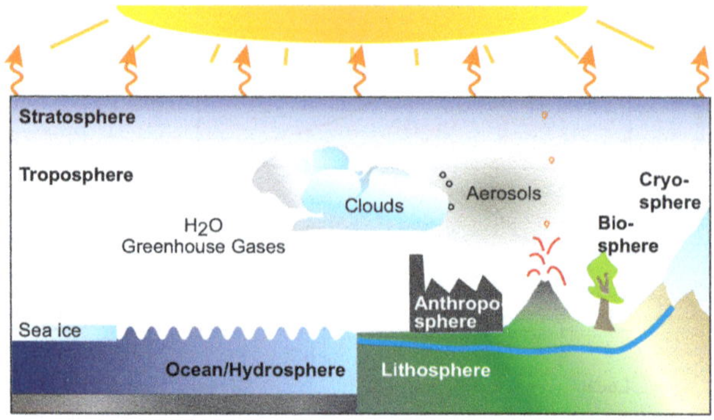

Fig. 1.5 Fluxes of (*top*) energy, (*middle*) mass, and (*bottom*) momentum in the climate system

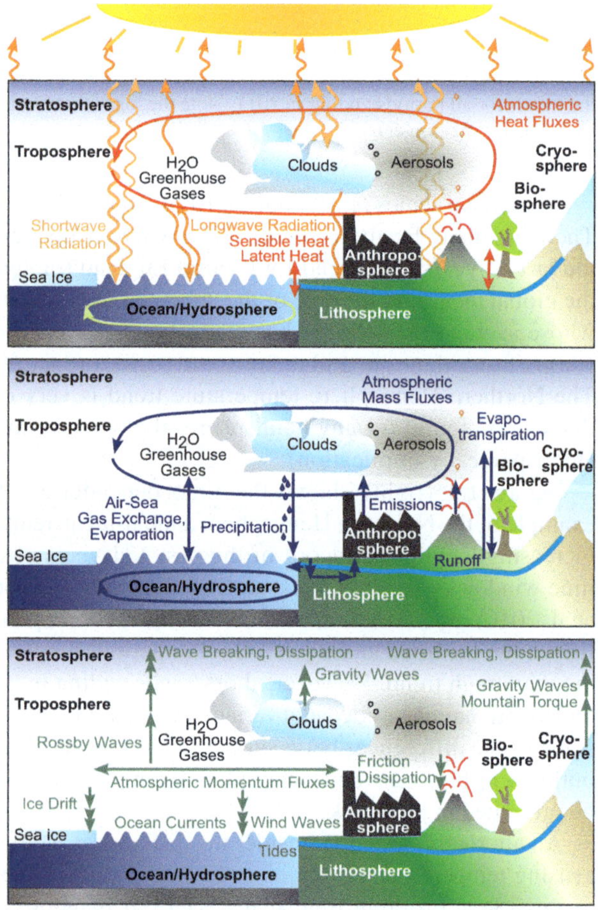

These properties are: energy, mass, and momentum. Here, we distinguish the atmosphere (with subspheres such as the troposphere, stratosphere, or the planetary boundary layer), ocean and hydrosphere, cryosphere, lithosphere (or pedosphere), and biosphere. To this system, at least when considering the climate since the 19th century, we have to add the anthroposphere. Specific constituents of the atmosphere such as clouds, water vapour and greenhouse gases, and aerosols play a crucial role in the climate system and deserve particular attention. This book will make use of this sketch of the climate system on numerous occasions.

Systems have boundaries. The upper boundary of the climate system is called the "top-of-atmosphere". Through this system boundary, the climate system receives energy from the sun that it eventually radiates back to space. At its lower boundary—the lithosphere—fluxes become irrelevant for the question studied. Thus, other than top-of-atmosphere radiation fluxes, the system as depicted here is a nearly closed system. Matter, energy, and momentum are quasi-conserved.

Climate science is all about understanding the fluxes and conversion of energy, matter, and momentum within the system. Solar radiation is converted to other forms of energy (sensible or latent heat) and transported through heat fluxes in the atmosphere and ocean in order to balance the differences in the energy budget. Evaporation and condensation link the energy balance (Fig. 1.5, top) with the mass balance for water (Fig. 1.5, middle). Apart from water, traces gases, aerosols, and nutrients are quantities that are often studied (in the system view, the term "cycle" is used for mass budgets of subsystems). Finally, the balance of momentum is important for the interaction of the solid earth with the flow or for describing the atmospheric flow on a rotating sphere.

The global temperature in Fig. 1.2 represents the combination of all processes in the system, but the fluxes of energy at the surface vary on small spatial scales, and the effects of these local fluxes for a given location such as Central England may be quite unique.

The unequal spatial distribution of energy fluxes is balanced by horizontal or vertical fluxes. Atmospheric processes are set in motion and carry out the fluxes. These processes (or mechanisms, because often it is a chain of processes) have typical temporal and spatial scales, which are often correlated (Fig. 1.6). Turbulence takes place on spatial scales of metres and temporal scales of seconds, while pressure systems and fronts have scales of hundreds of kilometres and time scales of days.

The processes, and therefore the scales, interact with each other. Often the term "energy cascade" is used for these interactions. For instance, meridional heat exchange between the subtropics and the subpolar regions proceeds "down scale" from large-scale waves all the way to dissipation at the molecular level. Conversely, organised convection in the tropics, through "up scale" mechanisms, is an important driver of tropical circulation. Section 3.1 will discuss these aspects in more detail.

As in the atmosphere, physical processes in the ocean show a relationship between time and spatial scales, but arguably a different relationship from that in

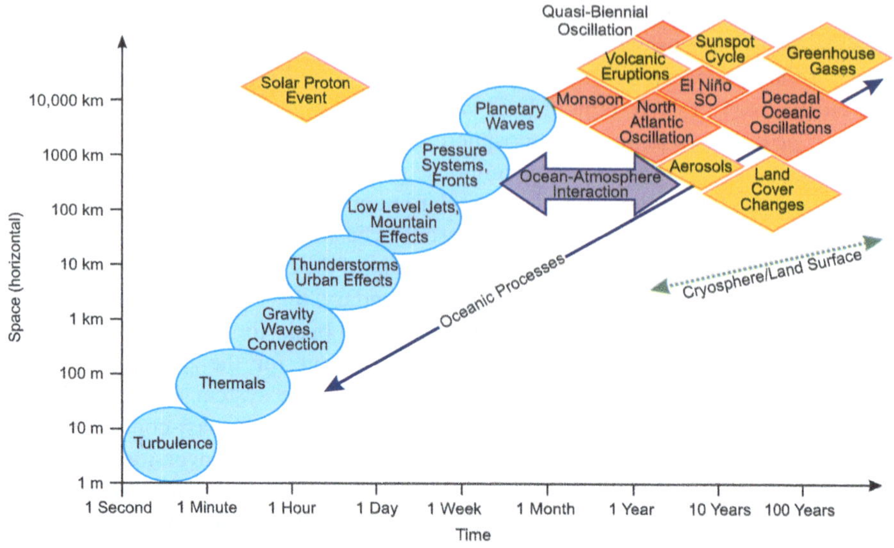

Fig. 1.6 Space–time scales of (*light blue ellipses*) meteorological phenomena, (*red diamonds*) climate phenomena, and (*yellow diamonds*) external climate influences (Adapted from Heinz Wanner)

the atmosphere. The same is true for cryospheric processes. Interactions among the spheres lead to climatic phenomena whose time scales are larger than those of atmospheric processes. The climate system may develop quasi-oscillatory variations such as El Niño–Southern Oscillation, internal modes of variability that vary on decadal time scales, or also seasonal variations such as monsoons due to the interaction with the annual cycle.

Forcing factors may also have very different space–time relations. For instance, they can have large spatial scales and long time scales (such as long-lived greenhouse gases), but there may also be short disturbances with near-global effects (such as solar proton events that affect the middle atmosphere) or very slow, local changes (such as land surface effects on climate). Forcings may be smooth and continuous (such as anthropogenic greenhouse gases), periodic (such as variations related to the 11-year sunspot cycle), or pulsing (such as volcanic eruptions).

External forcing contributions are usually small compared to variations in local energy fluxes, but they interact with regional processes such that different local systems may react differently to external forcings. As a consequence, the outcome will be different in Central England or at the global level, where globally integrated fluxes are most relevant.

The focus of this book is on the right side of Fig. 1.6—on external climate influences and internally generated variability on interannual-to-decadal time scales such as that due to ocean–atmosphere interaction.

Many of the mechanisms are relatively well understood, others are under research, and some are yet to be discovered. One main part of this book introduces the mechanisms of climate variability internal to the system (i.e., "the machinery"), followed by external influences (often subdivided into natural and anthropogenic), and feedback mechanisms that emerge through the reaction of the machinery to the forcing. A second main part of the book outlines global climate history since 1700 in the light of the mechanisms, that is, the machinery and its reaction to forcings. Before turning to the mechanism of climate variability, however, we will start with an overview of the very foundation of (almost) all research in this area: climate observations.

Chapter 2
The Basis: Past Climate Observations and Methods

With contributions and help from Renate Auchmann, Jörg Franke, Alexander Stickler, Petra Breitenmoser, and Silke Dierer

If you are worried about a thunderstorm forming near your town, if snowfall is imminent, if an El Niño event builds up in the Pacific, or if media reports that a heatwave strikes Australia, you can find a large amount of real-time information on weather and climate on the Internet. Just a few mouse clicks away, you will find observations, analyses, model simulations, satellite images, Radar data, and many other products (Fig. 2.1). Where does information on the atmosphere come from, what does it really tell us, and how can we today explore the weather patterns of the 18th century? In this chapter, we will cover these questions, starting with some general considerations of weather observations and measurements, the present day observing system, and historical climate observations. We cover uncertainties and problems in climate data and see how models can be used to learn about past climate and how they can be combined with observations. This chapter also covers climate proxies and the methods used to derive climate information from these proxies. Finally, this chapter concludes with a more detailed description of those datasets that form the basis of many of the analyses that follow in Chaps. 3 and 4.

Climate information has been the topic of many books and reviews. For a historical view of climate information that touches upon technological, institutional, and scientific aspects of climate data, data exchange, and data modelling, I strongly recommend the book "A Vast Machine" (Edwards 2010).

2.1 Observations of Weather and Climate

Research on climate variability and trends relies on atmospheric observations from the past and the present (see Brönnimann 2012, for more information about the following). Today, most of the observations are measurements, that is, a property of the atmosphere or ocean is quantified using instruments. Examples include

S. Brönnimann, *Climatic Changes Since 1700*, Advances in Global Change Research 55, DOI 10.1007/978-3-319-19042-6_2

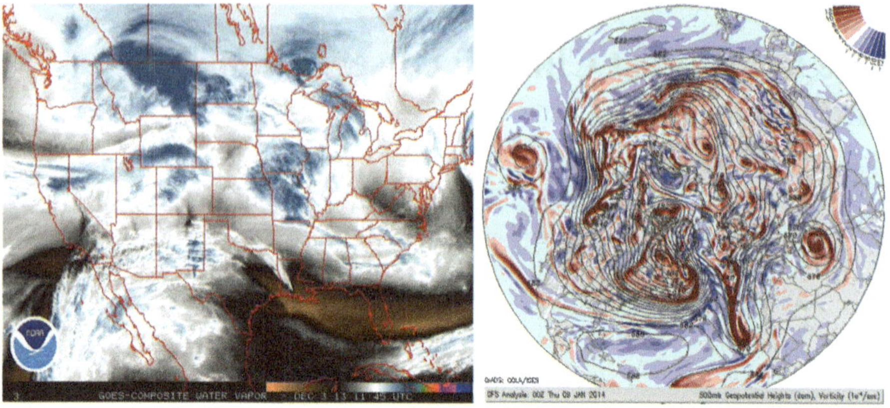

Fig. 2.1 Snapshots of atmospheric information available at our fingertips. (*left*) GOES Composite water vapour for 3 December 2013 1145 UTC (Data source: NOAA). (*right*) 500 hPa geopotential height (*contours*) and vorticity (*colours*) at 9 January 2014 0000 UTC, from the Global Forecast System analysis (Data source: http://wxmaps.org/pix/hemi.vort.html)

air temperature, ocean temperature, pressure, humidity, wind speed and direction, radiation, and the composition of the atmosphere. However, visual observations of cloud type and fraction, precipitation type, or weather type are also important, particularly further back in time.

It is beyond the scope of this book to give a technical description of instruments. The reader is referred to the overviews of Burt (2012) or, for procedures and practices, the World Meteorological Organization (WMO 2008). In the following, we focus on few principal properties of measurement systems.

In situ observing systems measure the physical or chemical properties of the surrounding air or water (Fig. 2.2). Remote sensing systems indirectly derive atmospheric properties from the perturbations of electromagnetic signals passing through the atmosphere. Platforms used for atmospheric observations include ground-based stations and towers on land, ships and buoys in the ocean, as well as weather balloons, aircraft, or satellites in the free atmosphere.

Measurements can be seen as partial descriptions of the atmospheric state at a given time. They are only helpful for science, however, because observing systems assure that measurements are meaningful and comparable. For instance, the WMO (1992) defines air temperature as

> "the temperature indicated by a thermometer exposed to the air in a place sheltered from direct solar radiation."

Furthermore, in order to achieve representative results when comparing thermometer readings at different places and at different times, a measurement protocol is required. This encompasses, for instance, a standardised exposure of the screen. According to WMO (1998) guidelines, the observed air temperature should be

> "representative of the free air conditions surrounding the station over as large an area as possible, at a height of between 1.2 and 2.0 m above ground level. Preferably measurements

Fig. 2.2 A pyranometer (*top*) for the measurement of shortwave radiation and (*bottom*) temperature and humidity measurements at a meteorological station of SwissMetNet in Zollikofen, Switzerland

> are performed over level ground, freely exposed to sunshine and wind and not shielded by, or close to, trees, buildings and other obstructions."

Measurements that do not follow such standards are equally "true" and perhaps helpful for some applications, but certainly not meaningful for comparisons.

For some observed variables (such as radiation, carbon dioxide (CO_2) concentration, or total column ozone), internationally maintained, traceable standards exist. Often standards refer to a world primary instrument (i.e., the true value is defined as that measured by a certain instrument) and are transferred to individual instruments using travelling standards (i.e., instruments that are regularly calibrated against the primary standard) and a chain of regional and national calibration centres. However, no traceable standards exist for many meteorological measurements.

Atmospheric observations are always performed with a certain intention, for instance, meteorological or climatological applications. The intention behind an observation is important as it affects not only the reporting, processing, and analysis, but also the data properties and quality. Meteorological applications (e.g., weather forecasting and analysis) call for a real-time availability, high precision, and high spatial resolution of the data. Important considerations for climatological applications (climate research, planning, or risk assessment) are high accuracy, representativeness, and long-term stability.

Systematic observations are expensive and require a high degree of coordination. At the same time, they can be of national interest (e.g., for warning systems or strategic considerations). Therefore, systematic observations are mostly performed by large, national organisations such as weather services, which are organised globally by the WMO. The WMO issues standards and recommendations concerning instruments, calibrations, measurement practices, reporting, and quality control/assurance. The WMO also regulates data exchange, promotes professional education and provides the link to the political world.

Only a subset of the global meteorological network is needed for climate observations, but with particular requirements for quality, comparability, and long-term stability. Ensuring the existence of suitable networks is one of the tasks of the Global Climate Observing System (GCOS), founded in 1992; an institution that is co-sponsored by the WMO and other intergovernmental organisations.

In 2009, the World Climate Conference established the "Global Framework for Climate Services (GFCS)", a global partnership of governments and organisations that produce and use climate information and services (http://www.gfcs-climate.org/content/about-gfcs). The GFCS ensures the development of science-based climate data products and their provision to users. Considering the history of meteorological measurements, GFCS and GCOS are very recent successes in a period of scientific development that started in the 19th century (see Edwards 2010). In Sect. 2.2, we will look back to this period and beyond in order to understand the basis of historical climatology: historical climate observations.

2.2 Historical Climate Observations

2.2.1 Documentary and Early Instrumental Data

We cannot rely on a global climate observing system for analysing climatic variations back to 1700. A global or even national observing system did not exist in the 18th century. Therefore, information from different sources must be combined. Figure 2.3 gives an overview of the observing systems and platforms used since 1700.

Weather, atmospheric phenomena, and climatic effects were observed at least since the classical epoch and were sometimes even written down. People noted the conditions they considered relevant, for instance, those impacting agriculture, extreme events affecting society (e.g., causing damage to common goods), or weather seen in a religious context. Since the middle ages, weather diaries were kept by individuals and weather events were reported in other documentary sources. Some observers took only sporadic notes, whereas others observed in a systematic way, sometimes providing semi-quantitative and verifiable information on snow, freezing, wind, precipitation, and other weather elements (see Glaser 2008; Pfister 1999). Weather information was particularly important, even essential for survival, at sea (Fig. 2.4). On ships, observations have been performed for a long time and noted meticulously in ship logs. Ship logs and weather diaries are important sources of information for studying climate of the 17th and 18th centuries.

Instrumental measurements started in the 17th century with the advent of the Enlightenment and the growing curiosity of humans to physically explore their environment. The longest continuous temperature time series is the Central England temperature, introduced in Chap. 1 (Fig. 1.2). Measurements that were taken before

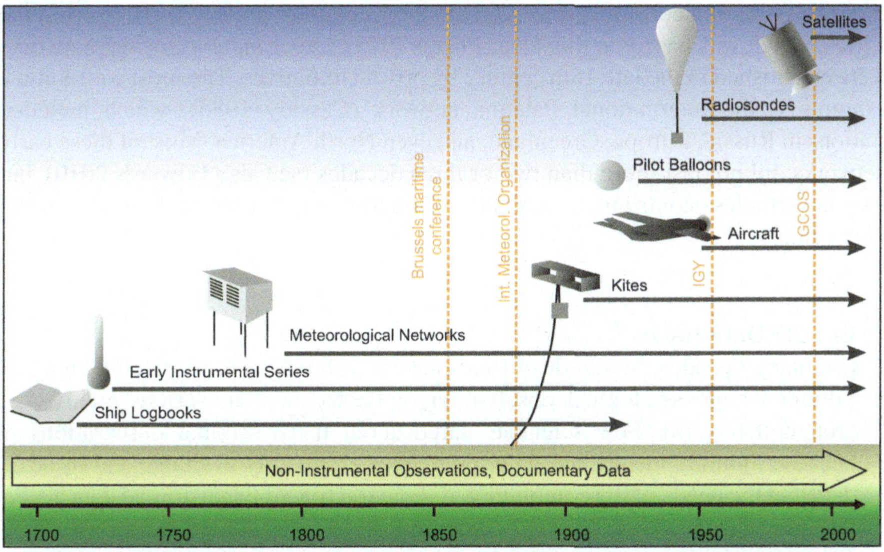

Fig. 2.3 Schematic overview of the observing systems and platforms for climate observations since around the 18th century. *Yellow dashed lines* denote important time lines in the development of the observing systems and (*IGY* International Geophysical Year, *GCOS* Global Climate Observing System)

Fig. 2.4 Sorely tried, the HMS Beagle off Cape Horn, 12 January 1833, by John Chancellor (Courtesy of Gordon Chancellor) (see also Fig. 2.26)

the start of (mostly national) meteorological networks are termed early instrumental measurements. They were typically taken by scientists (interested in the physics of the environment), physicians (e.g., trying to understand the causes of illnesses), colonial administrators (reporting the wealth of their assigned territory), or clerical persons. The efforts towards basing agricultural practices on science in the 18th century brought another source of demand for meteorological measurements. One example is the meteorological network of the Economic Society of Bern from the 1760s, which was based on this idea (Pfister 1975). Other meteorological networks were established in the late 18th century by private initiatives. The most well-known example is the international Palatina network (Cassidy 1985), which included stations in Russia, Europe, Greenland, and even North America. Most of these early networks did not last more than two or three decades (see also Edwards 2010), but they nevertheless contribute important information on weather and climate of the 18th century.

Box 2.1 Data rescue

For many decades, hundreds of thousands of volumes and boxes containing billions of meteorological observations collected dust in various archives around the world. Few scientists cared about these original observations because monthly means and other products were readily available and these data met the needs of the community. A large fraction of the historical data has never been digitised or, if it has been, only in the form of monthly averages of selected variables.

Today, original weather data are considered valuable in view of new research questions posed by society to science and as new tools become available. For instance, extremes have come into focus in recent years—climate change is no longer only about average climate. Regional detail is required for impact studies, and more variables than just temperature and precipitation are requested. Data assimilation techniques (see Sect. 2.6) allow us to exploit the full temporal information of, for example, air pressure, and to produce three-dimensional global weather datasets.

Fig. 2.5 Enormous amounts of historical data are still waiting to be digitised—citizen science promotes hope

(continued)

Box 2.1 (continued)

This requires going back to the archives and rescuing the original observations (Fig. 2.5). In recent years, several national weather services as well as large research projects have started data rescue activities. However, the amount of old weather data to be digitised is so great that other solutions are also being evaluated. Citizen science projects such as www.oldweather.org try to engage the public in digitising the data via the internet. Millions of weather measurements have already been digitised by the public and made available.

Data rescue activities on a global level are loosely coordinated through the Atmospheric Circulation Reconstructions over the Earth Initiative (ACRE; Allan et al. 2011) and the International Surface Temperature Initiative (ISTI; Thorne and Vose 2010).

2.2.2 *From National Weather Services to a Global Observing System*

The number of stations grew slowly during the first half of the 19th century and then increased rapidly during the second half of the century with the foundation of many national weather services. The latter emerged from a need for systematic observations for weather forecasts and storm warnings, new telecommunication technology enabling real-time data transmission (i.e., the telegraph), and political changes leading to new governmental responsibilities. National weather services brought a standardisation and professionalisation to observations. The new weather services also showed interest in compiling earlier historical observations and published them in their yearbooks. Moreover, meteorological instruments were refined (e.g., Assmann 1891); Fig. 2.6 shows several historical meteorological instruments.

During the same period, meteorology became international, although not yet at a governmental level (see also Edwards 2010). The Brussels Maritime Conference of 1853 and the International Meteorological Organization after 1873 (see Fig. 2.3) tried to establish standards for weather observations worldwide, although with moderate success. International scientific projects such as the International Polar Year 1882/1883 promoted scientific exchange at the multinational level. From the second half of the 19th century onward, the number and spatial distribution of meteorological stations was such that, from today's perspective, a global view became possible. Remote areas were targeted and high-altitude observatories were founded such as the Sonnblick Observatory in Austria and Mount Säntis in Switzerland. In fact, many of the widely used global climate datasets for the earth's surface reach back to the second half of the 19th century, comprising surface air temperature, sea-surface temperature, sea level pressure, and precipitation.

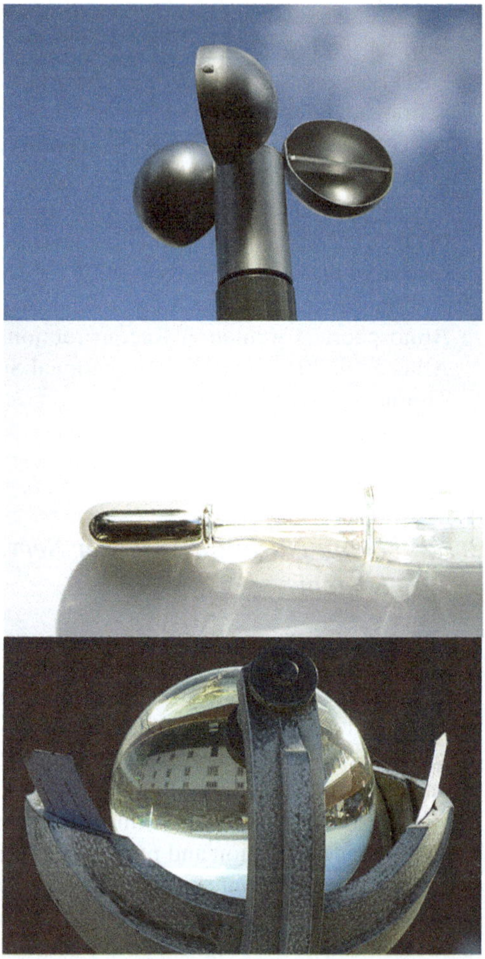

Fig. 2.6 Conventional meteorological instruments used in the 20th century. (*top*) Cup anemometer, (*middle*) Thermometer and (*bottom*) Campbell-Stokes sunshine recorder

The number of stations gradually increased during the first half of the 20th century. The quality of atmospheric observations was greatly improved during the International Geophysical Year (IGY) 1957/1958, a successor to the International Polar Year, endorsed by the International Council of Scientific Unions. The IGY led to the establishment of meteorological stations in Antarctica, improved and coordinated networks of weather balloon soundings, atmospheric ozone observations, and measurements of carbon dioxide in the atmosphere, among numerous other achievements. The data quality and coverage over both the terrestrial and marine domains further increased until the late 20th century (Fig. 2.7), although it was punctuated with dips during the World Wars and, for some regions and variables, a decline after the 1980s (Note that a substantial part of the decline is due to delayed reporting).

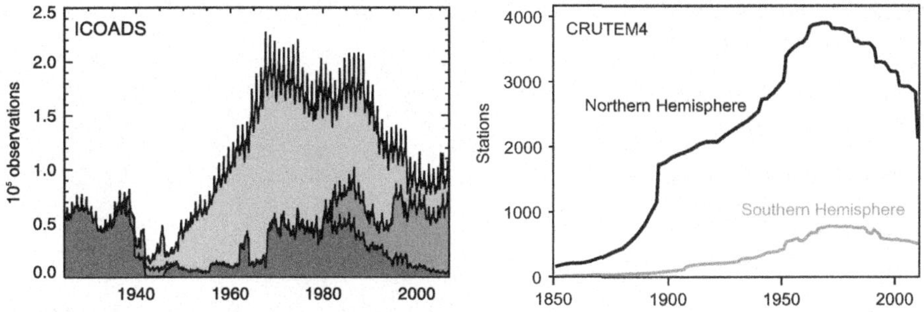

Fig. 2.7 Number of (*right*) land stations (Jones et al. 2012) and (*left*) sea-surface temperature observations. *Dark*, *middle*, and *light grey* denotes bucket, engine room intake, and unknown observation type (From Kennedy et al. 2011, John Wiley & Sons)

2.3 Upper-Air and Satellite Observations

Climate and weather data near the earth's surface only provide a limited view of the three-dimensional structure of the atmosphere. Upper-air measurements began in the late 19th and early 20th century (Fig. 2.3), but were only fully developed after the Second World War. In many countries, operational upper-air networks using registering balloons, kites, and pilot balloons (free-flying balloons tracked from the ground to derive wind profiles) were gradually established during 1900–1920s and were augmented in the 1920s by aircraft observations. Radiosonde networks (weather balloons measuring temperature, pressure, and humidity and transmitting the data to a receiver on the ground) only began in the late 1930s and 1940s—at first, rather experimentally (see Stickler et al. 2010). Figure 2.8 shows the launch of a radiosonde in the 1940s. The IGY finally brought a standardised global upper-air network of weather balloon stations, which remains the foundation of upper-air observations.

Another important change in the global climate observing system was the start of space-borne observations of the earth in the 1960s and 1970s (Fig. 2.9). Satellites carry instruments that measure irradiance emitted from the earth (infrared) or the atmosphere (microwave), solar radiation reflected at the earth or backscattered by the atmosphere, or solar or stellar radiation passing through the atmosphere in the earth's limb. Active systems emit signals and measure the reflected or backscattered radiance. All systems are based on the fact that electromagnetic signals interact with the earth's atmosphere. The physical properties of the earth's surface (e.g., reflectivity and temperature), the atmosphere (e.g., motion) and its constituents (e.g., trace gases, aerosols, and hydrometeors) alter the radiance intensities, spectral signatures, or signal travelling times. Deriving the atmospheric properties from the measured signal, however, is a mathematically ill-posed problem. Many possible atmospheric conditions could result in the same measurement. Complex algorithms, termed "retrieval", are necessary and involve incorporating prior information to obtain a unique solution.

Fig. 2.8 Launching of a weather balloon, ca. 1944. Women's first opportunities in meteorology occurred as a result of the Second World War (National Oceanic and Atmospheric Administration (NOAA) National Weather Service (NWS) Collection)

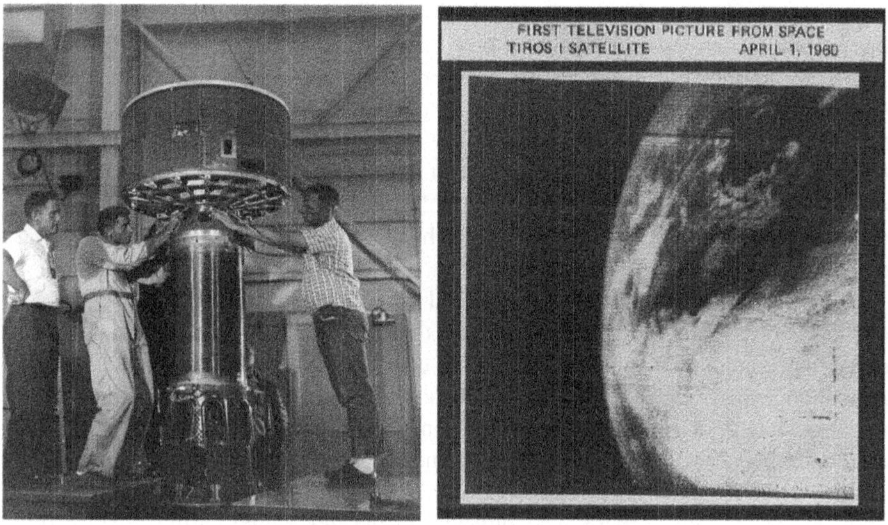

Fig. 2.9 (*left*) Television and Infrared Observation Satellite (TIROS) I and (*right*) first television picture from space, 1 April 1960 (Photos: National Aeronautics and Space Administration (NASA))

Satellites provide a near-global coverage of numerous climate variables, such as sea ice or snow coverage, cloud cover, the vertical temperature structure, or surface wind. For many sparsely observed regions such as the South Pacific, satellites brought a huge amount of additional weather information. In particular, satellites provide detailed information on the concentration of many reactive trace gases and on the amount and properties of aerosols. At least as important as for surface data, quality remains a fundamental issue for satellite data. The overlap between different sensors is often too short to obtain reliable transfer functions.

Ground-based remote sensing techniques supplement in situ observations of the free atmosphere. In fact, cloud motion was already systematically observed in the 19th century. Since the 1920s, spectroscopy was used to measure the total ozone column. Later, radar, wind profilers, and other devices were introduced to derive wind, precipitation, and aerosols, and time delays from Global Positioning System (GPS) techniques give atmospheric moisture. While these data sources are important for present day weather analysis, they are not useful when considering climate changes since 1700 and hence are only relevant for the last part of this book.

2.4 Data Dissemination in the Course of Time

Climate observations were made to be exchanged. Early meteorological measurements were disseminated through scientific journals such as the Philosophical Transactions of the Royal Society (Fig. 2.10) and extensive private correspondence. Scientists like the Bernese scholar Albrecht von Haller[1] (13,300 letters to him are preserved), were effective communicators and collected observations from their private networks of peers.

In the early 19th century, German meteorologist Heinrich Brandes[2] introduced a new way of using weather data by drawing the first weather maps (based on data from the year 1783), thus demonstrating the value of synoptic meteorology for diagnostic purposes (Brandes 1819). However, weather maps had no operational use before the invention of the telegraph. During the first half of the 19th century, more and more stations were starting to operate. The data were compiled by scientists and first quantitative overviews could be produced. James Pollard Espy,[3] Cleveland

January. Day. Hour.	*Thermoſcope.* inches.	*Baroſcope.* inches.	166$\frac{5}{6}$.	
19. 8. Morn.	14 $\frac{1}{16}$.	29 $\frac{1}{2}$.	Hard froſt.	Cloſe.
4. Even.	14 $\frac{1}{8}$.	29 $\frac{1}{4}$.	Hard froſt.	Cloudy.
9. Even.	14 $\frac{1}{4}$.	29 $\frac{1}{4}$.	Rain.	Wind
20. 8. Morn.	15 $\frac{1}{4}$.	28 $\frac{1}{4}$.	Sunſhine.	Wind.

Fig. 2.10 Temperature and pressure measurements for Oxford on 19 and 20 January 1665 (Gregorian dates: 29 and 30 Jan. 1666) (Wallis 1665)

[1]Albrecht von Haller, 1708–1777, was a Swiss physician (anatomist), scientist, and poet of the Enlightenment. He is said to have published ca. 50,000 pages of scientific works of consistently high quality.

[2]Heinrich Brandes, 1777–1834, was a German meteorologist, astronomer, and mathematician.

[3]James Pollard Espy, 1785–1860, was an American meteorologist and one of the first to use the telegraph for collecting meteorological observations.

Abbe,[4] and (for marine data) Matthew Maury[5] were pioneers in the United States from the 1830s to 1850s. In Europe, Heinrich Wilhelm Dove[6] (in the 1830s) and later Julius Hann,[7] Francis Galton,[8] Wladimir Köppen,[9] and Eduard Brückner compiled global datasets. Henry Blanford,[10] Gilbert Walker,[11] and Felix Exner[12] continued these efforts and brought in statistical concepts for the first time (see Box 4.1, p. 224).

With the start of national weather services, observations were published and made accessible in yearbooks. Brückner used such sources for his studies, arguably including the three volumes shown in Fig. 2.11. In the subsequent decades, attempts were made to systematically collect global weather data—a huge and extremely difficult undertaking. Most noteworthy in this context are the efforts of the Smithsonian Institution, whose "World Weather Record" contained a global compilation of climate data (see also Edwards 2010).

The data compilations enabled climatological products to be presented in graphical forms such as isotherms (pioneered by Alexander von Humboldt[13]). Graphical elements became more common in scientific publishing in the mid-19th century, and in 1852, Dove published the first global maps of temperature and precipitation (Dove 1852). Prior to the age of electronic computers, graphical methods were often used in science as efficient tools. Meteorological atlases

[4]Cleveland Abbe, 1838–1916, was an astronomer and meteorologist and founder of the U.S. Weather Bureau.

[5]Matthew Fontaine Maury, 1806–1873, was an American scientist (oceanographer and meteorologist). He published the first comprehensive book on oceanography.

[6]Heinrich Wilhelm Dove, 1803–1879, was a German physicist and meteorologist and one of the first to study global climate. He is known for his works on winds.

[7]Julius Hann, 1839–1921, was an Austrian meteorologist and climatologist and one of the founders of modern meteorology. Hann is known for the Föhn theory, among many other achievements. His "Handbuch der Klimatologie" (Hann 1883, 1897, 1908–1911) set the standard for five decades (see Stehr and von Storch 2000).

[8]Francis Galton, 1822–1911, was a British scientist, known for his work in statistics (correlation), meteorology (weather maps, anticyclones, and weather variability), and eugenics. Galton was a cousin of Charles Darwin.

[9]Wladimir Köppen, 1846–1940, was a German–Russian meteorologist, geographer, and climatologist. Among many other things, Köppen developed a classification of climates that is still used today (Köppen 1881).

[10]Henry Francis Blanford, 1834–1893, was a British meteorologist and palaeontologist and first director of the India Meteorological Department.

[11]Sir Gilbert Walker, 1868–1958, was a British mathematician and meteorologist. Walker worked for the India Meteorological Department from 1904 to 1924. He is best known for his work on climatic oscillations and statistics.

[12]Felix Exner, 1876–1930, was an Austrian meteorologist and director of the Austrian Meteorological Service (Zentralanstalt für Geodynamik). Exner made important contributions to dynamic meteorology.

[13]Alexander von Humboldt, 1769–1859, was a Prussian geographer, naturalist, and explorer. He was one of the most famous scientists of his time. His main work, "Kosmos" (von Humboldt 1845) is a compendium of science and nature.

Fig. 2.11 Data storage in the climatology group at the Institute of Geography in the University of Bern. (*top*) Printed volumes of Russian meteorological observations from 1879 published by Heinrich Wild,[14] these volumes may have been used by Brückner, (*middle*) punch cards, and (*bottom*) punch tape Facit 4070 (1968). The storage size of one tape is 120 KB. Currently, the group uses a 400 TB storage system

(e.g., Hann 1887) became important and popular for science and education. Most school atlases produced since the late 19th century contain climatological maps, an example of which is shown in Fig. 2.12.

Several technical developments in the early 20th century affected data handling and storage. Weather data could be stored on machine-readable punch cards (although the operations that could be performed were extremely limited; see Edwards 2010). The invention of the radio allowed wireless transmission of weather data to many receivers. However, a large fraction of the data was still written down and stored on paper. Only since the 1960s, has the acquisition, processing, and storage of meteorological data become fully electronic. Magnetic tapes came into

[14]Heinrich Wild, 1833–1902, was Swiss meteorologist. Wild was a physics professor in Bern and later founder of the Main Geophysical Observatory in St. Petersburg, Russia.

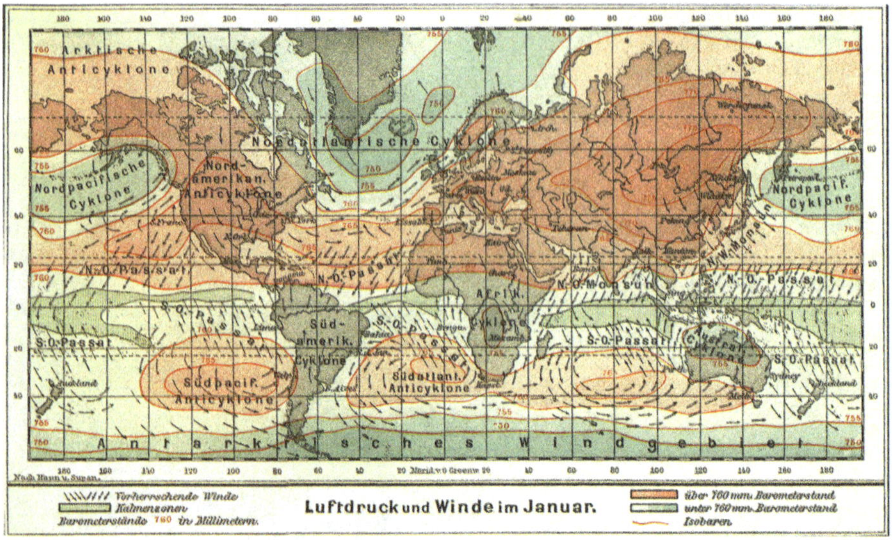

Fig. 2.12 Air pressure and winds in January from a school atlas (Diercke Schul-Atlas, Braunschweig, 1901) based on Hann's atlas

use as storage media, but most data were still stored on punch cards, computer code was often stored on punch tape (Fig. 2.11). An important role in the development of data dissemination was again played by the IGY, the first "big data project" (Aronova et al. 2010). The IGY required an organised form of data handling and established the World Data Center system, for the archiving and dissemination of scientific data through dedicated institutions in the U.S., Europe, the Soviet Union, and Japan. Although originally the World Data Centers did not yet store data on magnetic media and dissemination was limited to the IGY (see Edwards 2010), the system has played a key role in climate data dissemination to the present.

With the implementation of communication satellites, a system of terrestrial and satellite communication developed, enabling the global transmission of weather data, which today is known as the Global Telecommunication System (GTS). The GTS is an important component of WMO's World Weather Watch, a programme established in the 1960s that "combines observing systems, telecommunication facilities, and data-processing and forecasting centres" (WMO website). The challenge we face today is the sheer amount of data. Climate model, reanalyses, and satellite data worldwide add up to exabytes (10^{18} bytes). Doing climate research typically requires large storage capacities.

Box 2.2 When meteorology was not considered science

Although early climate observations were studied by the scientists of the Enlightenment, they were far from sufficient to form the basis of a physical

(continued)

Box 2.2 (continued)
theory of the atmosphere. Climate science in the early 19th century was not a discipline of its own, but an auxiliary science for several other disciplines such as botany, geology, or medicine. The task was to document, not to understand. Atmospheric scientists—if one can use this term—were aware of the large needs and gaps in their discipline. The following episode may elucidate this. In 1817, after the devastating "Year without a summer" of 1816 (see Box 3.2, p. 124 as well as Sect. 4.2.4), the Swiss Natural Sciences Society offered an award for answers to whether or not climate in the Alps had been deteriorating in recent years (see Bodenmann et al. 2011). The analysis of meteorological series was explicitly excluded from the call text. Rather, the call asked for indirect evidence (or proxies, as we would say today) and addressed scientists from other fields such as geology, natural history, or forestry. The president's speech (Usteri 1817) made clear why:

> "This branch of science [meteorology] has peculiar difficulties, and unless one wants to open a vaudeville, one has to refrain from seeking broad public attention for it."

Weather and forecasts clearly was the domain of astrology, not science.

2.5 Uncertainties in Climatic Data

2.5.1 Uncertainties and Errors in Measurement Series

Long climate records, which are required for most climate applications, have never been produced for the purpose of climate monitoring and reach back to a time when observing standards were different or inapplicable. Therefore, they are subject to several types of problems, ranging from measurement error to statistical uncertainty (Table 2.1). These should be considered for calculating climate trends and for assessing variability. In the 19th century, scientists were well aware of uncertainties and the danger of using low-quality data.[15] For instance, Brückner not only relied on expert judgement, but meticulously analysed differences between neighbouring stations to address possible breaks in the series. However, the statistical framework for data homogenisation was not yet established.

Before entering into details, it is necessary to define a number of terms; they are shown schematically in Fig. 2.13. Unfortunately, the terms "error" and "uncertainty" are not used consistently in the literature. In everyday language, error implies that

[15]"Barometer, pluviometer, and thermometer are the main instruments of the meteorologist and have been used for over 200 years. Yet one would be very mistaken to trust the old measurements. [. . .] And then the old temperature observations! Here, the inadequacy of the instruments pairs with insufficient placement of the instruments to render the observations completely useless." (translated from Brückner 1890, pp. 133–134).

Table 2.1 Errors and uncertainties in station time series for the example of surface air temperature (adapted from Willett et al. 2014)

	Problem	Cause	Detection	Solution
Reporting	Reported air temperature is not measured air temperature	Errors in reporting, units, data transmission, conversion, encoding, etc.	Analyse metadata, statistical outlier screening and testing	If possible, identify cause of error and correct. Otherwise, flag suspect values
Instrument	Measured air temperature is not true air temperature	Instrument error or malfunction, calibration error	Comparison with reference series or parallel measurements, analyse metadata	If possible, identify cause of error and correct. Otherwise, perform statistical homogenisation or flag suspect values
Station	True air temperature is not representative of ambient air temperature	Instrument shield or shelter, measurement practices, local (nonrepresentative) changes in station surroundings	Comparison with reference series or parallel measurements, analyse metadata	If possible, identify cause of error, use energy balance model of shield or model of microclimate to correct. Otherwise, perform statistical homogenisation
Surroundings	Representative ambient air temperature is affected by nonclimatic influences	(Representative) changes in station surroundings, urbanisation	Comparison with reference series, analyse metadata	Statistical homogenisation or no correction (depends on application)
Processing	Different ambient air temperatures are merged	Change in location	Comparison with reference series, analyse metadata	Statistical homogenisation or unmerge (depends on application)
	Changes in diurnal sampling affect statistics	Change in observation time	Comparison with reference series, analyse metadata	Statistical homogenisation or split (depends on application)

something is wrong, while uncertainty implies that something is unknown. The scientific meaning of the two terms is quite different. In applied mathematics, error is the difference between a measured value and the true value (x-axis in Fig. 2.13), while uncertainty is the statistics of the error (e.g., its probability density function, shown as the y-axis in Fig. 2.13, or a parameter such as the standard deviation). However, the term uncertainty is avoided by mathematicians and often only the term error is used. In meteorology, the true value is almost never known, and hence the error is not known either. It may, however, still be possible to quantify the uncertainty. Errors are sometimes subdivided into random and systematic components. The random error has an expectation value of zero for

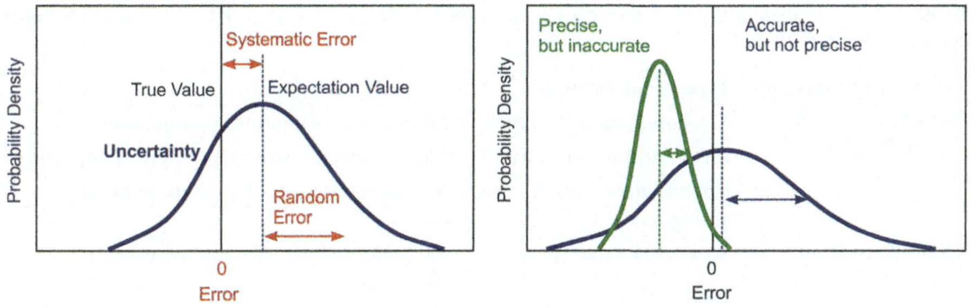

Fig. 2.13 Schematic diagram illustrating the terms error, uncertainty, systematic error, random error, accuracy, and precision

an infinite number of replications of the measurement, while the systematic error is the average of an infinite number of replications of the measurement. Measurements can be "accurate," which means that their systematic error is small. "Precise" measurements, conversely, have a small random error (Fig. 2.13).

Let us now look at the problem of data error and uncertainty using the example of air temperature. Recall that, according to the WMO, temperature should be measured in a sheltered place, freely exposed to weather, and should be representative of the free air conditions over as large an area as possible. These recommendations should ensure the comparability of measurements. There are many reasons why a measured temperature might not fulfil these conditions. Table 2.1 gives a summary of the possible sources of errors.

One common source of error in meteorological data is reporting. In other words, the number that is in front of us is not the measured temperature due to errors in units, decoding, processing, input-output, copying or digitising, or similar. If detected, it is sometimes possible to correct these errors. A second type of error refers to the instrument itself during observation (e.g., malfunction, power loss, or interference between measurements) or maintenance (calibration and handling). In this case, the measured air temperature is not the true temperature at the sensor. Errors may also occur due to the shielding of the thermometer. While there may be nothing wrong with the thermometer itself (it measures the true air temperature), this true air temperature represents the microclimate of an overheated shelter and thus is in conflict with the WMO definition of air temperature (see Sect. 2.1). Finally, a temperature reading may not fulfil the WMO recommendations because of the station environment (e.g., obstruction by trees, buildings, or paved roads). Thus, it is no longer representative for the surroundings. Identifying and correcting errors and reducing uncertainties are complicated by the fact that they change over time.

A different type of uncertainty is related to changes in the larger-scale station environment. When meteorological networks were built up in the late 19th century, stations were often erected at the outskirts of a city. Soon many of these locations became part of the urbanized area and therefore represent a more local climate that is subject to urban effects. Although real and in accordance with the above-mentioned WMO guidelines, this effect may be undesired. For studying global

Table 2.2 Examples of types of uncertainty relevant for climate research (based on draft by Peter Thorne)

Degree of processing	Type of uncertainty	Example
Station time series (see Table 2.1)	Measurement uncertainty	Instrument calibration, malfunction
	Procedural uncertainty	Change in standards (e.g., shelter), reporting
	Representativeness in time and space	Station relocation, change in observation times
Homogenisation	Structural uncertainty	All possible detection and correction strategies
	Parametric uncertainty	Choices and parameters in a selected detection and correction strategy
Averages and products	Coverage uncertainty	Gaps in time and space on scales for which averaging is attempted
	Representativeness in time and space	Over-/underrepresentation of urbanised areas or of specific local climates (valleys, coasts, etc.)
	Structural uncertainty	All possible averaging strategies
	Parametric uncertainty	Choices and parameters in a selected averaging strategy (e.g., treatment of gaps, adjustment for variance)

climate change, urban effects may receive undue weight as suburban and urban stations are overrepresented in the global station network by area. Here, a correction or downweighing of such stations is necessary. For a biologist studying changes in urban flora, however, a correction is obviously undesirable.

Further uncertainties come into play when considering sequences of measurement series (Table 2.2). For instance, stations may be relocated (e.g., from the city centre to an airport), or observation times may change. Each part of the series may be correct, but climate science often desires long time series and combining the sequences requires corrections that add further uncertainties. It is not clear, per se, for which purposes series from different stations can (or should) be combined.

Errors in temperature data can sometimes be identified (and in some cases subsequently corrected) by quality control and quality assurance (QC/QA) routines such as checks for physical plausibility, internal consistency, and statistical outlier tests. Data may also evolve due to reprocessing, for instance, if better sensor-response curves become available or if a new system allows a more accurate processing of the raw data.

2.5.2 Inhomogeneities and Homogenisation

Systematic errors (see Fig. 2.13) may change over time and lead to breaks or artificial trends in a time series, called inhomogeneities. Most temperature (or other climate variable) series are affected by inhomogeneities; breaks occur typically every 10–20 years on average. For climatological applications, all of these breaks need to be corrected. If an inhomogeneity is foreseeable (e.g., due to a change in instruments, shelters, or location), parallel measurements should be performed that can then be used for a detailed assessment of the inhomogeneity and subsequent correction. Otherwise, statistical methods are used and the procedure is called homogenisation.

Homogenisation of climate data ideally consists of three steps: (1) Detection of an inhomogeneity, (2) (if possible) attribution to a cause, and (3) (if possible and desirable) correction. In some cases, for instance if a station is relocated, inhomogeneities are obvious. Otherwise, detection is based on statistical methods (see Kuglitsch et al. 2012). Some tests are only based on changes in statistical properties within the series in question (Wang et al. 2007b). However, most methods use reference series—homogeneous series that are highly correlated (with a stable relation) with the series under question (e.g., Caussinus and Mestre 2004; Toreti et al. 2012). These may be series of the same variable at a neighbouring location, or series of a highly correlated variable at the same location. Break detection methods can be performed using daily, monthly, or annual data. Air temperature series are often tested on the basis of annual-mean values. Ideally, sufficient information on the station (i.e., metadata) is available to decide whether or not to accept a break.

If a break is accepted, in the ideal case it can be attributed to a specific problem. For instance, a break in a temperature series may by attributable to a change in shelter. If such an attribution can be made, it may be possible to find a more appropriate correction than possible through statistical methods. In most cases, however, statistical methods are used to homogenise the series once a break in a series is accepted. Conventionally, the part before the break is adjusted to the part after the break (see Fig. 2.14).

Often, climate data are homogenised with respect to a time average such as monthly means. However, in recent years climate extremes have come into focus, and consequently homogeneous meteorological records on a sub-daily scale are required. However, for daily or sub-daily data, applying an average correction may be insufficient as negative extremes might be affected differently by an inhomogeneity (such as a change in the shelter) than positive extremes. Some methods (e.g., Della-Marta and Wanner 2006) not only account for changes in the mean, but also changes in the higher moments of the distribution (see Fig. 2.14).

Physics-based correction approaches provide an alternative in some cases. The example in Fig. 2.15 shows corrections of temperature series for a change of shelter (from a Wild screen to a Stevenson screen) obtained with a physics-based approach that estimates the energy balance of the sensor-screen system. The corrections are different for each individual measurement and depend on solar elevation, cloud cover, and wind speed. In this case, parallel measurements were available to check the results and revealed a good agreement, specifically for heatwaves (Auchmann and Brönnimann 2012).

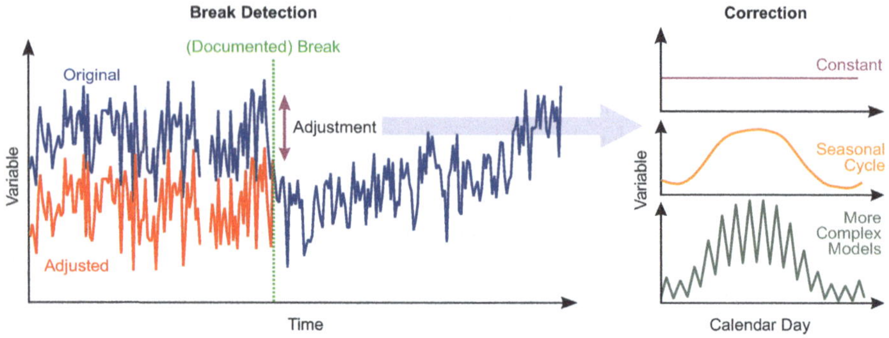

Fig. 2.14 (*left*) Schematic depiction of the homogenisation of a documented break (*green dashed line*). (*right*) Seasonality and complexity of the correction function

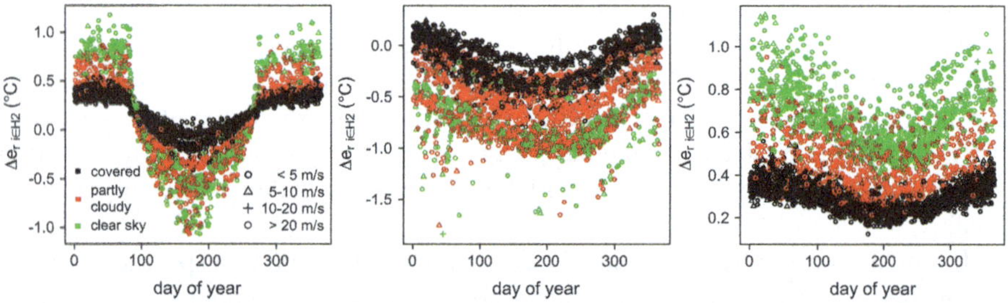

Fig. 2.15 Corrections for daily temperatures at Basel to adjust for the change from a Wild screen to a Stevenson screen around 1960 for (*left*) the early morning, (*middle*) noon, and (*right*) the evening for different ambient conditions (cloud cover, wind) as found with a physics-based, empirical approach (FromAuchmann and Brönnimann 2012, John Wiley & Sons)

While homogenisation is a necessary step for reaching a level of quality that allows trend calculations, it introduces additional uncertainties. Homogenisation involves the estimation of many parameters and many choices. Each of these estimates has its own uncertainties, often termed parametric uncertainties. The errors of the parameters can be quantified statistically, and a Monte Carlo approach can be used to estimate their effect on the results (i.e., the homogenisation approach is repeated a certain number of times, each with a different set of parameters). This approach has been used, for instance, to address uncertainties in sea-surface temperature data (Rayner et al. 2006) and is becoming increasingly common.

Similarly, the effect of different choices (e.g., how many years to use on both sides of the inhomogeneity, or the criteria to be used to qualify a reference series as good) can be assessed in a Monte Carlo framework. However, uncertainty already arises from the choice of a given method. Choosing different approaches will lead to different results (Venema et al. 2012). The related uncertainty is termed "structural uncertainty" and is very difficult to quantify as we only know a very small set out of an infinite number of possible homogenisation methods. Imagine two independent teams of engineers given the same task of building a satellite with some desired capabilities. The two teams will likely come up with different solutions.

2.5.3 The Chain of Uncertainties

In Sects. 2.5.1 and 2.5.2, we discussed the uncertainties involved in the measurement of temperature at a station, the process of homogenising temperature series, and the uncertainties introduced in the process. When producing gridded or averaged products from station observations (which will be discussed in Sect. 2.6) the same basic types of uncertainty apply. They are summarised in Table 2.2. An approach needs to be chosen (out of an infinite number), and within this approach there are additional choices and parameter estimations. Again, parametric uncertainty is the most easy to quantify, and current data products for surface air temperature are usually accompanied by a quantification of parametric uncertainty. With respect to the structural uncertainty, the only way of quantification is to have different groups produce the same product. A recent example for global surface temperature is the Berkeley Earth Surface Temperature (BEST) project (Rohde et al. 2013); an attempt by an independent group of scientists to create a new global record from raw station data. With respect to homogenisation, a recent benchmark exercise was very successfully applied to test different homogenisation algorithms (Venema et al. 2012).

Structural uncertainty is almost never incorporated into error estimations. As a consequence, published uncertainties are always lower-limit estimates. As our knowledge of uncertainties increases, estimates of uncertainty may grow (Fig. 2.16). Hence, increasing uncertainties (with respect to past publications) may reflect an increase in our understanding of uncertainties rather than a decrease in quality. A quantification of all errors along the chain from the measurement to the final

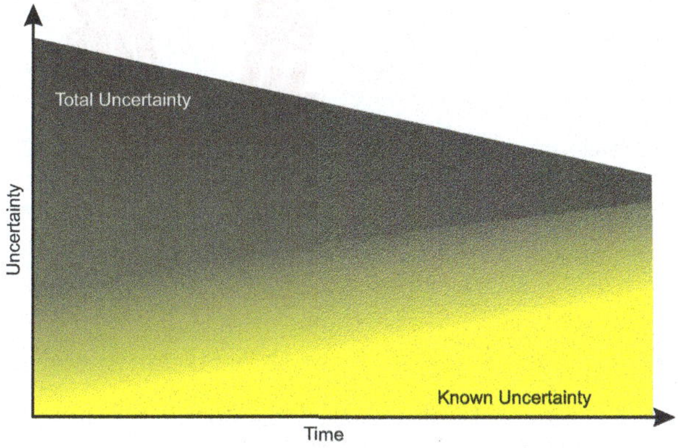

Fig. 2.16 Schematic view of the change in uncertainty over time. The decrease in total uncertainty is due to a better correction and homogenisation of datasets, while more of the full uncertainty is known and better quantifiable

statistical analysis is not possible. Therefore, it is particularly important to know the concept of uncertainties and to know which parts of it are quantified.

2.6 Data Products and Dynamical-Statistical Methods

2.6.1 *Spatial Information and Geostatistical Methods*

Every day, a large amount of meteorological information is gathered by numerous global observing systems (Fig. 2.17): surface-based stations, aircraft, radiosonde, buoys, and in particular satellites. This information is processed, disseminated, and archived, and can then be used for further studies. However, as the figure shows, the observations are irregular in time and space, with large gaps or, conversely, regions with extremely dense information. It is difficult to work directly with such data. For most applications, two- or three-dimensional datasets on a regular space–time grid are much more convenient. In the following, we will discuss methods used to produce such datasets.

Let us assume that we would like to analyse precipitation on 21 May 1999, when the northern part of Switzerland was affected by strong flooding due to severe precipitation and melting snow. Figure 2.18 shows different data products at different spatial resolutions that could be used. In Fig. 2.18 (top left), station data

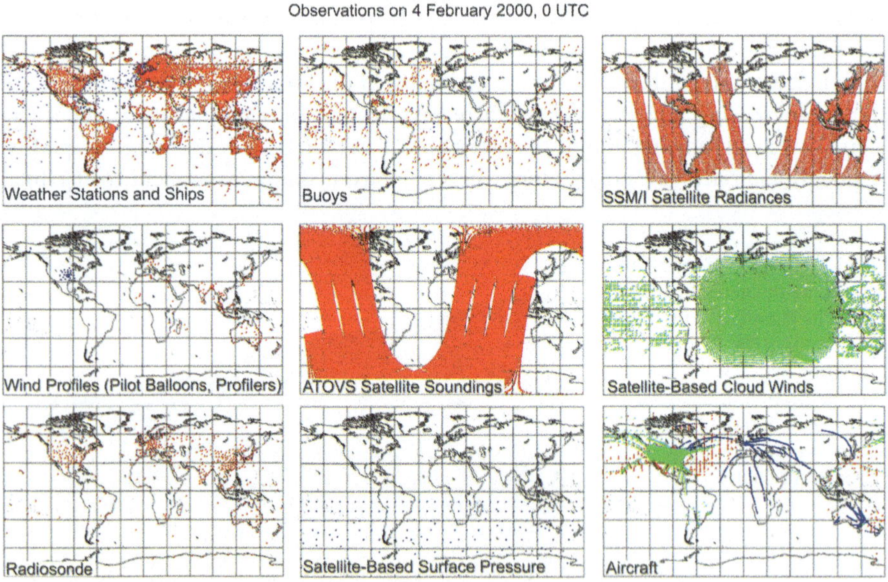

Fig. 2.17 Data gathered by the European Centre for Medium-Range Weather Forecasts (ECMWF) on one specific day and time, 4 February 2000, 0 UTC

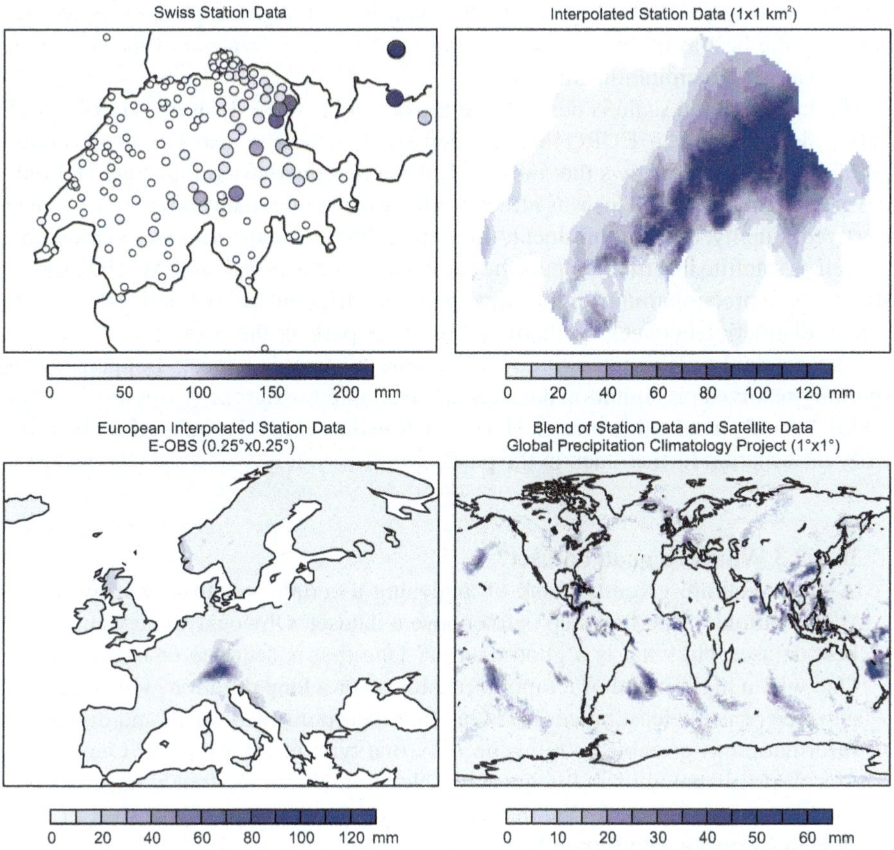

Fig. 2.18 Precipitation on 21 May 1999 in different datasets. (*top left*) Swiss precipitation station data and (*top right*) interpolated station observations (the typical inter-station distance and hence the effective resolution is ca. 15 km). (*bottom left*) Data interpolated from stations from the E-Obs dataset and (*bottom right*) blended precipitation data from stations and satellites (Data sources: MeteoSwiss) (*top*, RhiresD data and station data), Koninklijk Nederlands Meteorologisch Instituut (*bottom left*), and Global Precipitation Climatology Project (*bottom right*)

from Switzerland are given for this specific day. The precipitation network is dense and, after the quality assurance procedures described in the previous Sect. 2.5, the data provide detailed and direct information. However, this may not satisfy all of our needs. A hydrological model requiring a precipitation input value at each grid point of, for instance, a 2×2 km grid, could not be fed without further assumptions.

Using geostatistical methods, the station data can be interpolated to a regular, fine grid. Thereby, the effects of small-scale topographical features on precipitation can be incorporated statistically, while instrumental measurement noise can be filtered out. Such a dataset has been produced for Switzerland (Fig. 2.18, top right) and the Alps (Isotta et al. 2013) in the European Reanalysis And Observations

For Monitoring (EURO4M) project. The map is spatially complete and provides arguably the best available estimate of daily precipitation, but not every grid value is an observed precipitation amount.

If a larger spatial scale is desired, the European "E-Obs" dataset (Haylock et al. 2008, also updated in EURO4M) is often used, which is also based on station precipitation data. It shows that most of Europe remained dry on this day and only the northern part of the Alps was affected. However, individual valleys are no longer resolved. Finally, a global product (Adler et al. 2003) that incorporates station data as well as satellite information puts the local event into a global context. The product shows some precipitation over the Alps (note the different scale), but does not allow a clear distinction between northern and southern parts of the Alps.

The available daily datasets cover the past 60 years at most. Longer, highly spatially resolved precipitation datasets are available for Europe (Efthymiadis et al. 2006), but only as monthly means. Hence, station data are still required for obtaining daily information further back in the past.

Box 2.3 What is a good dataset?

Assume you are given the task of analysing a certain scientific question in "observations". The first step is to choose a dataset. Obviously, you want the best dataset, but what is a good dataset? One that is accurate or is precise? One with a high spatial or temporal resolution or a long duration? One that is gap free or is the least processed? One that incorporates as much and diverse information as possible or relies on only one type of information? One that is well documented or is freely accessible? One that you already have or is used by everybody else? One that occupies the least space on your disk or is in your favourite data format?

Applications have different requirements and methods to produce datasets have different strengths and weaknesses. Many of these properties are conflicting or even mutually exclusive. Moreover, technical or financial limitations need consideration. Ideally, all of these aspects need to be reconciled and the above questions need to be asked. In practice, however, inertia often takes over. Standard datasets are often used without question. In recent years, as for models, it has become common practice to perform the same analyses on several different observation-based datasets. This is a step forward in terms of addressing uncertainties. However, there is no theoretical framework for working with such "ensembles of observations" and it is not clear what uncertainty is expressed by the differences found.

Gridded global climate data products sometimes convey the image of a well-observed variable. However, prior to the satellite period, the underlying data can be quite sparse. Of the globe, 70 % is covered by oceans and vast areas are covered by deserts, ice sheets, pristine forests, and other uninhabited and sparsely observed

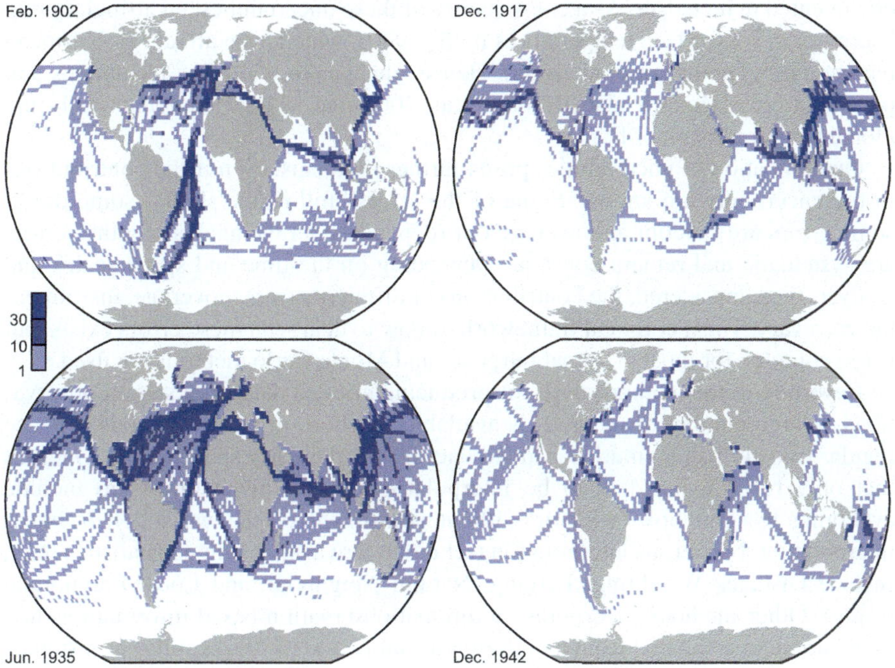

Fig. 2.19 The number of sea-surface temperature observations per month and the 2° × 2° grid cell in the ICOADS Release 2.5

regions. Figure 2.19 shows the coverage of sea-surface temperature observations in four specific months of the main historical ocean dataset (International Comprehensive Ocean-Atmosphere Data Set (ICOADS) Release 2.5, Woodruff et al. (2011)) on a 2° × 2° grid. Several features are noteworthy. First, most ocean grid cells have no observations. For those that do have observations, the number is rather small (mostly below 10). Second, sea-surface temperature at that time was measured exclusively from ships. Hence, the data coverage corresponds to ship routes. These routes change over time due to economic, political, or technical factors (such as the opening of the Panama Canal in 1914). In February 1902, the Atlantic Ocean was rather well covered, but not the North Pacific. In 1917, during the First World War, the coverage was much better over the North Pacific, but the Atlantic was sparsely covered. In 1935, most ocean areas had a comparatively good coverage, while the opposite was the case for 1942—during the Second World War. Even in recent years, spatial coverage of sea-surface temperature data is affected by economic and political factors such as piracy in the Indian Ocean (Smith et al. 2011).

Constructing gridded datasets in this situation is challenging. On top of the data scarcity, major changes took place in the observing system. Rather than measuring ocean temperatures with buckets, many ships started to measure temperatures in the engine cooling water. The changeover started around 1940, but is often not documented in metadata. As a consequence of these uncertainties, an artificial jump is present in sea-surface temperatures in the 1940s, which even affects global-mean temperature (Thompson et al. 2008). Hence, changes in observing techniques and poor data coverage (see also Brohan et al. 2009) led to high uncertainties in the global temperature record.

Quantification of uncertainties poses further problems, even if the contributions to the uncertainty are known. Some of the errors "follow the ships", some don't; some errors are random and average out (depending on the analysis), other errors are systematic and remain constant. Depending on the time and space scales and analysis type considered, the contributions of different errors (coverage, instrumental, etc.) vary. The theoretical framework of how to deal with these errors exists, but is too complex for most practical purposes and Monte Carlo methods are used.

Reverting to the methods used for producing gridded data from station observations, a large number of methods is available. In short, gridding methods produce regular (mostly rectangularly) gridded data from irregularly spaced data. Methods that only rely on the data to be interpolated and the given time step include averaging by grid box (with or without adjustment of the variance) or linear interpolation. Numerous interpolation techniques are available in geostatistics (e.g., Inverse Distance Weighting, Kriging, Natural Neighbour, and Local Polynomial fitting). Other methods incorporate additional observation-based information such as covariances estimated from a high-quality subperiod of the dataset. For instance, Reduced Space Optimal Interpolation employs principal components of a subperiod with dense observations. In the sparse part of the data, only the scores of the principal components are then estimated (similar to climate reconstruction techniques discussed in Sect. 2.8.4). Finally, data assimilation methods not only use observations and covariances, but in addition use a numerical model to obtain a state estimation.

Excellent technical literature (Atkinson and Lloyd 2011; Daley 1993; Hengl 2009) is available on these methods. In Sect. 2.6.2, I would like to focus on one of these methods in some more detail, as it plays an important role in atmospheric sciences and in the remainder of this book: data assimilation.

2.6.2 Data Assimilation and Reanalyses

The concept of data assimilation is to combine observations of the atmospheric state with a numerical weather prediction model to obtain an estimate of the true state of the atmosphere at a given time. Some also call it "physically consistent interpolation" as the resulting estimate is both physically consistent and consistent with the observations. However, the term "extrapolation" might be more appropriate as data assimilation provides the result in the model space (i.e., model variables, grid, levels, etc.), including non-observed variables, levels, and regions.

How does data assimilation work? Imagine a weather forecaster in Brückner's time, trying to draw a weather map from the sparse observations available (Fig. 2.20). First, the forecaster would draw points on a map showing the

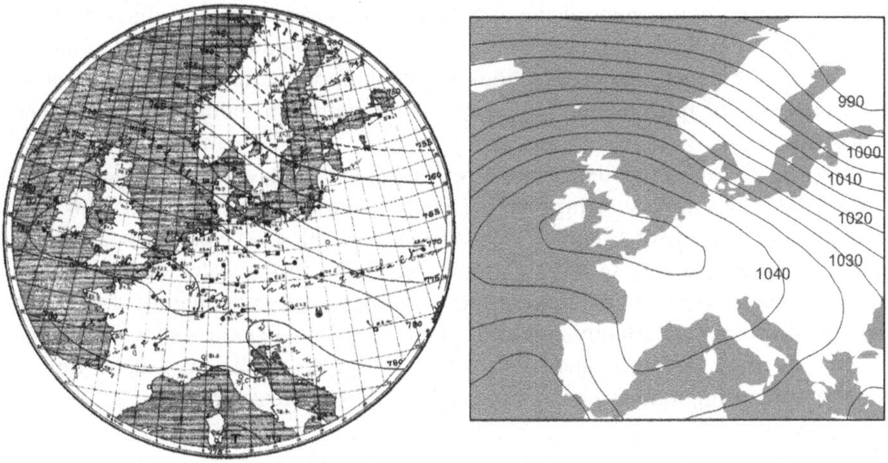

Fig. 2.20 Sea level pressure on 30 January 1896 in (*left*) the analysis of the Deutsche Seewarte (mmHg; 1 mmHg = 1.333 hPa) and (*right*) in the ensemble mean (hPa) of the Twentieth Century Reanalysis (20CR; Compo et al. 2011)

observations that were telegraphed to the weather service. Then, the forecaster would draw isolines of pressure and temperature and perhaps specific features such as fronts. The result for a specific day, 30 January 1896, is shown in Fig. 2.20. Note the high pressure system stretching from Ireland to Normandy, France.

In the process of drawing the map, the forecaster had to incorporate a lot of expert knowledge. First, the meteorologist was an expert on the quality of the observations. He or she knew which stations were reliable, whether they reflect local effects that are not relevant for the synoptic scale, and if so, under which conditions. Second, the forecaster knew the development of the current weather situation (e.g., yesterday's dominating pressure systems and their forecasted change). Third, the meteorologist ideally had long-term experience, encompassing many similar situations in the past, and a basic understanding of atmospheric physics. Only with this additional knowledge could the forecaster draw an accurate map (called "analysis") that was consistent with both the observations and the physics of the atmosphere.

Data assimilation shares the same concept, but is automated and objective. Again, the goal is to produce a map based on all available observations, and the same three sorts of information are necessary: information on the errors and representativeness of observations (in statistical form), the development of the weather situation from the previous analysis (in the form of a forecast with a numerical weather prediction model), and information on the statistics of a weather situation (i.e., how the atmospheric state at one location is related to that at another location; information in the form of the statistics of a weather model). Similar to a historical weather map, the result is called analysis. The weather map that results from this objective and automated procedure, for the same date, is shown in Fig. 2.20. Not only does the assimilation agree with the hand analysis, but the assimilation provides, in three dimensions, the full set of variables typically had by a model.

An essential component of the approach is the model (we will discuss models in more detail in Sect. 2.7), which needs to have high predictability at the time step of interest. Obviously the observations are also important, as was addressed in the previous parts of this chapter. How is the combination of model and observations achieved?

Statistically, the procedure can be regarded as a least-squares problem. Let us consider a very simple system: Imagine a tank filled with water and ice. How warm is the water? Theory (our "model", in the following called x_b) would suggest that the temperature is around 0 °C, but with some uncertainty as the water may not be pure. A thermometer (our observation, termed y) may give us a temperature of 0.1 °C, but again with an uncertainty. Let us assume the errors are Gaussian, with standard deviations of $\sigma_b = 0.2$ °C and $\sigma_y = 0.1$ °C , respectively. The optimal estimate x for the water temperature must minimise the following cost function:

$$J(x) = \frac{(x - x_b)^2}{\sigma_b^2} + \frac{(x - y)^2}{\sigma_y^2} \; . \tag{2.1}$$

Solving this problem is straight forward and leads (in this simple case) to the weighted average of the two estimates:

$$x = \frac{\dfrac{x_b}{\sigma_b^2} + \dfrac{y}{\sigma_y^2}}{\dfrac{1}{\sigma_b^2} + \dfrac{1}{\sigma_y^2}} = x_b + \frac{\sigma_b^2}{\sigma_b^2 + \sigma_y^2}(y - x_b) \; . \tag{2.2}$$

The best estimate for the water temperature is thus 0.08 °C. By a slight rearranging, we can express the estimate as an update to the model that is proportional to the deviation (Eq. 2.2, right-hand side).[16]

We have now found an optimal estimate for one point in time. In data assimilation, the system is propagated in time. Let us assume that the remaining ice is removed and the water slowly takes on room temperature. We can formulate a simple model and calculate temperature for a short interval into the future, but with some uncertainty. When the forecast time is reached, we can make a new measurement and estimate the actual temperature (i.e., we can correct our forecast). Using this estimate as starting point, we can make another short forecast and so on (illustrated in Fig. 2.21). Thus, the model is pushed towards the observations at each assimilation step.

The atmosphere is more complex than a tank of water. The state vector $\mathbf{x_b}$ of a numerical weather prediction model may have 10^9 variables and be confronted with a vector $\mathbf{y}$ of 10^6 observations (from the earth's surface, ships, aircraft, balloons, and

[16]This solution is equivalent to a Bayesian framework, where x_b would be called prior and x posterior.

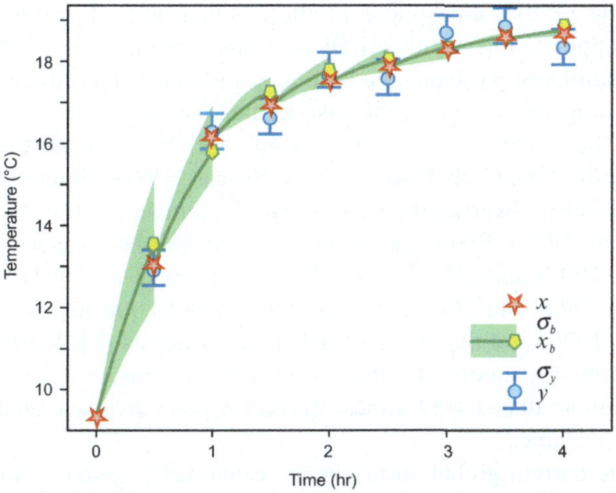

Fig. 2.21 Schematic depiction of data assimilation using the example of a relaxation of temperature in a water tank towards room temperature

satellites; an example is given in Fig. 2.17). The cost function and solution can be generalised to n dimensions (note the similarity to Eq. 2.1):

$$J(\mathbf{x}) = (\mathbf{x} - \mathbf{x_b})^T \mathbf{B}^{-1} (\mathbf{x} - \mathbf{x_b}) + (\mathbf{y} - H[\mathbf{x}])^T \mathbf{R}^{-1} (\mathbf{y} - H[\mathbf{x}]) \; . \tag{2.3}$$

$H[\mathbf{x}]$ is an observation operator that extracts the observations from the model space. Errors are now specified in the form of covariance matrices (**B** for the model and **R** for the observations)—they include the dependence of errors between elements. The solution to the cost function (Eq. 2.3) can be written as:

$$\mathbf{x} = \mathbf{x_b} + \mathbf{B}\mathbf{H}^T (\mathbf{H}\mathbf{B}\mathbf{H}^T + \mathbf{R})^{-1} (\mathbf{y} - H[\mathbf{x_b}]) \tag{2.4}$$

where **H** is the Jacobian matrix (i.e., the matrix of all first-order partial derivatives) of $H[\mathbf{x}]$. Note again the similarity to the right-hand side of Eq. 2.2. The background covariance matrix **B** is essential here as it determines how the information from the observations is spread into the unobserved part of the model space, as deviations in one variable at one location may be correlated with errors in other locations or other variables. Formulation Eq. 2.4 is used in the so-called "Ensemble Kalman Filter", a widely used assimilation technique.

Data assimilation is used heavily for global weather analysis and forecast. Here, $\mathbf{x_b}$ is typically a short-term forecast over 6 hours. However, given the size of the vectors and matrices, the problem can only be approximated. In recent years, numerical techniques such as 4D-Var (four-dimensional variational data assimilation) or the Ensemble Kalman Filter have been developed, providing efficient solutions to this problem.

Analysis datasets are attractive for many applications as they provide a complete (in time and space), comprehensive (a large range of variables), and physically consistent view of the atmosphere in three dimensions. In large projects, the procedure is applied consistently to all past observations using a state-of-the-art model and assimilation system. The output is a global, three-dimensional, 3 to 6-hourly meteorological dataset for the past, termed "reanalysis".

Reanalyses have been extremely successful since the first short reanalysis performed during the First Global Atmospheric Research Programme (GARP) Global Experiment (FGGE) covering the period from December 1978 to November 1979 (Bengtsson et al. 1982). Two longer reanalyses, the ECMWF Re-Analysis ERA-15 (Gibson et al. 1997) and NASA's GEOS-1 (Schubert et al. 1993), built on these efforts. In the 1990s and 2000s, the two most popular reanalyses, NCEP/NCAR (Kalnay et al. 1996) and the 40-yr ECMWF Re-Analysis (ERA-40; Uppala et al. 2005), were arguably among the most widely used datasets in earth science and were cited far more than 10,000 times. In recent years, diverse reanalysis products have become available.

A list of the current global atmospheric reanalyses is given in Table 2.3. Only few institutions or centres are able to undertake such projects and the production cycle is rather long, but several generations of products are available (see also Hartmann et al. 2013, for more information about the following). Substantial development has taken place over the past two decades. These developments involve the assimilation algorithms, the models, the model resolution and complexity, but also the observations and the treatment of satellite data or biases. In addition to the global reanalysis datasets listed in Table 2.3, there are regional reanalysis datasets (e.g., the Arctic Systems Reanalysis, North American Regional Reanalysis, or the EURO4M) and the number of products is growing rapidly. One particular new product that is used widely in this book is the Twentieth Century Reanalysis (Compo et al. 2011), which provides 6-hourly global data back to 1871. It is introduced in more detail in Sect. 2.9.

The strength of reanalyses products lies in providing precise and comprehensive information on scales ranging from weather systems to interannual variability. However, they may not be accurate and may have erroneous long-term trends. Model biases, changes in the observations (e.g., errors and inhomogeneities, coverage, and changes in platforms such as the introduction of satellite data), and errors in the boundary conditions may lead to step changes in time or gradual changes. Even the latest generation of reanalyses suffers from such problems. Although some trends may be well depicted, others may be erroneous and trend suitability needs to be assessed on a case-by-case basis (Dee et al. 2011; Thorne and Vose 2010).

2.7 Climate Models

2.7.1 Characteristics of Climate Models

What will the weather be like on the weekend? Will extreme precipitation events increase in a future climate? How much more CO_2 can we emit? Today, we try to answer many of these scientific questions by using computer models. This is also

Table 2.3 Overview of global dynamical reanalysis datasets after FGGE (see also Blunden and Arndt 2012; Hartmann et al. 2013; www.reanalysis.org), resolution is lat × lon

Institution	Reanalysis	Model	Obs	Assimilation method	Period or start year	Resolution at equator (km)	Reference
NASA	GEOS-1	GEOS, 2° × 2.5°, L20	All	SI	1980–1995	270	Schubert et al. (1993)
ECMWF	ERA-15	IFS T106/L31	All	SI	1979–1993	190	Gibson et al. (1997)
NCEP and NCAR	NCEP/ NCAR R1	NCEP/ MRF, T62/L28	All	SI	1948	320	Kistler et al. (2001)
NCEP and U.S. Dept. of Energy DOE	NCEP/ DOE R2	NCEP/ MRF, T62/L28	All	SI	1979	320	Kanamitsu et al. (2002)
ECMWF	ERA-40	IFS, T159/L60	All	3D-Var	1957–2002	125	Uppala et al. (2005)
JMA	JRA-25	JMA, T106/L40	All	3D-Var	1979	190	Onogi et al. (2007)
NASA	MERRA	GEOS-5, 0.5° × 0.66°, L72	All	Gridpoint SI	1979	75	Rienecker et al. (2011)
ECMWF	ERA-Interim	IFS, T255/L91	All	4D-Var	1979	80	Dee et al. (2011)
NCEP	CFSR	CFS, T382/L64 (coupled)	All	3D-Var spectral SI	1979	50	Saha et al. (2010)
CIRES and NOAA	20CR, vers. 2	NCEP/ GFS, T62/L28	*p*	Ensemble Kalman filter	1871–2011	320	Compo et al. (2011)
ECMWF	ERA-20C	IFS T159/L91	*p* wind	4D-Var	1900	125	Poli et al. (2013)
JMA	JRA-55	JMA, T319/L60	All	4D-Var	1957	60	In preparation

20CR Twentieth Century Reanalysis, *3D-VAR* Three-dimensional variational data assimilation, *4D-VAR* Four-dimensional variational data assimilation, *CFSR* Climate Forecast System Reanalysis, *CIRES* Cooperative Institute for Research in Environmental Sciences, *ECMWF* European Centre for Medium-Range Weather Forecasts, *NASA* National Aeronautics and Space Administration, *GEOS* Goddard Earth Observing System Data Assimilation System, *ERA* ECMWF Re-Analysis, *NCEP* National Centers for Environmental Prediction, *GFS* Global Forecast System, *IFS* Integrated Forecast System, *JMA* Japan Meteorological Agency, *JRA* Japanese Reanalysis, *MERRA* Modern-Era Retrospective Analysis for Research and Applications, *MRF* Medium Range Forecasting model, *NCAR* National Center for Atmospheric Research, *NOAA* National Oceanic and Atmospheric Administration, *p* air pressure, *SI* Statistical Interpolation the case for past climates. Computer models of weather and climate have become very powerful tools. It is therefore important to recall what a model is, how it is used, and how we can learn from it.

At the very basic level, a climate model is an executable computer code (see Müller 2010, for more information on the following). It numerically solves a set of equations. Models reflect our physical understanding of the system, captured in basic physical laws. These laws are expressed as prognostic differential equations, such that a forecast of the state of the system can be made. The equations are evaluated on a discrete space–time grid (or similar discrete representation). Numerical algorithms need to be formulated to solve the equations, and these algorithms need to be converted into computer code, which then is run and produces output. Because the model incorporates our physical understanding of the system, the output can be understood as an interpolation or extrapolation of our knowledge.

For the case of weather forecasts or climate projections, the output of a model is directly the final product. For other applications, it is the analysis of the model output by scientists that generates knowledge in that models provide the scientists with arguments (see Beisbart 2012, for a more detailed discussion). In this sense, it may be helpful to think of a model not as an approximation to reality, but as an abstraction and simplification of reality. This allows us to design experiments using a set of control parameters and established diagnostics, although the applicability to the target system—the real climate—remains uncertain. Thus, simplicity is an important asset of a model.

The basic equations of a weather forecast model were already known during Brückner's time. Vilhelm Bjerknes[17] published the concept of numerical weather modelling in 1904 (Bjerknes 1904):

> "The necessary and sufficient conditions for a rational solution of the problem of meteorological prediction are the following:
>
> 1. One has to know with sufficient accuracy the state of the atmosphere at a given time.
> 2. One has to know with sufficient accuracy the laws according to which one state of the atmosphere develops from another."

Bjerknes then derived the seven basic equations, which are based on the basic conservation laws (conservation of momentum, conservation of mass, and conservation of energy) as well as the ideal gas law. However, he could not find a convenient way to solve these equations. Based on the ideas of Bjerknes, Lewis Fry Richardson (Richardson 1922)[18] undertook the first numerical forecast (of a past weather event) by manual computation, though his result was grossly wrong. He foresaw a "forecast factory" and calculated that 64,000 "computers" (i.e., humans doing computations) would be necessary to keep pace with the actual weather.

---

[17]Vilhelm Bjerknes, 1862–1951, was a Norwegian Meteorologist. He is known for his work on synoptic meteorology and was the founder of the Geophysical Institute in Bergen. Vilhelm Bjerknes is the father of meteorologist Jacob Bjerknes.

[18]Lewis Fry Richardson, 1881–1953, was pioneer in numerical analysis in meteorology, numerical weather prediction, and modelling of conflict (Hunt 1998).

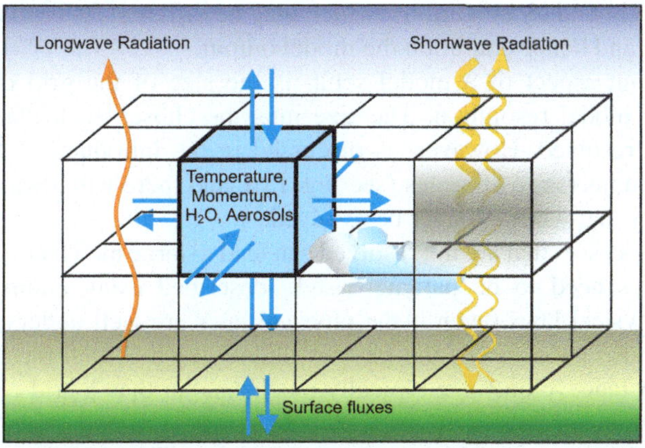

Fig. 2.22 Schematic depiction of a climate model consisting of boxes in a three-dimensional grid

When numerical computers became available after the Second World War, weather forecasting was one of the first applications. In the early 1950s, a team of scientists succeed in the first prediction of the atmospheric state with a simple one-layer model (Charney et al. 1950). Model and, in parallel, computer development rapidly progressed during the following decades (see Edwards 2010). Only in the 1980s, however, did numerical forecast outperform expert analyses. The basic equations of Bjerknes, which can be expressed and solved in various ways—including in a spectral representation[19]—still form the dynamical core of each climate model.

To understand what a numerical model does, Fig. 2.22 shows in a schematic way how the atmosphere is discretised into grid boxes, each of which is attributed quantities of certain properties (e.g., water vapour). These quantities are exchanged over a discrete time step between the boxes, horizontally or vertically. The properties within a box change due to processes within each gird box (e.g., condensation) or due to fluxes between grid boxes (e.g., advection). In addition to the dynamical core, more processes need to be accounted for (e.g., radiation). The figure also shows that a model, like any system, has boundaries. In this case, because the schematic shows an atmospheric model, the land surface is not part of the model. However, it influences the lowest model layer across the model boundary (e.g., due to evapotranspiration). Thus, these fluxes need to be prescribed. Processes that take place outside the boundary and hence need to be specified are called "boundary conditions".

As pointed out by Bjerknes, the model also needs values for each property to start with. These are called "initial conditions". For a given model setup, initial

[19]The spectral representation characterises the horizontal structure with a series expansion of spherical harmonics, whose coefficients are then determined. Truncation (mostly triangular truncation, denoted "T") indicates the highest wave number represented.

conditions and boundary conditions (which may be different for each time step) are the only input and hence condition the model output.

An important aspect of a model setup is the size of the grid cells, or more generally, the model resolution. The size must be chosen such that the relevant processes are resolved. However, as discussed in the introduction, the processes cover a continuous range of scales (see Fig. 1.6), and there will always be smaller-scale processes that will no longer be resolved.

Ignoring processes that are not resolved can lead to large model errors. Therefore, these processes need to be parameterised—described using empirical relations rather than physical laws (even if the physical laws are well understood). Parameterisation is also necessary if the physics underlying a certain mechanism is not known. For instance, clouds are typically much smaller than a grid cell. Saturation may be reached in one area of the grid cell but not nearby. Clouds may therefore start to form before the entire grid box reaches saturation. The model needs to account for that.

Typical resolutions for global climate models that are used for studying the past decades or centuries are shown in Fig. 2.23. The model simulations used in this book are mostly either T30 or T63. The resolution of T30 (denoting a truncation at wavenumber 30 in spectral models, see above) reproduces the large-scale flow, but regional detail is not expected as orography is only crudely depicted, as is the land–sea mask. Note that the actual resolution of a model is lower than the distance between two grid cells because typically more than one cell is needed to characterise a feature (see Laprise 1992). In rough terms, T30 corresponds to a resolution at the equator of ca. 700 km.

We use T30 in simulations that encompass the entire troposphere, stratosphere, and mesosphere with 39 levels. These simulations focus on stratospheric processes

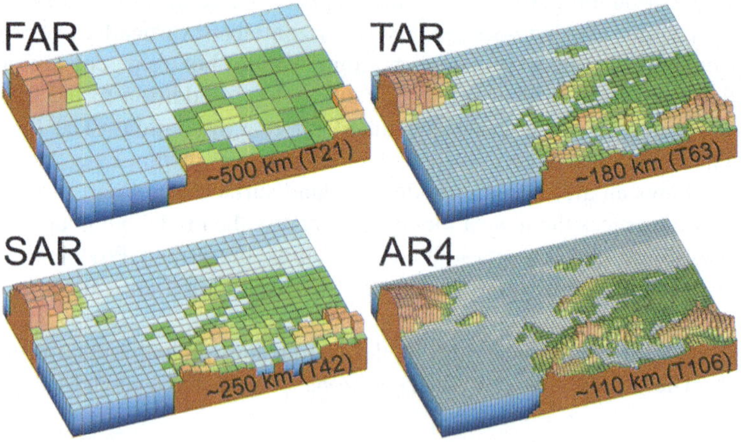

Fig. 2.23 Schematic depiction of climate model resolutions typically used in the first four IPCC Assessment Reports (FAR, SAR, TAR, and AR4) (From Le Treut et al. 2007)

(including chemistry), thus requiring a good representation of the stratosphere. The resolution of T63 (corresponding to around 320 km; Fig. 2.23, top right), with 17 levels in the vertical (model top at 10 hPa), is used in most of the model-based figures shown in the remainder of this book. This setup better captures tropospheric dynamical processes, but still allows multi-century simulations. The finer resolutions are more typically used for simulations on decadal-to-century time scales or for reanalysis datasets.

The vertical resolution plays an important role in, for instance, vertical wave propagation. It is generally believed that a high resolution around the tropopause will more accurately depict dynamical coupling processes and produce better results. Near the surface, terrain-following coordinates are mostly used. However, this requires a smooth orography.

Which aspects of a model setup can limit the insight we get from models? Is it the initial condition, the boundary condition, the model physics, the resolution, or the number of simulations? Obviously this depends in the application.

Bjerknes saw weather forecasting as an initial condition problem. This is certainly true for short forecast times of a few days. After a certain period, the weather can no longer be predicted deterministically. Even the most tiny change in initial conditions can change the outcome of the model completely. Edward Lorenz[20] showed this in the 1960s (Lorenz 1963); a concept that is known by the name "butterfly effect". When attempting to model long time scales, model boundary conditions become important. The result is no longer a deterministic forecast of weather, but is a forecast of the statistics of weather (i.e., climate). The boundary conditions include solar radiation, greenhouse gases, and volcanic aerosols. Depending on the model, sea-surface temperatures or tropospheric aerosols are also boundary conditions—or they are modelled. Thus climate projections, in contrast to weather predictions, are sometimes seen as a boundary condition problem.

As described in Fig. 1.6, oceanic and atmospheric processes proceed on different time scales. Hence, climate may be a boundary condition problem with respect to the atmosphere but an initial condition problem with respect to other components of the climate system such as oceans or the cryosphere. By providing realistic initial values for the ocean in forecast models, scientists try to make seamless predictions from the weather to the climate scale.

Model resolution was long thought to be the most important limitation of model studies. However, a high resolution may not be necessary for all research questions. Recall that models also can be seen as an abstraction or simplification of reality. They should be sufficient to capture the process under study, but not more. This is specifically true for palaeoclimate applications where long periods need to be simulated and computing resources are a limiting factor. Thirty simulations with slight changes of the initial conditions, allowing us to address the randomness of

---

[20]Edward Lorenz, 1917–2008, was an American mathematician and meteorologist, and a professor at the Massachusetts Institute of Technology. Lorenz is the founder of chaos theory.

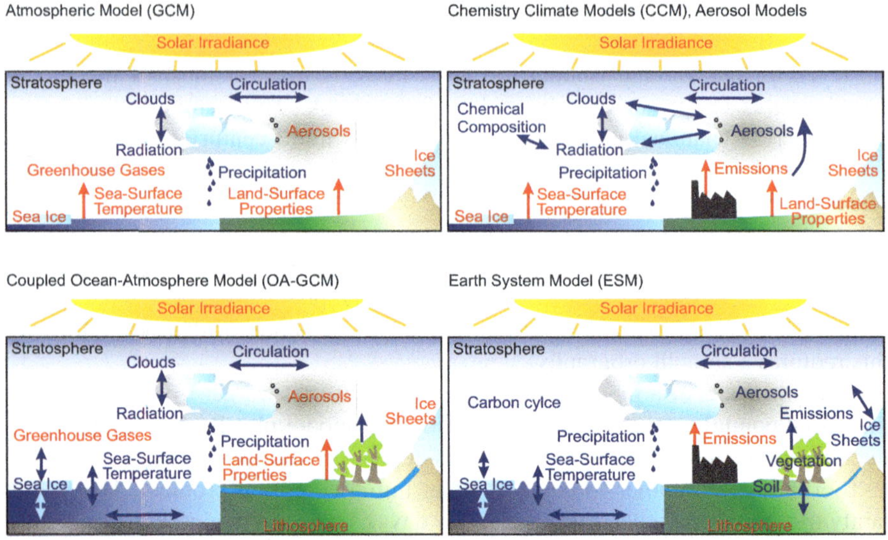

Fig. 2.24 Types of climate models. *Red arrows* and terms denote prescribed properties, *blue arrows* and terms denote model-generated properties

a result, may be more useful than one high-resolution simulation. This strategy is called an "ensemble simulation" and is described in more detail in Sect. 2.7.2.

2.7.2 *Types of Climate Models and Experiments*

Models are tools and should be catered for their purpose. As a consequence, many different model setups have been developed as well as different simulation strategies. The models differ in area covered, resolution, and complexity. Here, I summarise those that cover the atmosphere in three dimensions and that are relevant for this book.

Models can cover the entire globe (global models) or only a part (regional models); the latter need boundary conditions at their lateral boundaries, which typically come from global models. In this book, we mostly deal with General Circulation Models (GCMs) of the atmosphere, shown schematically in Fig. 2.24 (top left; also in Fig. 2.22). In these models, sea-surface temperatures and sea ice are prescribed as boundary conditions. Because 70 % of the world is covered by oceans, these boundary conditions are very determinant. If sea-surface temperatures and sea ice are known, the model output may be comparable on a year-by-year basis to the real climate. However, in reality there is a two-way interaction between the atmosphere and sea-surface temperatures or sea ice. The cost of treating this as a one-way interaction is difficult to quantify. Ocean–Atmosphere GCMs, often just called coupled models (Fig. 2.24, bottom left), resolve the two-way interaction, but the simulated oceans may evolve in a different fashion than in the real world. In other words, we lose the year-to-year comparability of model and real climate. Then, comparisons only make sense statistically, process based (e.g., if specific events such as El Niño are selected from the model and analysed), or on much longer time scales.

Sea-surface temperatures and sea ice are only known from instrumental measurements for about the last 150 years. Therefore, GCMs are only used for this time period or shorter (e.g., Scaife et al. 2009). For palaeoclimatological simulations going further back in time, coupled models are mostly used. Boundary conditions in these models include external forcings from greenhouse gases, volcanic eruptions, solar irradiance changes, volcanic aerosols, tropospheric aerosols, and land surface properties. This book is an exception as we heavily rely on GCM simulations of the atmosphere (i.e., using prescribed sea-surface temperatures and sea ice) back to 1600. They are described in more detail in Sect. 2.9.

In addition to the atmosphere and ocean, other components of the climate system can be added to a model. These models are often modular, and the term "Earth System Models" is often used (Fig. 2.24, bottom right). The modules usually encompass land surface processes or dynamic vegetation, the carbon cycle, or ice sheets. Thus, the models allow us to address feedbacks among these components. Models with a very low resolution of the atmosphere (Earth System Models of Intermediate Complexity (EMICs)) are sometimes used in order to focus more efficiently on slower parts of the climate system (e.g., ice sheets or vegetation changes).

Finally, atmospheric chemistry and aerosol microphysics can be added (Fig. 2.24, top right). Chemistry–climate models (CCMs) incorporate gas phase and heterogeneous chemistry of the atmosphere and are able to reproduce, for instance, changes in the ozone layer (CCMVal SPARC et al. 2010). Aerosol models (e.g., HAM2, which is coupled to the GCM ECHAM6; Stier et al. (2005)) incorporate the basic chemical and microphysical processes of aerosols and cloud formation, which, at least until a few years ago, were missing from most coupled climate models. Simulations with chemistry–climate models and aerosol models are also used in this book to explore the past ca. 100 years.

With these different models, a number of experimental setups can be used. One of the first simulations done by all models is called a "control simulation", that is, a simulation with constant boundary conditions. This simulation provides a quality check. Does the model explode or drift? Is it realistic? If the model passes this test, the control can be used as a reference for all other simulations. The control simulations also allow us to study the internal climate variability of the model. However, in many cases we are less interested in the model's internal variability. Therefore, we apply forcings and the boundary conditions are no longer held constant. For instance, all boundary conditions can be specified realistically and in a time-dependent manner (transient simulations). These simulations are sometimes called "all-forcings simulations". For atmospheric models, the term "AMIP-type simulations" is also used and refers to the Atmospheric Model Intercomparison Project (Gates 1992) which developed protocols for this type of simulation. Similarly, the Coupled Model Intercomparison Project (CMIP) developed protocols for coupled ocean–atmosphere models, and the Paleoclimate Modelling Intercomparison Project (PMIP) assisted with palaeoclimatological simulations (Braconnot et al. 2012).

All-forcings simulations may produce a climate that is more realistic and comparable to the real climate than the control. However, we do not know which of the forcings was responsible for this. Therefore, some of the external forcings are varied individually and compared to the control or to simulations in which all forcings are specified. These simulations are often termed "sensitivity simulations". Detection and attribution approaches (e.g., Hasselmann 1993) can then be used to relate the observed climate variable to a linear combination of the modelled climates.

Sensitivity runs can also be performed for individual regions. For instance, sea-surface temperature may specified according to observations in one ocean basin, but set to a climatological seasonal cycle elsewhere, or they may be specified in one area and a coupled ocean used elsewhere (e.g., Kosaka and Xie 2013). This can help to identify the origin of the forced variability (conversely, some argue that sensitivity simulations may be subject to inconsistencies in energy fluxes).

To quantify the effect of internal atmospheric variability on the outcome of a climate simulation, an ensemble of simulations can be run—each simulation with a slightly different initial condition. The simulations can then be compared to the real climate. If the real climate lies outside the ensemble spread for a certain feature, we can say that the model does not reproduce this feature. If it lies inside the spread, the model is consistent with the real climate. If the ensemble mean also shows this feature (or differs from an ensemble with other boundary conditions), then it is said to be "forced" by the boundary conditions. For the sake of completeness, it should be mentioned that ensembles can also be performed to sample uncertain parameters or boundary conditions rather than initial conditions.

In this book, we will heavily use an initial condition ensemble of 30 atmospheric model simulations of the past 400 years from an all-forcings setup. In addition, several sensitivity experiments have been performed with this model. The simulations are further introduced in Sect. 2.9.

For complex models consisting of several modules, many other types of setups and experiments can be used and compared. For instance, different parts of the model can be run coupled or uncoupled ("offline"). For instance, the chemistry module of a chemistry–climate model (termed "chemistry–transport model") can be run offline using data from a climate model or a reanalysis. These simulations obviously do not include feedbacks. For example, the modelled trace gases will not affect the radiation in the climate model. So, comparing the uncoupled simulations with coupled simulations may give hints as to the feedbacks involved.

"All models are wrong, but some are useful" is a common expression. Indeed, models can produce many misleading or wrong results. Most models have biases that need to be corrected before they can be analysed. Therefore, all models need to be evaluated against observations. The evaluation targets the model climate, its variability, and the representation of relevant processes (process-oriented validation). For palaeoclimatological studies, the evaluation of variability is often restricted to the comparison with proxies.

2.7.3 Downscaling, Nudging, and Other Techniques

Global models and reanalyses correctly represent large-scale circulation and synoptic weather systems, but only depict topography in a very crude form (see Fig. 2.23). In the Twentieth Century Reanalysis, for example, the Alps are a smooth, 1000-m high hill. However, variables such as precipitation and wind at a given weather station depend strongly on local topography. In order to derive local weather or climate information, downscaling is used (often in the context of climate scenarios, but the same methods are also applied for past climate). Statistical downscaling uses transfer functions calibrated against observations in order to obtain information on a small scale. Many different methods are used, some of which are similar to those discussed in the context of interpolating precipitation in Sect. 2.6, except model biases also need to be considered.

Alternatively, regional weather models can be used in a similar way as in operational weather forecasting to downscale a global model output to a smaller region in high resolution. The regional model is nested into a larger domain—it takes its lateral boundary conditions from another model. This method is termed "dynamical downscaling". The same methods can be used to obtain local weather from the past. It often requires several steps to bridge the gap from the global scale to the that of weather stations. The example shown in Fig. 2.25 is wind speed at 10 m above ground for a storm in Switzerland, 23 February 1935 (see also Sect. 4.5.4). Few topographic details are depicted in the global field, which is from the Twentieth Century Reanalysis. The model depicts the general flow over the Atlantic and the land–sea differences in wind speed due to changes in surface roughness. The Alps as an obstacle in the flow are clearly visible in the first downscaling step (here using the Weather Research and Forecasting (WRF) model; Skamarock et al. (2005)). The second step already shows high winds in the Swiss Plateau, between the Alps and the Jura mountains. The third nesting (right) depicts individual valleys at a 3-km

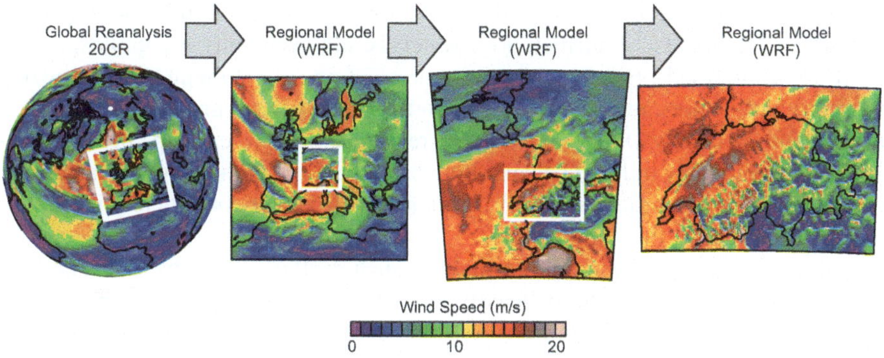

Fig. 2.25 Dynamical downscaling of a global model to the local scale using a regional model. Here, a storm in 20CR at 12 UTC 23 February 1935 is downscaled using the WRF numerical model, the variable shown is 10 m wind speed. WRF simulations were computed by Silke Dierer

resolution. Other than topography and land surface information, downscaling does not add additional information. However, the model output is much more realistic. Downscaling will be used later in this book to quantitatively study past storm events with a high resolution back into the 19th century (Sect. 4.5.4).

In the previous paragraph, we discussed offline simulations and how, for instance, chemistry–transport models can be driven by climate model outputs. When trying to mimic observed chemical composition, chemistry–transport models can also be used with reanalysis data, although no feedbacks are taken into account. One way of allowing such feedbacks, at least to some extent, is *nudging*. Here, the flow in a climate model is relaxed towards the circulation of a reanalysis or another model simulation by adding small tendencies at each time step (see a further discussion in Sect. 2.8.5). Only the most basic flow fields are nudged, otherwise the model is free and thus can produce feedbacks, for instance, with cloud formation and radiation. In this book, we will use simulations with an aerosol-enabled climate model that was nudged to a reanalysis (see Sect. 2.9 for a detailed description).

2.8 Palaeoclimate Information and Techniques

2.8.1 Climate Proxies

To study the climate of the early 18th century or to investigate multidecadal variability, instrumental series are usually insufficient. Long climate records are required that reach beyond the limited spatial and temporal availability of measurements. In 1890, when Brückner wrote his book, this was even more the case than today because observed series were short. Even the longest temperature series were only ca. 130 years long and precipitation series were even shorter. To establish the existence of a 35-year climate cycle, Brückner had to consult additional sources of climate information such as lake levels. Lake levels are related to the regional water balance and hence precipitation and evaporation. However, the relation is indirect (i.e., other factors also contribute to lake levels and eventually a complex system needs to be analysed). Brückner also used descriptions of cold winters, the freezing of rivers, and—after carefully checking neighbouring series—grape harvest dates, a method introduced by Dufour (1870). The relationship between grape harvest dates and climate is highly indirect because both biological and anthropogenic factors are contributors. Using these sources of information, Brückner constructed time series that reached back to the 16th century.

Indirect sources of climate information, such as those used by Brückner, are called "proxies" and are used heavily in climate research. In this chapter, we describe proxies and products based on proxies that are used in the study of past climate variability.

Climate proxies can come from "archives of nature" (see Table 2.4, adapted from Christian Pfister). They record and preserve climate-related signals through biological, chemical, and physical mechanisms. Well-known archives of biological

Table 2.4 Types of proxy data (Adapted from Pfister 1999)

	Archives of nature				Archives of society		
Direct					Weather observations, event descriptions, instrumental measurements		
Indirect	*Biological*		*Non-biological*		*Biological*	*Non-biological*	*Cultural and religious*
	Archive	Proxy	Archive	Proxy	Phenology (plant phases), animals, harvest dates and yields	Lake levels, frozen rivers and lakes	Religious ceremonies, archaeological sources
	Tree rings	Ring width, late wood density, $\delta^{18}O$	Ice cores	Air composition, $\delta^{18}O$ of ice, snow chemistry			
	Lake sediments	Pollen, microbial life	Lake sediments	Mineralogy, geochemistry, grain size			
	Terrestrial sediments	Fossil plants	Terrestrial sediments	Mineralogy, geochemistry, grain size			
			Speleothems	$\delta^{18}O$			

processes are tree rings, peat bogs, and lake sediments. Archives of non-biological processes include ice cores, speleothems, and terrestrial sediments. Another source of non-instrumental climate information are "archives of society", mostly consisting of written sources or archeological evidence. Written sources can contain direct observations of weather elements, or indirect information such as the grape harvest dates. This special type of proxy data is often called "documentary data" (Fig. 2.26).

The use of documentary and other proxy data has a long tradition. Before the era of instrumental data, they were the only source of climate information, not only for chronists, but also for science. In the 18th century, climatic changes (or lack thereof) in the Mediterranean since the classical epoch were analysed based on descriptions in classical writings (Mann (1790); see also Ideler (1832)).

Reported changes in land use, the length of glaciers, or the presence of erratic blocks were used in the early 19th century to address the recent and historical climatic deteriorations in the Alps (see also Box 2.2, p. 22). Astronomer William Herschel[21] published a paper in 1801 relating sunspots and climate using wheat prices as a proxy for climate (Herschel 1801).

[21] William Herschel, 1738–1822, was a British–German astronomer and composer. He discovered Uranus and several moons.

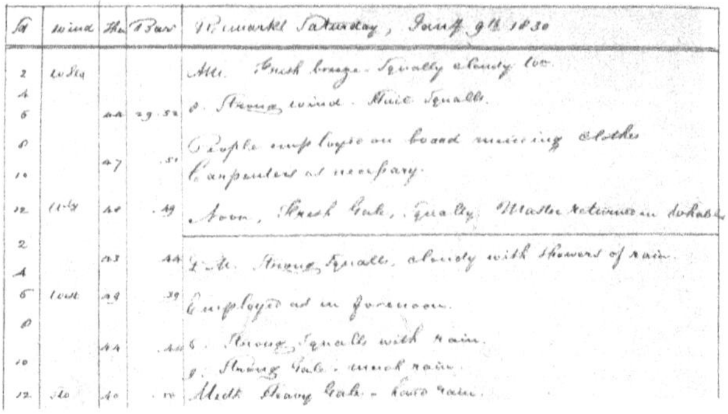

Fig. 2.26 Excerpt from the Captain FitzRoy's log of the HMS Beagle, 9 January 1830 (see also Fig. 2.4) (Source: the CORRAL Project: http://badc.nerc.ac.uk/data/corral/index.htm)

Today, documentary data have become an important source of information for climate research (e.g., Brázdil et al. 2005; Glaser 2008; Pfister 1999). Their strength lies in their exact dating and—in some cases—their abundance. In previous centuries, people were specifically interested in, and therefore documented, extreme events and climate impacts. These areas have returned to the focus of climate science. Hence, documentary data deliver information for those situations in which we are most interested. On the other hand, they often need to be transformed into quantitative information on climate variables before comparisons with other datasets or models or statistical analyses can be undertaken. Moreover, they may often not be suitable for trend analyses.

Documentary sources often concern the water cycle, which was relevant for the predominantly agrarian societies in the preindustrial epoch. They also concern plant life, such as grape harvest dates or wheat yields. Plant phases were often systematically observed and reported. These phenological data can be used for climate and climate impact research (Rutishauser et al. 2008). Even today, their collection finds increased popularity not only for science, but also for educational purposes and awareness raising (e.g., OpenNature, http://www.opennature.ch/).

Plant species distribution in unmanaged environments also is a source of information. Natural archives such as lake sediments preserve pollen and macrofossils from the surrounding region and can allow for palaeoecological reconstructions (e.g., Wick and Tinner 1997). Lake sediments can also contain other proxies that more directly record temperature changes. For instance, remnants of chironomids (non-biting midges) and diatoms in high-altitude Eurasian lakes have been used as a proxy for past temperature (Trachsel et al. 2008). Mass accumulation rates and turbidites can give information on floodings or glacier changes (e.g., Stewart et al. 2011).

Probably the most widely known natural biological archive are tree rings—the annual growth layers of tree stems (Fig. 2.27). Tree rings are also widely used in the remainder of this book. Several proxies can be derived from tree rings. These include ring width, maximum late wood density, and isotopic composition (e.g.,

Fig. 2.27 The width of annual growth rings of trees is one of the most widely used climate proxies

Fritts 1976). Several climate variables—such as temperature, precipitation, and light—may influence tree growth. In addition, there can be non-climatic limiting factors. These include pests and diseases, nutrients, species competition, and fire.

The response of a tree is often assumed to depend on the most limiting factor (Liebig's principle). Therefore, field sites are chosen such that the most limiting factor is clearly known and non-climatic signals are suppressed. For instance, at a high-altitude location such as Bergün in the Swiss Alps, growth is assumed to be always limited by temperature. Conversely, in a semi-arid basin, moisture availability limits growth. However, the growth response is still that of a complex biological system, with many other factors such as pests, species competition, or soil playing a role. Tree rings usually do not have dating errors. With respect to climatic changes since 1700, tree rings are a key source of information because hundreds of chronologies are available worldwide.

An example of non-biological archives are ice cores from mountain glaciers, Arctic glaciers, or ice sheets (for a historical overview, see Jouzel (2013)). Many different proxies can be derived from ice cores. First, ice cores are archives of snow with all its properties, ranging from its isotopic composition to its abundance. Second, ice cores record contamination of snow (e.g., aerosols). Third, air bubbles trapped in the ice preserve the composition of the past atmosphere. Historical concentrations of trace gases such as CO_2, nitrous oxide (N_2O), or methane (CH_4) can be determined from ice cores by sampling the bubbles. Many polar ice cores were drilled with the aim of reconstructing conditions as far back as possible. These are typically low-accumulation sites with low time resolution, which makes them less useful for studying climatic changes since 1700. Other ice cores from high-accumulation sites give a more detailed view of our period.

Ice cores can also be drilled from mountain glaciers, which give a more regional view of past climate, albeit not back to glacial times. Ice cores are studied in the

Alps (e.g., Eichler et al. 2000; Wagenbach 1989), Svalbard (e.g., Isaksson et al. 2001), the Rocky Mountains (Moore et al. 2002), the tropical Andes (Ramirez et al. 2003), the Altai (Olivier et al. 2003), and even Kilimanjaro in Africa (Thompson et al. 2002). However, dating to the exact year is often not possible.

Long time series from a large range of proxies were recently used to compile continental climate reconstructions within the initiative PAGES2k of the "Past Global Changes" programme (Ahmed et al. 2013). These reconstructions provide a first global view of continental climatic changes over the past two millennia.

2.8.2 Problems Arising When Working with Proxies

There are many pitfalls when working with climate proxies, and one has to be aware of their limitations. Arguably the most important limitation is our incomplete understanding of the proxies. If we do not understand the underlying mechanisms, we may misunderstand or misinterpret the signal we find in proxies. Another problem that is common to many proxies is dating. Dating adds an additional dimension of errors to those listed in Table 2.2. For documentary data, we need to consider human perception and the social system in which the observer was involved—work that needs to be done by trained historians.

Other areas of uncertainty are similar to those listed in Tables 2.1 and 2.2. Obviously the measurement itself (e.g., ring width) introduces an error. Further errors are introduced by the pre-processing of the data, which aims at removing non-climatic signals. For instance, tree ring width depends on the age of the tree, and removing this age trend can be done in various ways. Similar processing steps apply to other proxies (e.g., thinning of ice layers with depth).

Another area of uncertainty is related to the signal preservation. For instance, even if $\delta^{18}O$ measured in an ice core[22] was a perfect proxy for local air temperature, it might only contain limited information about annual-mean temperatures because the ice core only preserves information when it snows. The more snow that a weather system delivers, the more weight that it has in the $\delta^{18}O$ record. Heat waves are unnoticed, as are bitter cold but dry continental winters. As a consequence, changes in the seasonality of precipitation may be confounded with temperature changes (e.g., Laepple et al. 2011; Persson et al. 2011; Sime et al. 2011; Werner et al. 2000). In an example from our own work, we found excellent agreement between $\delta^{18}O$ in quasi-annual layers of an ice core and corresponding (quasi annual) precipitation-weighted temperature averages from a meteorological station at 50 km distance (Fig. 2.28, top). However, no correlation with annual-mean temperature was found. In fact, a global map of the correlation between annual-mean temperature and precipitation-weighted temperature (Fig. 2.28, bottom; calculated over calendar

[22] A temperature-dependent fractionation of water molecules with light or heavy oxygen isotopes (^{18}O and ^{16}O) occurs during evaporation and condensation. This effect can be used for temperature reconstruction.

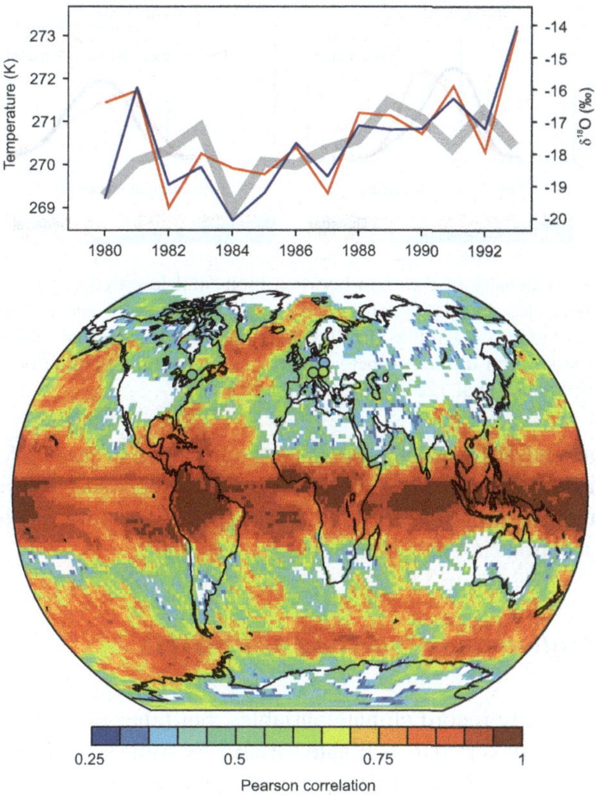

Fig. 2.28 (*top*) Quasi-annual values of δ^{18}O in the ice core of the glacier Grenzgletscher (*blue*), corresponding quasi-annual precipitation-weighted temperature averages (*red*), and annual-mean temperature averages (*thick grey line*) from the Great St Bernard station. (*bottom*) Map of the Pearson correlation coefficient between annual-mean temperatures and precipitation-weighted temperatures (calendar year) over the period 1979–2012 calculated from the ECMWF Interim Re-Analysis (ERA-Interim). *Circles* denote correlations between annual-mean station temperatures and δ^{18}O in precipitation. Only correlations that are significant ($p < 0.1$), accounting for autocorrelation, are shown (From Brönnimann et al. 2013c)

years) shows a rather bad agreement in the extratropical continents. Similar issues apply to other proxies. Note that these problems are not errors, but are due to the fact that proxies do not always record what we would like them to record.

Another uncertainty is related to the time scales resolved. Archives may react differently on different time scales. For instance, trees may be able to cope with a single cold or dry year, but may suffer much more during dry or cold decades. Thus, tree rings may exhibit a higher ratio of low- to high-frequency variability than climate itself (see Fig. 2.29; Franke et al. (2013)). The spectrum becomes more "red".[23] Moreover, observed differences between the spectral characteristics

[23] A spectrum in which all frequencies have the same power is called white, if low frequencies dominate it is called red, and if high frequencies dominate it is called blue.

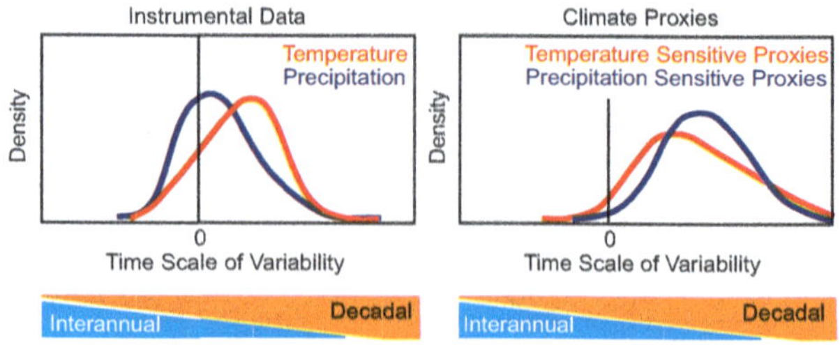

Fig. 2.29 Probability density of the ratio between high- and low-frequency variability in (*left*) instrumental temperatures and precipitation and (*right*) temperature- and precipitation-sensitive proxies. Proxies tend to have a red bias (i.e., they exhibit too much decadal relative to interannual variability), and the spectral difference between temperatures and precipitation is lost (idealised schematic based on Franke et al. (2013))

of precipitation and temperatures (usually temperature is more "red"—this will be discussed in more detail in Sect. 3.2.1) are not reproduced when comparing temperature- and precipitation-sensitive tree rings. Accounting for these factors in climate reconstructions is very challenging.

2.8.3 Calibration and Modelling of Proxies

Proxies do not directly record climate variables, but other properties such as $\delta^{18}O$ that first need to be converted to a climate variable by some sort of calibration. In the ideal case, if the proxy–climate relation is well understood and reproducible in the laboratory, it may be possible to use an experimentally derived calibration. In most cases, however, calibration is a statistical procedure and involves the comparison of the proxy time series with times series of observed climate data (calibration-in-time).

Many different statistical concepts and techniques are used, but the basic goal of them is to find a function that translates the signal in the proxy p to a signal in the climate data x. The classical approach is regression:

$$x = f(p) + e_x, \tag{2.5}$$

where the residual e_x can be autocorrelated. In reality, climate is not a function of the proxy but vice versa. Reverse regression models reflect this:

$$p = f(x) + e_p, \tag{2.6}$$

where the inverse of f provides the desired function. The functions relating proxies to climate are termed "transfer functions". A least-squares regression or robust regression methods are often used to fit the functions empirically, but many other statistical concepts can also be used, including neural networks or Monte Carlo techniques.

Different methods result in different properties of the reconstructed series. A least-squares regression may provide a best estimate of the mean, but it will underestimate the variance by construction. Other methods may provide a realistic variance, but larger errors for the individual time step.

Evaluating the skill of a reconstruction method within the calibration period is always optimistic because some over-fitting inevitably occurs. Thus, the transfer function needs to be validated in an independent period. Often, the period of overlap between proxies and climate data is subdivided into a calibration and a validation interval. In other words, function parameters are estimated in period 1 (the calibration) and then a reconstruction is performed for period 2 and compared with climate data (the validation). Cross-validation techniques (e.g., the leave-one-out method) are also often used. Other calibration approaches make use of spatial (e.g., altitudinal) gradients in place of temporal statistics ("calibration-in-space").

The time window for calibration and validation is often limited by the availability of instrumental data. As a consequence, the resulting model can only reconstruct the range of variability that is found within the calibration period. Furthermore, the quality of instrumental data is essential and inhomogeneities in the observations may affect the results (see Sect. 2.5.2).

Application of a transfer function, derived using a certain calibration period, to a past period implies the assumption of stationarity of this function (i.e., that the proxy–climate relationship is stable in time). However, limitations of a proxy may change over time, which would contradict the stationarity assumption. Splitting the available record into a calibration and validation period is of limited help, especially if the two periods are climatically similar. Stationarity issues may still occur because during reconstruction the transfer function is often extrapolated to unobserved parts of the phase space (e.g., very cold climates). Many statistical approaches suffer from a bias of the reconstructed series towards the mean of the calibration period.

A different approach of comparing proxies with climate data is to use proxy forward models, which have been developed in recent years for various proxies such as tree rings or corals (e.g., Tolwinski-Ward et al. 2011). In these approaches, the proxy is modelled from climate data taking into account the most important mechanisms (ranging from fully physical or physiological models to semi-empirical models). Forward models require the calibration of some parameters and hence there is also a stationarity assumption here, but it is much less restrictive than a purely statistical model because the calibrated parameter is close to the actual mechanisms. Although proxy forward models do not provide climate reconstructions, they can be used, for instance, in data assimilation approaches (see Sect. 2.8.5). A different way of forward modelling is the explicit representation of processes in GCMs. Some GCMs have oxygen isotope schemes or aerosol schemes that then allow comparison with ice core proxies (see Sturm et al. 2010).

An example of forward proxy modelling is given in Fig. 2.30 (from Breitenmoser et al. 2014). In this case, tree ring widths from 2287 chronologies were modelled with the Vaganov-Shashkin Lite model (Tolwinski-Ward et al. 2011) using global monthly climate data (temperature and precipitation) from an observation-based dataset covering 1901–1970. The model requires the calibration of five growth

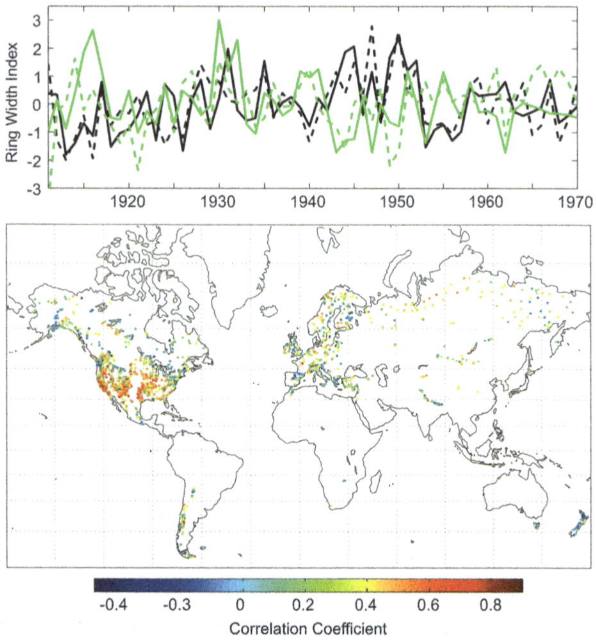

Fig. 2.30 (*top*) Observed (*solid*) and modelled (*dashed*) tree ring widths for the Swiss sites (*black lines*) Bergün Val Tuors (European larch, 2065 m a.s.l., temperature limited, $r = 0.69$, $p < 0.01$) and (*green lines*) Krauchthal (Scots pine, 550 m a.s.l., moisture limited, $r = 0.44$, $p < 0.01$). (*bottom*) Correlations between modelled and observed tree ring widths for the 1901–1970 period for 2287 chronologies (see Breitenmoser et al. (2014), for details)

parameters (see Breitenmoser et al. 2014). Results for two sites in Switzerland are shown in Fig. 2.30 (top) and a map of correlations between modelled and observed series of tree ring widths is shown in Fig. 2.30 (bottom).

Correlations are in a similar range as those found by direct comparison of tree ring widths with seasonally averaged temperature or precipitation. However, forward models have several advantages. For instance, they at least semi-explicitly represent the underlying mechanisms. The estimated parameters have a physical meaning (such as the lower temperature threshold for growth initiation), whose stationarity can more safely be assumed. They may even capture a change in the dominant limitation of growth. A subset of these 2287 chronologies is used later in this book.

2.8.4 Climate Reconstructions

Calibrated proxies provide estimates of one climate variable at one location. However, for studying climatic changes since 1700, we may be interested in a large-scale view of the climatic state. How can we reconstruct Northern Hemisphere–mean temperatures or even spatial fields of climate variables back to 1700? Traditionally, statistical reconstruction techniques have been used to derive past states. The problem is similar to the calibration problem for individual proxies described above, except that now we deal with many proxies and possibly many climate variables.

For the case of the Northern Hemisphere–mean temperature series, the most simple reconstruction technique is called "composite plus scaling". Standardised proxy time series are averaged and the resulting time series (composite) is scaled to have the same variance as the Northern Hemisphere temperature series. Another way is regression, similar to proxy calibration (see Eq. 2.5):

$$x = f(p_1, p_2, p_3, \ldots, p_n) + e_x. \tag{2.7}$$

Again, different methods, estimators, and noise structures (etc.) are in use. Northern Hemisphere temperatures can also be derived by averaging climate field reconstructions, which is discussed in more detail below. There is abundant literature about the methods used for climate reconstruction, especially in the case of Northern Hemisphere–mean temperatures (e.g., Frank et al. 2010; Juckes et al. 2007; Ljungqvist 2010; Mann et al. 1999; Moberg et al. 2005; Smerdon 2012; von Storch et al. 2004). In fact, reconstruction methods have become the spotlight of the recent palaeoclimate debate.

One specific and often-discussed methodological aspect refers to the preservation of variability on different time scales (e.g., Moberg et al. 2005; von Storch et al. 2004). We have already seen that proxies themselves might not be able to represent the variability at all time scales in the same way (Franke et al. 2013). Reconstruction methods add a further complication. Because the climate signal and proxy noise may have different spectral characteristics that may be inadequately captured in the statistical approach, systematic biases may occur, for instance, when calibrating a statistical approach based largely on interannual variability but then analysing centennial variability. This is illustrated schematically in Fig. 2.31. We assume that the proxy is recording a climate signal plus a noise. The climate signal is assumed "white" in the case shown on the left but "red" in the case on the right and vice versa for the noise. Most reconstruction methods essentially perform a linear scaling of the time series (green lines). The figure shows that if the spectral characteristics of proxy and climate are not the same, scaling will lead to an over- or underestimation of low-frequency variability. Traditional reconstruction methods are thought to suffer from an underestimation of low-frequency variability (von Storch et al. 2004) and more complex methods may need to be used.

To test reconstruction techniques, climate model information is increasingly used. Proxies are produced from climate model output by extracting a times series and perturbing it with a known noise structure (termed "pseudo proxies"). Because the truth is known (i.e., the unperturbed, complete model state), this allows us to test the behaviour of our methods (Smerdon 2012).

For climate field reconstructions, the above applies but many climate variables are predicted. Often, the number of proxies and particularly the number of variables to be reconstructed (e.g., temperature at different grid points) is large compared to the number of proxy observations (time steps) overlapping with instrumental observations. Therefore, Principal Component Analysis (PCA) is often used as a first step to reduce the number of variables and the number of proxies (see also Reduced Space Optimal Interpolation introduced in Sect. 2.5). PCA provides

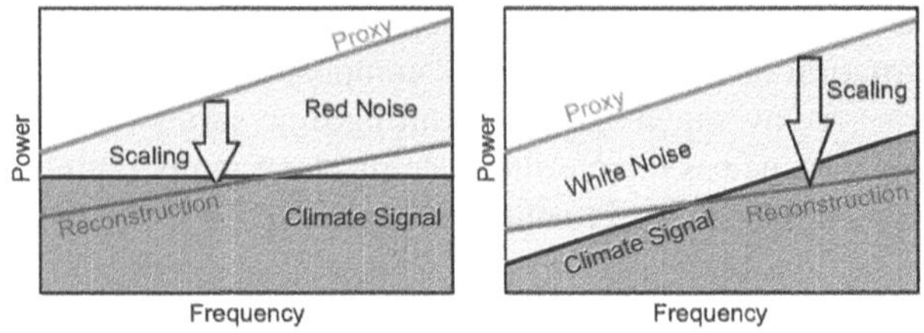

Fig. 2.31 Schematic of how the calibration of reconstructions can bias the spectrum of the reconstructed series (Adapted from Osborn and Briffa 2004)

transformed variables (z in the following) that are independent from each other (uncorrelated) and where each is a linear combination of all original variables. If the original variables are grid cells in a field, the coefficients of the linear combination can be interpreted as a spatial pattern a pertaining to each transformed variable z. The first principal component (PC) explains the largest fraction of the total variance and each subsequent PC explains the maximum fraction of remaining variance, such that usually a few PCs are sufficient to capture most of the variability. The calibration of the reconstruction then becomes a set of regression equations (where subscripts x and p refer to climate variables and proxies, respectively):

$$\begin{aligned} z_{x.1} &= f_1(z_{p.1}, z_{p.2}, z_{p.3}, \ldots, z_{p.n}) + e\ , \\ z_{x.2} &= f_2(z_{p.1}, z_{p.2}, z_{p.3}, \ldots, z_{p.n}) + e\ , \\ &\ldots\ , \\ z_{x.m} &= f_m(z_{p.1}, z_{p.2}, z_{p.3}, \ldots, z_{p.n}) + e\ , \end{aligned} \tag{2.8}$$

Note that m and n are much smaller than the original dimensions.

Once calibrated, the functions f are used to calculate all z_x in the reconstruction period, for which $z_{p.1},\ z_{p.2},\ \ldots,\ z_{p.n}$ can be derived from the proxies. The field reconstruction (here expressed as a state vector $\hat{\mathbf{x}}$) for a given time can then be expressed as a linear combination of a vectors containing the reconstructed transformed variables, $\mathbf{z_x}$, and the matrix of the patterns $\mathbf{A_x}$ from the calibration period:

$$\hat{\mathbf{x}} = \mathbf{z_x A_x}\ . \tag{2.9}$$

As for the calibration and time series reconstruction, many other methods can be used, and there is abundant literature on these methods. Again, the problem of stationarity applies. It is even more relevant here, because not only the regression coefficients, but also the PCA patterns are assumed constant. This assumption is crucial—and may not hold. Lehner et al. (2012) found that reconstructing a climate variability mode such as the North Atlantic Oscillation from two proxy time series may easily fail if variability patterns change slightly.

Climate field reconstructions have become widely used and are also presented in this book. In addition to global annual-mean temperature field reconstructions (Mann et al. 2008, 2009b), several regional products are available for the period since 1700 covering Europe (Luterbacher et al. 2004; Pauling et al. 2006), Asia (Cook et al. 2010), North America (Cook and Krusic 2003), and South America (Neukom et al. 2011)—some providing monthly or seasonal resolution and multiple variables. Climate field reconstruction techniques were also used by our group to reconstruct monthly global three-dimensional circulation back to the 19th century (Brönnimann et al. 2012b; Griesser et al. 2010). Some of these reconstructions are used later in this book.

2.8.5 *Data Assimilation in Palaeoclimatology*

Data assimilation methods, that is, the method of combining observations with model forecasts to obtain an estimate of the atmospheric state, are heavily used in atmospheric sciences and weather forecasting (Sect. 2.6.2). Can they also be used for state estimation in palaeoclimatology? The Twentieth Century Reanalysis (Compo et al. 2011) extends the "traditional" reanalysis approach as far back as 1871, but there is a limit to further extension because state observations are needed. Proxies do not capture atmospheric states, but time-integrated functions of states (averages, in the most simple case). This requires other assimilation techniques that allow assimilation time steps to be very long. Moreover, models must have predictability over this time step, which is much longer than the predictability of atmospheric weather.

Efforts are currently underway in the area of palaeoclimate data assimilation (see Hakim et al. 2013) that may change or at least supplement our way of working in the future. The approaches that are currently developed can be divided into three categories: "Classical" covariance-based approaches such as the Kalman Filter or variational techniques (similar to the techniques shown in Sect. 2.6.2), approaches based on analogues such as Particle Filters, and nudging techniques see (see Brönnimann et al. 2013b, for more information about the following). A schematic view is given in Fig. 2.32.

Recall that the assimilation problem can be formulated as a cost function, assuming Gaussian probability distributions (Eq. 2.3, see Sect. 2.6.2):

$$J(\mathbf{x}) = (\mathbf{x} - \mathbf{x}_b)^T \mathbf{B}^{-1} (\mathbf{x} - \mathbf{x}_b) + (\mathbf{y} - H[\mathbf{x}])^T \mathbf{R}^{-1} (\mathbf{y} - H[\mathbf{x}]) \,. \quad (2.10)$$

The solution to which can be written as (Eq. 2.4):

$$\mathbf{x} = \mathbf{x}_b + \mathbf{B}\mathbf{H}^T (\mathbf{H}\mathbf{B}\mathbf{H}^T + \mathbf{R})^{-1} (\mathbf{y} - H[\mathbf{x}_b]) \quad (2.11)$$

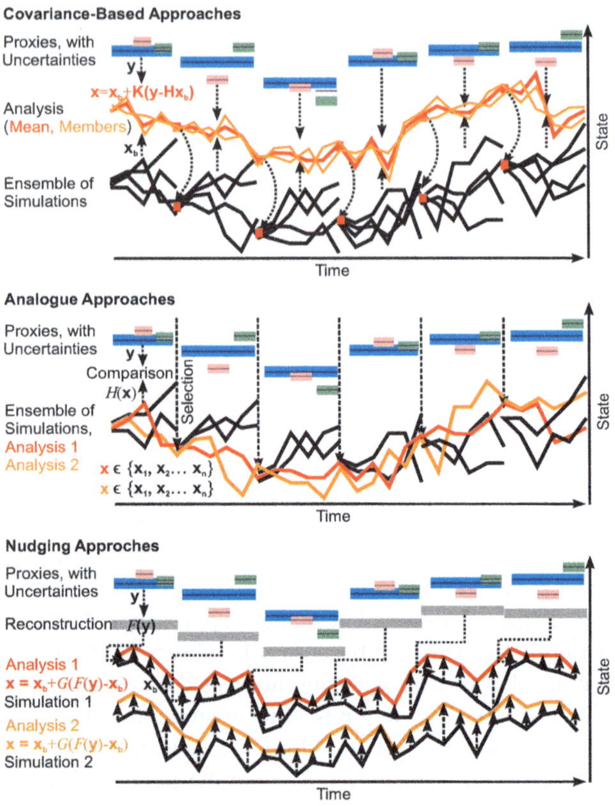

Fig. 2.32 Schematic overview of assimilation approaches. *Arrows* denote steps in the procedure (From Brönnimann et al. 2013b)

Normally **x** is a vector containing the atmospheric state. However, Dirren and Hakim (2005) have successfully extended the concept to time averages.

Data assimilation implies that the models have predictability on the time scales considered. In numerical weather prediction or reanalyses applications (described in Sect. 2.6.2), assimilation time steps are short (hours) and predictability mainly comes from initial conditions. In other words, the model is used to generate a short-term weather forecast, which is then corrected according to observations. The corrected state **x** is then used as a new initial condition for the next step. On the time intervals given by the resolution of proxies, there is hardly any predictability from initial conditions. However, simulations may have predictability from the ocean or from the boundary conditions (sea-surface temperatures, volcanic aerosols, etc.). In the latter case, Eq. 2.4 can be used in an "offline" approach, that is, without updating the initial conditions (Bhend et al. 2012). This allows us to use already existing simulations, which is convenient. Furthermore, because **x** is not used as an initial condition, there is no need for **x** to contain the entire state. It can be small and a vector of averaged model states (or other statistics). Function H can be a simple proxy forward model, that is, a time-integrated function of the elements of $\mathbf{x}$. This approach, which we call Ensemble Kalman Fitting (EKF), is used for the main product presented in this book. It is discussed in more detail in Sect. 2.9.

Covariance-based approaches are powerful but computationally intensive and can be sensitive to assumptions (e.g., of Gaussian distributions), to the treatment of covariance matrices, or to the behaviour of the observation operator H.

Reverting to cost function (Eq. 2.3), we can also look for an existing $\mathbf{x}$, for example by choosing among different ensemble members. This is the concept behind analogue approaches. The cost function then reduces to:

$$J(\mathbf{x}) = (\mathbf{y} - H[\mathbf{x}])^T \mathbf{R}^{-1} (\mathbf{y} - H[\mathbf{x}]) \qquad \text{for } \mathbf{x} \in \{\mathbf{x}_1, \mathbf{x}_2, \dots \mathbf{x}_n\} \,. \tag{2.12}$$

As for the covariance-based approaches, there are "online" and "offline" versions. In "online" approaches, the successful simulation is continued and non-successful ones are terminated. New ensemble members are then generated for the next time step by adding small perturbations to the successful $\mathbf{x}$ and the final analysis is a quasi-continuous simulation. Particle Filters (Goosse et al. 2010) use a distribution of $\mathbf{x}$ to calculate a weighted sum of cost function contributions to Eq. 2.12.

Offline analogue approaches include the "proxy surrogate reconstruction" approach (Franke et al. 2011) and the "best ensemble member" approach (Brönnimann et al. 2013b). Here, existing simulations are used with $\{\mathbf{x}_1, \mathbf{x}_2, \dots \mathbf{x}_n\}$ denoting either different slices of a long simulation (proxy surrogate reconstruction) or an ensemble of simulations for the same time step (best ensemble member), respectively. The "analysis" in both cases is a sequence of short, discontinuous simulations. The latter approach ("best ensemble member") is used later in this book and is outlined in Sect. 2.9.

Analogue approaches have some advantages over covariance-based approaches. Function H may be nondifferentiable (e.g., H can be a complex forward model driven by the full simulation output). Matrix $\mathbf{R}$ may be nondiagonal, and $\mathbf{x}$ may be very large (e.g., 6-hourly model output over a 6-month period). However, to reconstruct the state of systems with a large number of degrees of freedom, these approaches require a huge pool of possible analogues (Annan and Hargreaves 2012).

A third group of techniques is nudging (Widmann et al. (2010); see also van der Schrier and Barkmeijer (2005)). These approaches do not explicitly minimise a cost function. The distance between the model state and observations is reduced by adding tendencies to (usually a subspace of) the model state at each time step, similar to an additional source term in the tendency equations. Following our notation:

$$\mathbf{x} = \mathbf{x}_b + G(F[\mathbf{y}] - \mathbf{x}) \,, \tag{2.13}$$

where $F[\mathbf{y}]$ represents the target field, which is derived from the proxies $\mathbf{y}$ using method F (which can be a statistical reconstruction). Parameter G is a relaxation parameter.

Palaeoclimatological applications are much more disparate than atmospheric sciences in terms of time, time scales, systems analysed, and proxies used. Therefore, a plurality of data assimilation approaches is a logical way forward. However, all approaches still suffer from problems and uncertainties. Ensemble approaches provide some information on spread, which, however, only represents one (difficult to characterise) part of the uncertainty. Further uncertainties are related to model biases, limited ensemble size, errors in the forcing, and proxies. Therefore, evaluation of the approaches using toy models and climate models (pseudo-proxies) and evaluation of the results using independent proxies is particularly important. Any approach, however, fundamentally relies on a good understanding of the proxies.

2.9 Datasets Used in This Book

2.9.1 Overview

Over the last several decades, many datasets of past climate have been produced. Together, these offer a detailed view of climatic variations during past centuries and form the basis of palaeoclimatology. Although Chap. 4 opens with an overview of many reconstructions, the remainder will focus on just a few datasets to preserve the traceability of this book. Hence, this book does not do justice to the tremendous palaeoclimate efforts.

The chosen datasets comprise observations, model simulations, and reconstructions. The latter are based on statistical or data assimilation techniques. Some of these datasets were produced by our group. Although they are not better than other datasets, I use them because I know their strengths and weaknesses. Furthermore, the chosen datasets provide the specific perspective taken in this book: the climate as a three-dimensional, physical-chemical system within which changes occur. The datasets are described to the extent necessary for their use in Chaps. 3 and 4. Table 2.5 presents an overview of the most frequently used datasets. More details can be found in the literature.

2.9.2 Observations and Reanalyses

Instrumental observations are the most direct source of climate information. However, they also have many uncertainties as described in this chapter. This book uses early instrumental observations from a number of locations such as Geneva, Switzerland (a data series with which we have worked in the past). For the greater Alpine area, I use the Historical Instrumental Climatological Surface Time

Table 2.5 Main data products used in the remainder of this book. *T* refers to temperature, *P* to precipitation, *SLP* to sea-level pressure and *GPH* to geopotential height

Name	Type	Period	Temporal Resolution	Area	Spatial Resolution, levels	Variables	Reference
Observations							
HadCRUT4	Gridded observations	1850–present	Monthly	Global	5° × 5°	*T* (Land and ocean)	Morice et al. (2012)
HISTALP	Gridded observations	1780–2005	Monthly	Alps	5 arc min	*T* and *P*	Auer et al. (2007)
Reanalyses							
20CR v2	Reanalysis (see Table 2.3)	1871–present	6-hourly	Global 3D	2° × 2°, 28 l.	All	Compo et al. (2011)
ERA-Interim	Reanalysis (see Table 2.3)	1979–present	6-hourly	Global 3D	0.5° × 0.5°, 91 l.	All	Dee et al. (2011)
Model simulations (all forcings)							
CCC400	ECHAM5.4 (30 members)	1600–2005	6-hourly	Global 3D	2° × 2°, 19 l.	All	Bhend et al. (2012)
Aerosol	ECHAM5.5-HAM nudged with ERA-40	1960–1990	6-hourly	Global 3D	2° × 2°, 31 l.	Aerosols	In preparation
CASTRO	SOCOL (9 members)	1900–1999	12-hourly	Global 3D	3.8° × 3.8°, 39 l.	All (incl. chemistry)	Schraner et al. (2008), Fischer et al. (2008)
Reconstructions and offline assimilations							
EKF400	Offline assimilation of observations and proxies into CCC400	1600–2005	Monthly	Global 3D	4° × 4°, 3 l.	*T, P, SLP, GPH*	In preparation
REC1	Statistical reconstruction	1881–1957	Monthly	Global 3D	2.5° × 2.5°, 6 l.	*T* and *GPH*	Griesser et al. (2010)
REC2	Statistical reconstruction	1918–1957	Monthly	Global 3D (gaps)	2.5° × 2.5°, 6 l.	*T, GPH*, and wind	Brönnimann et al. (2012b)
HISTOZ/BDBP	Offline assimilation of ozone observations into CASTRO; satellite data after 1979	1900–2007	Monthly	Global 2D	5° lat, 30 l.	Ozone concentrations	Brönnimann et al. (2013a), Bodeker et al. (2013)

Series of the Greater Alpine Region (HISTALP) dataset. This monthly, gridded instrumental dataset goes back to the late 18th century. Global-scale analyses are only possible after 1850. I rely on the monthly, gridded HadCRUT4 dataset (Hadley Centre/Climatic Research Unit, version 4), which covers both land and ocean. Various other datasets are also used. They are introduced at appropriate places in this book.

In addition to observations, this book heavily uses reanalyses (see Sect. 2.6.2). For the period since 1871, I use an atmospheric reanalysis (20CR, version 2) that only assimilates surface and sea-level pressure information (i.e., the distribution of atmospheric mass). Sea-surface temperatures and sea ice are used as boundary conditions (Compo et al. (2011); also see Table 2.3). This information is sufficient, at least at midlatitudes, to provide reasonable estimates of the three-dimensional atmospheric state. Since 20CR is an ensemble product, 56 equally likely and physically consistent members are available. This provides some information on uncertainty.

Many validation studies have shown that 20CR provides a realistic depiction of climate in the northern midlatitudes, but has discrepancies in the tropics (Compo et al. 2011), Arctic (Brönnimann et al. 2012a), and other poorly monitored regions. For the last part of this book, I will use ERA-Interim.

Several data products have been derived from 20CR, including a classification of weather types. This provides a simple statistic of the daily synoptic circulation for a given region. I use the Grosswetter-Types 18 (GWT18) classification for Switzerland, which is based on sea-level pressure (Weusthoff 2011). GWT18 partitions the regional sea-level pressure field into 18 types (cyclonic- or anticyclonic-curved flow from the north, northeast, east, southeast, south, southwest, west, or northwest; plus pure cyclonic or anticyclonic types) by correlating the sea-level pressure field for the Alpine region with prototype gradient fields. Before 1870, local weather types can be defined based on station data.

2.9.3 Model Simulations

This book only considers atmospheric model simulations (i.e., no coupled ocean–atmosphere simulations). The basic simulations are known as "all forcings". They include greenhouse gases, solar irradiance, volcanic aerosols, tropospheric aerosols, and land surface properties in a time-dependent manner, as well as sea-surface temperatures and sea ice.

The simulations stem from the ECHAM family, but the model versions differ. CCC400 (Chemical Climate Change over the past 400 years) model simulations (Bhend et al. 2012) encompass a 30-member ensemble of atmospheric simulations that span from 1600 to 2005 and are performed with ECHAM5.4 at a resolution of T63/L31 (i.e., model top at 10 hPa). The model is forced with reconstructed sea-surface temperatures (Mann et al. 2009a), augmented to by adding monthly ENSO-related variability to obtain a more realistic depiction of variability (see

Bhend et al. (2012) for details). Total solar irradiance, stratospheric aerosol optical depth, tropospheric aerosols, and greenhouse gases are also prescribed. Unfortunately, trends in land-surface properties were erroneous, which has some effects on summer temperatures at northern midlatitudes.

The CASTRO (Climate and Stratospheric Ozone) simulations (a 9-member ensemble that spans from 1900 to 1999, described in Fischer et al. (2008)) use an older model version (middle atmosphere ECHAM4) with a coarser horizontal resolution but with a high model top (1 Pa) and full stratospheric (and rudimentary tropospheric) chemistry (the Model for Evaluating Ozone Trends (MEZON)). This model configuration, termed SOCOL (Solar Climate Ozone Links), is described in Schraner et al. (2008). In the CASTRO simulations, the Quasi-Biennial Oscillation is nudged using our reconstructions (Brönnimann et al. 2007a). With the same setup, sensitivity experiments were performed with individual boundary conditions held constant. These are sea-surface temperatures (fixed to a climatology around 1901), concentrations of greenhouse gases (fixed to 1901 values) and ozone-depleting substances (fixed to 1951 values), or emissions of NOx and CO (fixed to 1951 values).

Finally, we analysed a simulation performed with ECHAM5.5 (T63/L31) coupled to the Hamburg aerosol model (HAM, Stier et al. (2005)). This simulation was nudged to ERA-40 (Uppala et al. 2005). Thus, the coupled model sees the transient emissions of aerosols or their precursors at the surface as well as the winds from the reanalysis, but produces its own fields of aerosol properties, radiation, and clouds as a response.

2.9.4 Reconstructions and Offline Assimilation

Several statistically reconstructed climate fields are used in this book, including two of our own data products: REC1 and REC2. These monthly, three-dimensional, global temperature and geopotential height reconstructions use historical upper-air (Stickler et al. 2010), land temperature, and sea-level pressure data. Reconstructions were performed using a principal component regression as outlined in Sect. 2.8.3. In the case of REC1 (Griesser et al. 2010), the principal components were calculated for all fields and over large regions (15°–90 °N, 20 °S–20 °N, and 90°–15 °S) to provide spatially complete reconstructions. Locations that are far from observations are constrained by their teleconnections to other regions. In contrast, REC2 (Brönnimann et al. 2012b) calculates principal components for vertical columns with each grid column reconstructed independently. Grid columns are reconstructed from predictors within a "cone of influence" of ~1500 km around the column, and a minimum number of upper-air variables are required. In this sense, REC2 is close to actual observations. However, it is spatially incomplete.

Our main dataset for addressing atmospheric circulation variability from the 17th to 19th century is termed EKF400. This dataset is an offline assimilation (Ensemble Kalman Fitting; see Sect. 2.8.4) of early instrumental, documentary, and

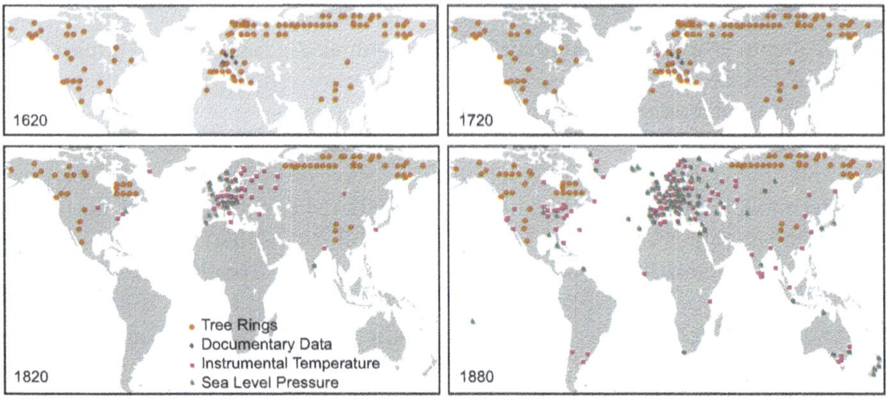

Fig. 2.33 Information assimilated into EKF400 in the years 1620, 1720, 1820, and 1880

tree ring data into the CCC400 model simulations. In other words, CCC400 is used as $\mathbf{x}_b$ for every second grid point (i.e., the resolution is 4° × 4°). The state vector $\mathbf{x}$ of the assimilation contains monthly averaged fields of sea-level pressure, precipitation, land surface air temperature, and several upper-level fields (850 and 200 hPa wind, 500 hPa temperature, geopotential height, and vertical velocity). Although $\mathbf{x}$ contains monthly data, the assimilation step is seasonal. Thus, $\mathbf{x}$ contains monthly data for 6 months (from April to September or from October to March).

The observations and proxies used by EKF400 are shown in Fig. 2.33. From tree rings, we used maximum late wood density and ring width. Tree rings and documentary data (for Europe) are available since the beginning of the period. Instrumental observations are available for Europe since the late 17th century, for some colonies for the 18th century, and for North America since the early 19th century. Note that no precipitation measurements and only a small selection of the more than 2000 tree ring chronologies shown in Fig. 2.30 were assimilated. Therefore, there are ample independent data for comparison and validation.

In the assimilation, tree rings are expressed with a regression approach. Thus, H is a set of regression coefficients for monthly temperatures (maximum late wood density) or monthly temperatures and precipitation (tree ring width) during the growing season. The regression model was calibrated using the Climatic Research Unit surface temperature dataset CRU TS3.21 (Harris et al. 2014) from 1901 to 1970. Likewise, monthly or seasonal documentary data were calibrated using CRU TS3.21 data from the 20th century. Instrumental temperatures or sea-level pressure were incorporated by using the closest grid point.

A first estimation of errors for the (diagonal) $\mathbf{R}$ matrix was obtained by analysing the residuals of the regression in the 20th century. However, expert judgment was also needed to set the appropriate errors for the different types of data.

The assimilation was performed using Eq. 2.4. Because the bias of observations and proxies can change over time, we corrected the proxies relative to the model background $\mathbf{x}_b$ by subtracting $H(\mathbf{x}_b)$ averaged over a 70-yr moving window. Thus, on longer (from multidecadal to centennial) time scales, EKF400 becomes almost identical to CCC400. Meanwhile, information from proxies and observations is used to constrain variability on time scales from months to decades.

To localise the covariance matrix **R**, the method of Bhend et al. (2012) was used. For each field, a typical decorrelation distance was estimated. For localisation across variables, the smaller decorrelation distance of the two variables was chosen. The full description of the dataset is in preparation (Jörg Franke, Univ. Bern).

Here, we show the performance of the method for a limited set of predictors (see Brönnimann et al. (2013b) for more information). Figure 2.34 shows the April–September 1810 average surface air temperatures relative to 1801–1830. The unconstrained ensemble mean (CCC400; Fig. 2.34, top left) shows the effect of boundary conditions: cooler-than-average summer temperatures following the volcanic eruption in 1808/1809. Anomalies are small and smooth, which is typical for an ensemble mean. The EKF analysis (Fig. 2.34, bottom left) suggests a more pronounced cooling over northern Europe (constrained by observations) and a slightly westward-shifted warming in northern Asia (constrained by covariances).

From the CCC400 simulations, we can determine the best ensemble member (see Sect. 2.8.4). For this purpose, we used high-quality tree rings from 35 locations. A tree growth model (Tolwinski-Ward et al. 2011) was used as H and Eq. 2.9 was minimised with a non-diagonal **R** matrix. The best member (Fig. 2.34, top right) exhibited strong anomaly centres over Alaska and western and eastern Siberia. These locations were not represented by observations in EKF400. In contrast, European cooling is less than the EKF400 ensemble mean.

Due to the small ensemble size, the best member may not fit each region. In other words, it does not have high statistical skill. However, we can combine these approaches. Figure 2.34 (bottom right) shows the EKF400 assimilation for the best

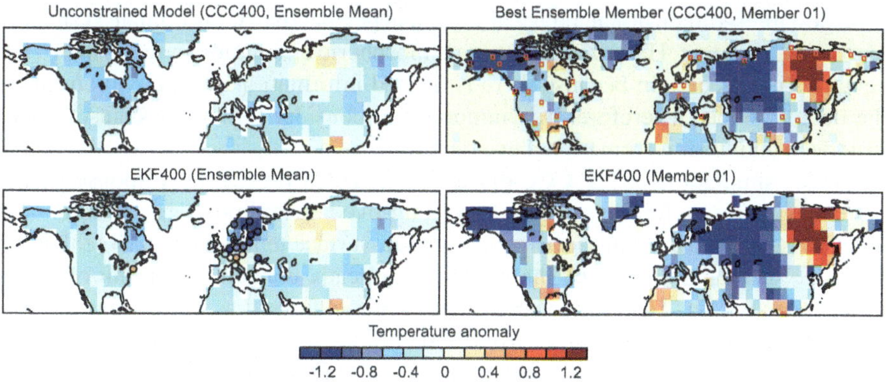

Fig. 2.34 Northern Hemisphere temperature anomalies from April to September 1810 (relative to 1801–1830) from the unconstrained ensemble mean, the EKF 400 mean, the best ensemble member (member 01), and the EKF 400 analysis for member 01. *Circles* indicate locations and anomalies of the assimilated instrumental measurements; *red squares* give the locations of tree ring proxies used to define the best ensemble member (From Brönnimann et al. 2013b)

member. It is constrained by assimilated observations (and reproduces the stronger cooling over Europe) but, away from the assimilated information, it retains the characteristics of the "best ensemble member".

In this book, I often show the ensemble mean and the best ensemble member of EKF400 (for April–September). The ensemble mean arguably is the product with the highest skill; the best ensemble member has a more realistic variance.

It is difficult to provide a measure of skill for EKF400. Skill varies by region, with time, and according to variable. Pseudo-proxy experiments (Bhend et al. 2012) with a very sparse proxy network reveal similar skill to conventional (statistical) reconstructions. However, the examples and comparisons shown in Chap. 4 of this book provide a better impression than statistical skill measures.

The historical ozone dataset (HISTOZ), described in Brönnimann et al. (2013a), was generated in the same way as EKF400. It is a global, two-dimensional (longitude and altitude) dataset of ozone concentrations. Historical ground- and (after 1970) satellite-based total column ozone observations were assimilated into the 9-member ensemble of CASTRO simulations. As for EKF400, 6 months were incorporated into the same state vector. The determination of errors and the localisation of the covariance matrix **B** is described in Brönnimann et al. (2013a). After 1979, the dataset continues with the satellite-based BDBP (Binary Database of Profiles) dataset (Bodeker et al. 2013), which is also used for the de-biasing step.

2.9.5 Conclusions

To enhance the readability of the book, we use few, selected data products in our analysis of climatic changes since 1700 . These datasets allow us to discuss climatic variability in the context of atmospheric circulation. After the 1930s, these data even provide a three-dimensional, physical-chemical perspective of the climate system, at least to some extent. However, analysing atmospheric circulation in the 17th or 18th centuries is less certain because there is limited information available to constrain the data products. Therefore, it is important to consider how these datasets were generated and what information they contain.

Model simulations (e.g., CCC400 and CASTRO) reflect the response of the climate system to forcings exerted by long-lived greenhouse gases, solar and volcanic activity, tropospheric aerosols, and sea-surface temperatures. Note that the latter are seen as a forcing by the model, but already contain an imprint of atmospheric circulation. Thus, they cannot be strictly addressed as an oceanic forcing. EKF400 assimilates information from observations and proxies into the CCC400 simulations and can be seen as a best guess estimation of the climate system of the past, combining forcings, proxies, and observations. Early instrumental data and statistical reconstructions, in contrast, are independent of models and forcings.

For the period after 1870, atmospheric circulation is addressed in 20CR and three-dimensional reconstructions (e.g., REC1 and REC2). The former is based on surface pressure, sea-surface temperatures, and a numerical model. The latter relies on upper-air data, land surface temperatures, sea-level pressure, and a statistical model. Although not fully independent, these datasets are rather different. Atmospheric ozone is studied in the CASTRO simulations and corresponding HISTOZ assimilations. The nudged simulations with ECHAM5.5-HAM are used for analysing aerosols and their effect on radiation.

Our climate data sets follow a tradition. Brückner was one of the first scientists to reconstruct climate time series extending several hundred years into the past. Based on early instrumental data, he also drew maps of pressure anomalies for the first half of the 19th century. Since Brückner's time, many new proxies have been discovered. Statistical reconstruction techniques have been used since the early 1990s to derive field reconstructions for climate research. This constitutes a major step forward. Another major step forward is currently underway and involves the use of numerical techniques such as data assimilation. Consequently, the datasets used in this book allow a much broader view of climatic changes since 1700 than possible 10 or 15 years ago. Together with climate model simulations, another new pillar in palaeoclimatology, they allow us to explore new aspects of climatic changes since 1700.

Chapter 3
The Machinery: Mechanisms Behind Climatic Changes

With contributions and help from Florian Arfeuille, Yuri Brugnara, Martin Wegmann, Jörg Franke, and Matthias Röthlisberger

> "I have attempted to depict climatic changes that have affected the globe during the last centuries. Like cogwheels in a clockwork, different meteorological elements interact to produce climatic changes. We see the wheels spin and the hand move in a certain rhythm, yet the driving force remains hidden." (translated from Brückner 1890, p. 322)

Brückner's depiction of climate as a clockwork[1] is remarkable. During the last 125 years, science has unravelled many of the hidden mechanisms. Before discussing climatic changes since 1700, it is useful to start with a tour of the "machinery" of the climate system. Basic physical concepts and mechanisms operating in the climate system will be briefly introduced—the motor, transmission, and basic mode of the operation of the machinery. Then, we will analyse how these mechanisms produce variations over time. We will look at changes in the general circulation, how they are expressed in systematic form in circulation variability modes, and how they affect local climate in the form of teleconnections. Special attention is devoted to how the machinery reacts to external forces and how feedback processes evolve in the climate system.

One aim of Chap. 3 is to provide necessary background for Chap. 4, where climatic variations since 1700 will be addressed. Chapter 3 is also a synthesis of Chap. 4, because many processes and mechanisms have been uncovered based on the analysis of climate data from the past 400 years. Therefore, this chapter contains introductory material as well as statistical analyses of the climate of the past 400 years. As for all parts of the book, Chap. 3 is biased by my own research interests, putting emphasis on some mechanisms while neglecting others and using my own examples, for which I feel more competent. The book does not replace—but

[1]I prefer the image of a machinery, as a clockwork implies precision, predictability, and deterministic behaviour.

S. Brönnimann, *Climatic Changes Since 1700*, Advances in Global Change Research 55, DOI 10.1007/978-3-319-19042-6_3

hopefully complements—standard textbooks. For a more comprehensive introduction to basic atmospheric physics the reader is referred to the excellent text books of Hartmann (1994), Wallace and Hobbs (2006), and numerous others.

3.1 Basic Climate Physics: The Mean State

3.1.1 Energy Budget and Heat Transport

The climate system is a heat engine that is driven by spatial energy differences and that performs the transport and conversion of energy. Therefore, a fundamental perspective of climate physics is the energy budget. This chapter starts with the energetics of the climate system, first at the level of global-mean values, then considering spatial differences and the transport of energy within the system. Then, I outline the general circulation of the atmosphere and emphasise the role of the stratosphere (Sect. 3.1.3). Ocean–atmosphere interaction is briefly touched on in Sect. 3.1.4. The book provides only an extremely limited introduction. The interested reader is referred to more comprehensive reviews of individual aspects (e.g. Fasullo and Trenberth 2008a,b; Schneider 2006; Schneider et al. 2010; Stephens et al. 2012). The advanced reader may wish to skip Sect. 3.1 altogether.

Relatively little of what is presented in this chapter was known to late 19th century scientists such as Brückner. Although they started to identify the main patterns of climate variability and had established the physical basis (fluid dynamics, thermodynamics), there was no theory of atmospheric circulation that related the two. Physical theories of atmospheric states and processes only developed during subsequent decades. In fact, in terms of the quantification of energy fluxes—the starting point of this chapter—a consolidated quantitative view only emerged in the past two decades with the help of satellite data.

3.1.1.1 Global-Mean Fluxes

A schematic depiction of fluxes in the climate system is given in Fig. 3.1 (Hartmann et al. 2013; Wild et al. 2013). The only relevant source of energy in the climate system on a global scale is the sun. Solar radiation at the top of the earth's atmosphere (termed "solar constant", despite it varying) amounts to about 1361 W m^{-2}.[2] On average, each square metre of the earth's surface receives a quarter of that amount, corresponding to the ratio between the surface area and cross section of the earth. About 30 % of this shortwave radiation is reflected on clouds or at the earth's surface, and 25 % is absorbed by atmospheric constituents (ozone, water vapour,

[2]This number was only recently quantified, see Kopp and Lean (2011); for a long time it was estimated higher.

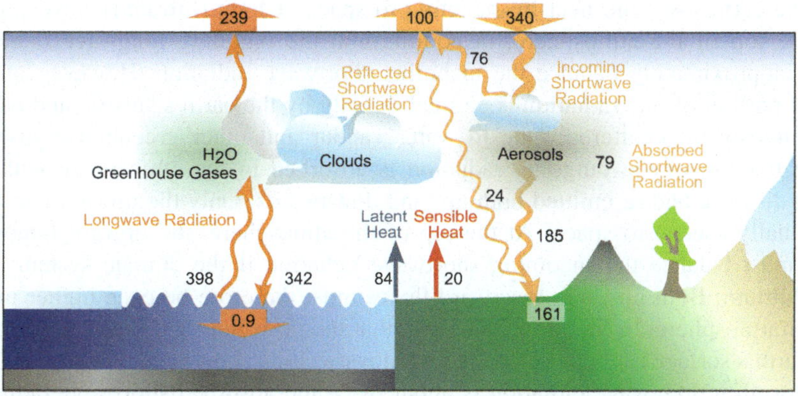

Fig. 3.1 Globally averaged radiation fluxes in the climate system (Adopted from Wild et al. 2013 and Hartmann et al. 2013)

aerosols). The remaining 45 % is absorbed by the earth's surface (see Hartmann et al. 2013). Here, a smaller fraction of radiation is transformed into heat that warms the atmosphere (sensible heat). A larger fraction is used to evaporate water, releasing the energy during condensation in the atmosphere (latent heat). This is the main mechanism that compensates for the net radiative energy surplus at the surface and net radiative energy deficit in the atmosphere. From the heated surface, heat can also penetrate downwards into land and ocean. There are very large energy fluxes between atmosphere and ocean or land in both directions depending on the time of day, season, and region; on a global average, a small net flux of 0.9 W m^{-2} into the ocean remains. This flux corresponds to the net radiation imbalance at the top of the atmosphere due to the negligible heat capacity of the atmosphere.

The large proportion of latent heat flux is crucial not only for understanding the basic mode of operation of the machinery, but also its response to warming. The latter comes from the fact that the saturation water vapour pressure increases by about 7 % per 1 °C of warming (this follows from the Clausius-Clapeyron relation).[3] In a warmer world, air will thus contain more latent energy that is transported horizontally and vertically by atmospheric circulation (which itself must balance energetic differences) before it is released and eventually radiated to space. To some extent, water vapour changes thus constrain the response of atmospheric circulation to global temperature changes in the future as well as the past (see Schneider et al. 2010).

[3]The Clausius–Clapeyron relation describes the line of coexistence of two phases of matter (in our case liquid water and water vapour) in a temperature–pressure diagram. For atmospheric conditions: $de_s/dT = L_v(T)e_s/R_vT^2$, where e_s is the saturation water vapour pressure, T is temperature, $L_v(T)$ is the temperature-dependent latent heat of vaporisation, and R_v is the gas constant for water vapour.

The earth loses the heat by radiating to space at long (infrared) wavelengths. At the top of the atmosphere, in equilibrium, the outgoing longwave radiation must approximately balance the incoming shortwave radiation. However, only a small portion of this radiation is emitted directly by the earth's surface and passes the atmosphere unaltered. Due to water vapour and other greenhouse gases, a large fraction of the outgoing radiation is absorbed in the atmosphere within a short distance and re-emitted both up- and downwards. Only the upward part may eventually escape to space. At the top of the atmosphere, the upward longwave flux must balance the incoming shortwave radiation if the climate system is in equilibrium. Both up- and downward fluxes must therefore be even higher in the lower atmosphere. In fact, Fig. 3.1 shows that the upwelling longwave radiation at the earth's surface is about twice the total downwelling shortwave radiation and the downwelling longwave radiation is about twice the absorbed shortwave radiation (see Sect. 3.3.3 for the effect of well-mixed greenhouse gases).

Radiation emission depends linearly on the fourth power of temperature. Thus, to provide the necessary radiation fluxes, the earth's surface will warm by absorbing the downwelling longwave radiation. If the vertical transport of energy was only achieved through radiation, superadiabatic temperature gradients would have to be established in the troposphere to maintain the fluxes (in the stratosphere, the absorption of ultraviolet (UV) radiation by ozone alters the temperature gradient and leads to a warming). Heat will therefore be transported by convection upwards through the troposphere in the form of sensible and latent heat. The resulting vertical temperature profile is called the radiative–convective equilibrium, where the troposphere adjusts to a convective equilibrium (with approximately a moist adiabatic lapse rate) and the stratosphere in a global mean is in radiative equilibrium (locally, though, it may be far from equilibrium). Hence, one important task of the climate system is to transport energy upwards in the atmosphere, from where it can eventually be radiated to space.

This basic concept of the one-dimensional mean temperature profile was outlined in the 1960s by Manabe and Wetherald (1967), although the first work on the radiative–convective equilibrium already started in Brückner's time (Emden 1913), triggered by the discovery of the stratosphere, observations of longwave radiation at different altitudes (Ångström 1916), and laboratory work on the infrared emissivity of gases.

3.1.1.2 Spatial View of Heat Transport

Let us now turn to the horizontal distribution of energy. Figure 3.2 (top left) shows the top-of-atmosphere net radiation (short- and longwave). For geometric reasons, the tropics receive more solar radiation than the extratropics. They also lose more energy due to radiation (the dry subtropical deserts are even a net energy sink), but the former dominates the energy budget. Temperature differences are more smoothed out, reflecting a quasi-steady state where poleward transport of energy balances the built-up temperature differences. However, the temperature field shows additional features that do not arise from radiation differences such

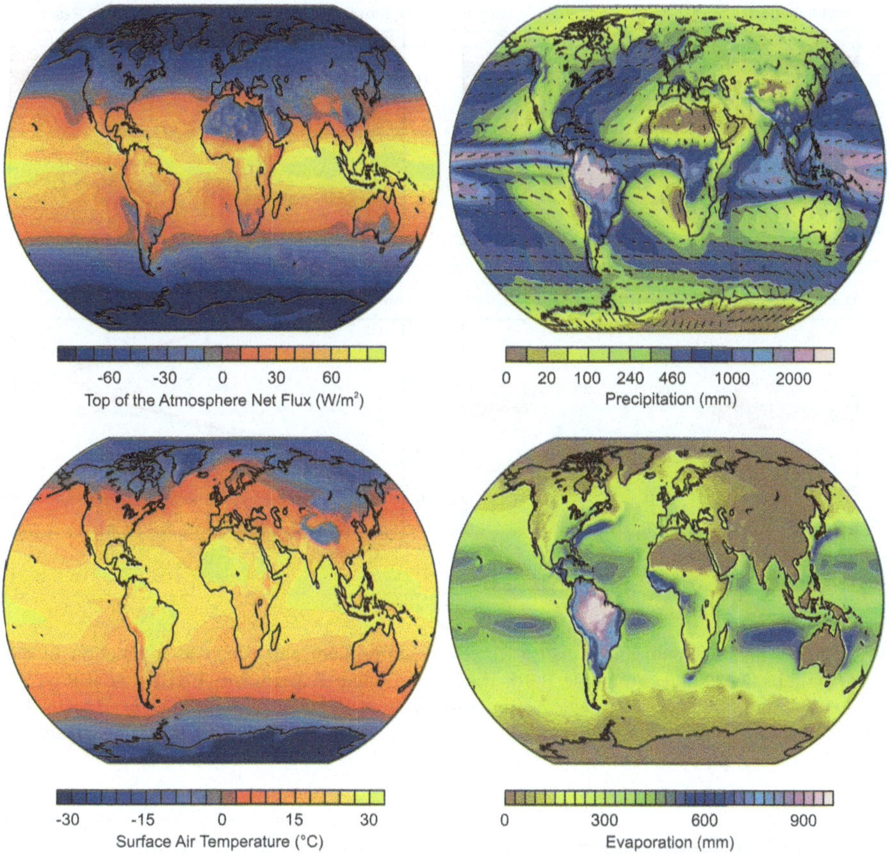

Fig. 3.2 Annual means for the period 2000–2011 of (*top left*) top-of-atmosphere net radiation (short- and longwave), (*bottom left*) surface air temperature, (*top right*) precipitation annual sum and 10-m wind, and (*bottom right*) evaporation. Note the nonlinear scale for precipitation. Data sources are CERES Level 3B (radiation), ERA-Interim (all other variables)

as the longitudinal structure over the North American and North Atlantic sector. Precipitation shows sharp gradients and a strong link to atmospheric circulation. It falls in zones of convergent winds such as the Intertropical Convergence Zone (ITCZ) or the midlatitude planetary frontal zones (see Fig. 3.2). Zones of divergent winds near the surface indicate the cloud-free high pressure areas of the subtropics. Evaporation maximises over the subtropical oceans and Amazon basin (highlighting the role of vegetation) and is minimal over the deserts and the polar regions.

The driver is the unequal spatial distribution of incoming solar energy. Thermal gradients and pressure differences build up and set the atmosphere in motion. The resulting circulation counteracts the imbalance and transports energy poleward. This heat transport is accomplished by both oceans and atmosphere. Although the oceans store large amounts of heat, their contribution to heat transport to the polar regions rapidly decreases outside the tropics, though it may still be regionally important. Poleward heat transport mainly takes place in the atmosphere in the form of sensible

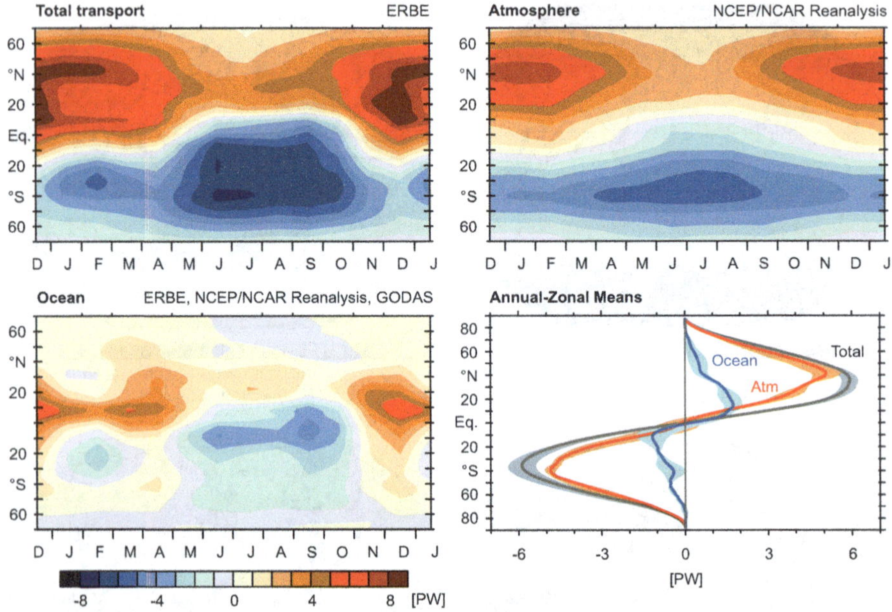

Fig. 3.3 Northward zonal-mean energy fluxes in atmosphere and ocean as a function of calendar month. Data sources include Earth Radiation Budget Experiment (ERBE), NCEP/NCAR Reanalysis, and Global Ocean Data Assimilation System (GODAS). (*bottom right*) The median annual-mean transport by latitude with the associated $\pm 2\sigma$ range (*shaded*) (Fasullo and Trenberth 2008b, © Copyright 2008 AMS)

and latent heat fluxes (see Fig. 3.3). The poleward heat flux is stronger during the winter season (Fasullo and Trenberth 2008a,b), when the gradients in the energy balance and surface temperature are strongest. Therefore, atmospheric circulation is more vigorous in winter.

Energy fluxes are different over land and ocean regions. Oceans absorb more shortwave radiation (Fig. 3.2). There is a net atmospheric heat flux from the ocean to the land areas. A substantial part of this heat flux occurs in the form of latent heat, which is balanced by the hydrological cycle. In terms of mass flux, river runoff balances the net moisture transport from the oceans to the land.

3.1.2 The General Circulation of the Atmosphere

3.1.2.1 The Zonally Symmetric Circulation

Climatic changes at a given location are rarely just the product of an altered local energy balance. Usually they are closely related to changes in atmospheric circulation—monsoons, jet streams, trade winds, or storms—which in turn may be induced by large-scale radiative forcing.[4] Therefore, before addressing climatic changes since 1700, we need to understand the large-scale, general atmospheric circulation and its driving factors based on the energetic considerations from the previous chapter.

Circulation is initiated by pressure differences, which are related to density changes (due to temperature or humidity changes). In the tropics, the latent heat released during condensation in regions of deep convection is the main driver of the atmospheric circulation. The released energy is partly radiated to space and partly transported poleward in the outflow of the convection areas, thus contributing to the poleward heat transport in the atmosphere (see Webster 2004). As a zonal average, convection within the ITCZ forms the inner, upwelling branch of the tropical meridional overturning circulation—the Hadley circulation (Schneider et al. 2014a). The outflow from the ITCZ is deflected towards the east due to the Coriolis force and does not reach beyond the subtropics, where it subsides. The circulation is closed through the return flow along the surface—the trade winds. The Hadley cell is a "thermally direct" circulation; it is caused by warmer air that rises and colder air that sinks.

A zonal-mean view of the meridional circulation in the form of the meridional mass streamfunction is shown in Fig. 3.4 (top left) for boreal winter. The northern Hadley circulation appears prominently as a clockwise circulation cell ascending equatorward of 10°N and descending between 20° and 30°N. Poleward of the Hadley cell, a circulation cell with an opposite flow direction appears, which often is called the Ferrel cell. This is more variable and only appears as a closed circulation cell in a zonal-mean depiction. It is a thermally indirect circulation; it cannot be maintained by thermal processes (nor transport energy poleward). The vertical velocity ω at 500 hPa (Fig. 3.4, top right), to which the precipitation distribution (Fig. 3.2) is closely related, shows that the up- and downwelling in the Hadley cell is far from zonally uniform. For instance, over the western and central tropical Pacific, the ITCZ shows a branching off. The ITCZ is a well-defined narrow band over the Atlantic, but becomes a broad upwelling zone over the Amazon and Congo basins.

The causes of trade winds and upper-level "anti-trade winds" have been a matter of scientific research since the 17th century (Halley 1686). During Brückner's time the Hadley cell was well known and a theory, although not fully correct, was established (Ferrel 1856; Hadley 1735). However, only surface winds were directly measured, and more complex circulations such as the monsoons could not be understood before upper-air measurements became more widespread.

The most outstanding feature of the longitude–height structure of the zonal-mean zonal wind (Fig. 3.4) is the subtropical jet around 30°N and at an altitude of around 10 km (200 hPa). The jet also clearly appears on the map of zonal wind at 200 hPa (Fig. 3.4, bottom right), leading from the North Atlantic across Asia to the

[4] Radiative forcing denotes a change in the net radiative flux at the tropopause or the top of the atmosphere that is due to an external driver such as greenhouse gases.

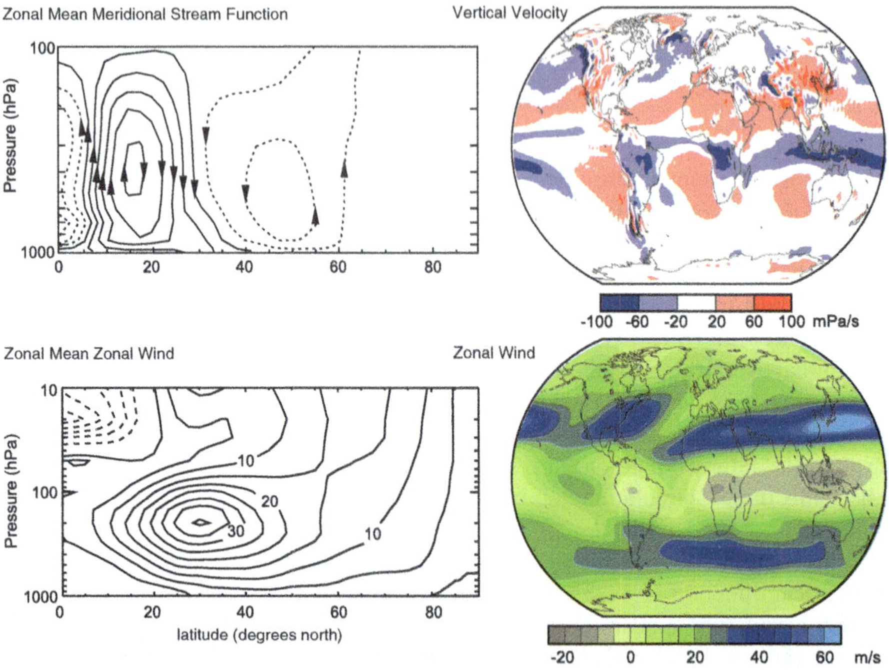

Fig. 3.4 Long-term boreal winter averages of (*top left*) zonal-mean meridional mass streamfunction (i.e., the vertically integrated meridional mass flux above a level, here given in arbitrary units; *contour lines* and *arrows* can be understood as trajectories of air parcels in steady flow), (*top right*) vertical velocity at 500 hPa, (*bottom left*) zonal-mean zonal wind ($m\,s^{-1}$), and (*bottom right*) zonal wind at 200 hPa. Shown are December–February averages from 1961 to 1990 (Data source is NCEP/NCAR Reanalysis)

Pacific and over North America. In the equatorial region, easterlies dominate over Africa, the Indian Ocean, and the Maritime Continent.[5] Westerlies dominate over the Pacific.

While latent heat is important in the tropics, the large-scale circulation of the extratropics is largely controlled by the temperature difference between tropics and the polar regions on the one hand and the angular momentum balance on the other hand. The structure of the zonal-mean zonal wind can be understood from the thermal wind relation, which describes the change of geostrophic wind Δu_g between two pressure levels p_1 and p_2 due to horizontal gradients in virtual temperature dT_v/dy within that layer:

[5]Meteorologists refer to the region of Indonesia, Philippines, and Papua New Guinea as the "Maritime Continent". It consists of many islands, peninsulas, and shallow bodies of water. Sea-surface temperatures are among the highest on earth, and the interspersed islands with their topography enhance convection.

$$\Delta u_g = -R/f \, dT_v/dy \, ln(p_2/p_1) \qquad (3.1)$$

where R is the gas constant and f is the Coriolis parameter. Because the equator-to-pole temperature gradient is positive up to the tropopause and negative above, westerly winds increase with height in the troposphere, peak near the tropopause, and then decrease in the stratosphere.

Alternatively, the subtropical jet can be understood from the conservation of angular momentum in the upper-tropospheric outflow of the ITCZ. However, wind speeds calculated in this way would be far larger than those observed. In reality, interaction of the subtropical jet with extratropical eddies slows down the jet and allows the exchange of energy and momentum. Due to the stronger meridional temperature gradient in the winter hemisphere, the subtropical jet is stronger in winter than in summer.

Changes in the Hadley circulation on an interannual scale are related to climate variability modes such as El Niño–Southern Oscillation (see Sect. 3.2.6), variability in the Atlantic Ocean, or external forcing. Changes in the Hadley circulation are an important component of past climatic changes in the tropical and subtropical belt; this concerns not only variations in the strength, but also in the position and poleward extent of the Hadley circulation. A widening or contraction of the Hadley cell is expected to lead to a pole- or equatorward migration of the subtropical dry zones, respectively; a delicate change in the global climate system.

Poleward of the subtropical jet and up to 75°N, westerly winds dominate at all tropospheric levels (Fig. 3.4). In winter, strong westerlies occupy the entire troposphere and stratosphere in a consistent manner, building the so-called polar vortex poleward of 60°N. Westerlies bring rainfall and storms from the midlatitude oceans to the western parts of the continents. They are the dominating circulation system for European climate, which forms a special focus of this book and of Brückner's work. Changes in the midlatitude westerlies and therefore in the polar vortex, which will be further discussed in Sect. 3.1.3, are a prominent expression of interannual-to-decadal variability in surface climate since 1700.

3.1.2.2 The Zonally Asymmetric Circulation

A zonal-mean view of circulation is insufficient for understanding the general circulation of the atmosphere. The earth's land masses are not zonally symmetric, the absorbed energy is not zonally symmetric, and the Hadley circulation is not zonally symmetric (see Figs. 3.2 and 3.4). Over Asia during the summer, the convergence zone shifts poleward by a substantial amount and becomes part of the Asian monsoon system (see Sect. 3.1.4).

In the tropics, there are zonal circulation cells. The Pacific Walker circulation consists of upwelling over the western tropical Pacific (Fig. 3.4, top right) eastward flow along the equator (Fig. 3.4, bottom right), downwelling over the eastern tropical Pacific (Fig. 3.4, top right), and return flow as trade winds (Fig. 3.2, see also Sect. 3.2.6). There are also Walker cells over the Indian Ocean, with ascent over the eastern Indian Ocean and descent over East Africa, and over the Atlantic Ocean, with ascent over South America and descent over the eastern tropical Atlantic. The Hadley circulation is not zonally symmetric but is more pronounced in the upwelling regions of the Walker cell. In addition to the tropical circulation, subtropical jets and the westerly winds poleward of the jets are zonally asymmetric. The latter show longitudinal structures, which are to some extent fixed in space and to some extent variable.

A strictly zonal flow would not allow meridional exchange of heat (apart from diffusion and radiation), and a large temperature gradient would develop. However, above a certain threshold of that gradient, slight disturbances grow rapidly and the flow becomes unstable. Waves develop and grow and it is mainly through these waves and associated disturbances that heat exchange takes place.

Waves play an important role in atmospheric sciences and I will revert to waves at several instances, but even a basic introduction is beyond the scope of this book. Table 3.1 summarises different types and characteristics of atmospheric waves. For more details the reader is referred to Holton (2004).

Statistical analyses of pressure or wind fields shows that waves do not occur randomly in space and time. Figure 3.5 shows long-term means of 500 hPa geopotential height in the winter of the respective hemisphere. Shadings indicate anomalies from the zonal average, orange indicates ridges, blue indicates troughs. Although these maps are based on 30-yr averages, the waves do not average out but show preferred patterns. In the Northern Hemisphere, we find two or three trough–ridge patterns (wavenumbers 2 and 3). In contrast, in the Southern Hemisphere, wavenumber 1 dominates.

These large-scale waves are long Rossby waves (see Table 3.1), which are also called planetary waves. They are quasi-stationary and exhibit preferred patterns. Changes in the patterns of these waves are another important ingredient for understanding climatic changes of the past. Patterns of temporal changes in the wave structure, climatic consequences, and relations to other climate factors are addressed in Sect. 3.2. In the following, I briefly analyse some of the underlying processes.

What anchors the quasi-stationary planetary waves? The main reason is that their sources are spatially fixed, and this is due to two important contributions: land–sea heating contrasts and orography. Mountain ranges that need to be overflown—a prominent example are the Rocky Mountains—can induce waves. During the overflow, air parcels are vertically compressed but horizontally extended. In order to balance potential vorticity (which is conserved), absolute vorticity[6] must increase to compensate for the vertical compression. The only way of establishing this is

[6]Vorticity describes the spinning motion in a fluid. In a two-dimensional field, vorticity ζ can be written as $\zeta = dv/dx - du/dy$, where u and v are the flow components in the x and y directions, respectively. In the atmosphere, this term is also called relative vorticity as it describes vorticity relative to the earth reference system. Absolute vorticity $\zeta + f$ also includes the planetary vorticity f induced by the earth's rotation. Potential vorticity relates absolute vorticity to static stability and is a conserved property, from which follows that $(\zeta + f)/H = constant$ (H is the thickness of the air parcel).

Table 3.1 Types and characteristics of atmospheric waves

Wave	Examples	Restoring mechanism	Source	Wavelength, propagation
Sound	Sound	Compression (oscillating pressure gradient force)	Many	Longitudinal waves (displacements in the direction of wave propagation)
Gravity	Kelvin-Helmholtz waves	Density gradient (buoyancy)	Topography, convection, wind shear (jet streams, thunderstorms)	Vertical displacements along slanted paths, propagate vertically to the stratosphere and mesosphere (amplitude grows with decreasing density), wavelengths of tens of kilometres
Inertia gravity	Midlatitude gravity waves, mountain waves	Buoyancy/Coriolis parameter	Hurricanes, strong vortices in general	Horizontal (>100 km) and vertical (ca. 2 km) displacements, low frequency (hours), upward propagating
Mixed Rossby/gravity	Equatorial waves (incl. Kelvin waves)	Sign change of Coriolis parameter		Horizontal and vertical displacements, east- or westward propagation
Rossby	(see below)	Latitudinal variation of Coriolis parameter, gradient of potential vorticity	(see below)	Horizontal displacements, propagates westward relative to the mean flow
	Forced stationary planetary waves		Orography, land–sea contrast	Wavenumbers 1–4, upward propagation to stratosphere possible, westward phase speed may equal mean westerly flow (stationary)
	Free travelling planetary waves		Orography, diabatic heating, baroclinic instability, extratropical transition of tropical cyclones	Wavenumbers 5–8, upward propagation to stratosphere damped, westward phase speed usually smaller than mean westerly flow
	Synoptic waves		Diabatic heating, baroclinic instability	Wavenumbers 9–12, upward propagation to stratosphere impossible

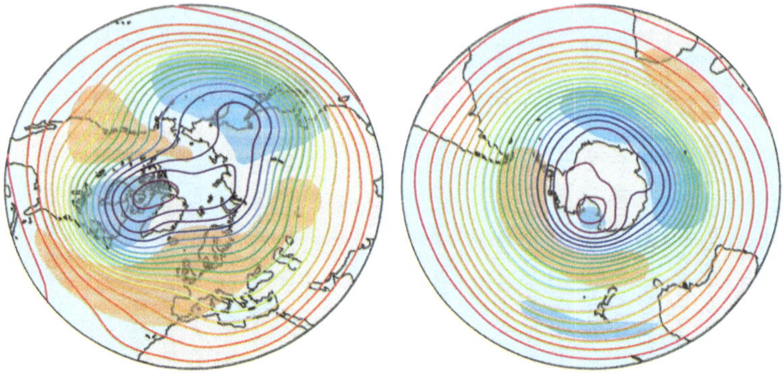

Fig. 3.5 Long-term mean geopotential height at 500 hPa for (*left*) the Northern Hemisphere in December–February and (*right*) the Southern Hemisphere in June–August. Isoline distance is 50 gpm, *red* and *blue* shadings show positive and negative anomalies from the zonal mean that exceed 50 gpm (Data source is NCEP/NCAR reanalysis, 1981–2010)

by displacement to the north. After crossing the Rocky Mountains, the opposite occurs. The air parcel stretches, the air moves south, and a trough forms. As a consequence, lee troughs develop downstream of mountain ranges. The wave structure in the Northern Hemisphere is thus strongly influenced by the Rocky Mountains, Greenland, and the Tibetan Plateau. Other factors also contribute to establishing and maintaining the wave. Off the coast of North America, the air masses encounter the warm North Atlantic Current. Off the coast of Asia, the warm Kuroshio affects the flow. The sharpened east–west thermal contrast induces a meridional component to the mean flow.

The Atlantic is downstream of the Rocky Mountains, and the quasi-stationary wave over eastern North America and the North Atlantic is partly explained by these factors. Note that Europe is on the front side of this wave and thus, on average, receives warm and moist subtropical air masses whereas eastern North America receives cold continental air. The fact that Europe is so much warmer than North America on the same latitude is partly due to this wave (Seager et al. 2002).

The sharpened land–sea thermal contrast also destabilises the atmospheric flow, particularly in winter. Weather systems develop and interact with the background flow. Therefore, in the next chapter, we will change from a statistical to a process view and take a look at the weather scale.

3.1.2.3 The Weather Scale

Cyclones preferentially develop and grow in regions of high baroclinicity, that is, large temperature gradients and significant wind shear. The presence of an upper-level jet stream—the polar front jet—helps in the development of a cyclone. In fact, the above-mentioned regions of sharpened land–sea heating contrasts are the beginnings of the North Atlantic and North Pacific storm tracks, respectively. As the disturbances develop, they interact with the mean flow and may sharpen the trough and strengthen the jet stream. Thus, they may contribute to anchoring the quasi-stationary wave pattern even though each disturbance is not stationary, but is steered by the mean flow and has a limited lifetime.

The polar front jet—not to be confounded with the subtropical jet—owes its existence to this interaction. It is much more variable in space and time compared to the subtropical jet and its imprint in climatological fields is thus smaller (e.g., Fig. 3.4). In the midtroposphere, blocking highs may develop in the westerly flow, kept in place by one (dipole or "Rex" pattern), two ("Omega" pattern), or more low pressure systems. The polar front jet is then strongly deviated from its path, or it even splits.

In the mid- and high latitudes, the sea level pressure field on an individual day shows the volatile weather, with highs or lows changing rapidly from day to day—the weather systems are "transient". Sea level pressure is the surface imprint of the atmospheric circulation; it reflects troughs and ridges, regions of convergence and divergence.

Similar as for the 500 hPa height fields, there is also a stationary component in the locations and paths of these weather systems, which is apparent when averaging over long time periods or analysing the spatial pattern of variance. The high and low pressure systems found in the long-term average are called quasi-permanent pressure systems. Because they play an important role for month-to-month or winter-to-winter variations of weather of the entire extratropics, they are also called centres of action. Examples of centres of action (see Fig. 3.6) are the Icelandic low or the Azores high, which determine weather and climate in Europe. Similarly, the western North American climate is strongly influenced by the Aleutian low and the North Pacific high. Subtropical highs are also found over the three Southern Ocean

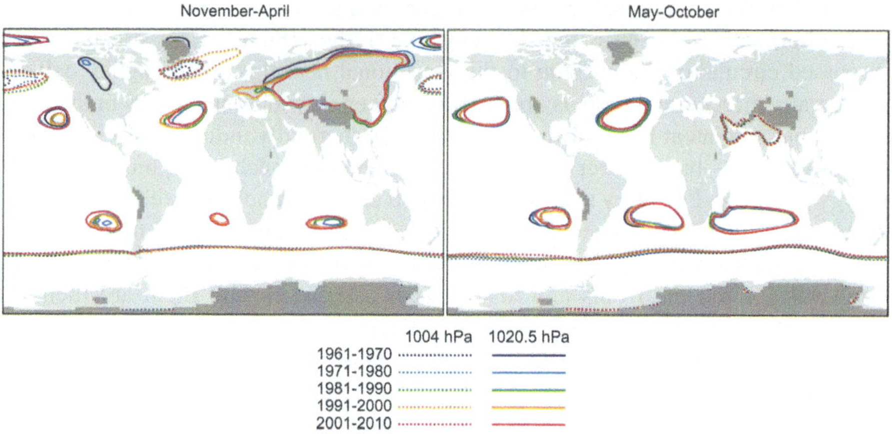

Fig. 3.6 Decadal averages of mean sea level pressure in (*left*) November–April and (*right*) May–October from the 20CR dataset (from Hartmann et al. 2013). Each colour shows a sea level pressure map for a given decade, depicted in the form of two selected contour lines (see also Fig. 2.12)

basins, as are low pressure systems such as the Amundsen–Bellingshausen Seas low. Climatologically, however, the southern high latitudes appear as one large pressure centre (Fig. 3.6).

In addition to these dynamical systems, thermal high and low pressure systems develop over the continents and affect the large-scale flow (Fig. 3.6). They are most pronounced over Eurasia, with a strong Siberian high in winter and the monsoon low in summer.

These weather systems or centres of action were discovered a few years before Brückner's book appeared (e.g., Teisserenc de Bort 1881; Galton 1863), and climatologies of pressure and atmospheric circulation were quickly established (see also Fig. 2.12 based on Hann's atlas published in the 1887). Strong or mild winters in Europe were related to changes in the Icelandic low and Siberian high, and some authors speculated about oceanic influences on the pressure centres (Meinardus 1898).

Today we know that not only interannual, but also decadal changes in the centres of action can be large. Even just during the past 50 years (displayed in Fig. 3.6 in terms of the position and spatial extent of isolines; see Hartmann et al. 2013), there were decades when the Azores high and Icelandic low in boreal winter were very large (e.g., in the 1990s) or very small (1960s). The North Pacific high in boreal summer extended farther westward in the 1960s and 1970s than in the 2000s. In recent decades, the subtropical highs in the Southern Hemisphere were larger in austral summer than in earlier decades. Decadal changes in the pressure centres are important for understanding weather and climate changes since 1700 and will be revisited in Chap. 4.

3.1.2.4 Energy Transport

Recall that the meridional gradient in energy received by the climate system is equilibrated to a large extent by atmospheric heat fluxes. How do the above discussed circulations contribute to the meridional heat flux? The Hadley cell, as a thermally direct circulation, is able to transport energy to around 30°N in the form of sensible and latent heat. Farther poleward, the thermally indirect Ferrel cell also transports heat—but equatorward. Therefore, another mechanism must transport energy poleward. Planetary waves can transport energy without an exchange of mass provided that there is an exchange of energy (e.g., heat uptake from the surface at low latitudes and radiative cooling at high latitudes). The transient weather systems (or "eddies") and the fronts and disturbances associated with them are then the ultimate location where energy exchange between the tropics and the polar regions takes place through dissipation and exchange of air.

Let us now switch from a process view to a statistical view of heat flux and waves. The meridional heat flux at given location and level can be written as the product of meridional wind and temperature vT. Its zonal mean can be split into the contribution of the zonal-mean flow, time-averaged stationary waves (mostly the expression of planetary waves), and transient eddies (i.e., weather systems). This is achieved by expressing v (and the same for T) in the following form:

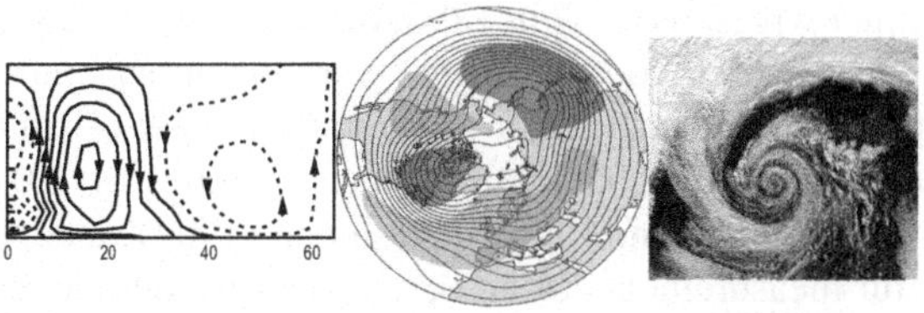

$$\overline{[vT]} = [\overline{v}][\overline{T}] + [\overline{v}^*\overline{T}^*] + [\overline{v'T'}]$$

Meridional Heat Flux | Mean Meridional Circulation | Stationary Waves | Transient Eddies

Fig. 3.7 Decomposition of the zonal-mean meridional heat flux over a season into contributions from the mean meridional circulation, the stationary waves, and eddies

$$v = [\overline{v}] + \overline{v}^* + v' \tag{3.2}$$

where square brackets and asterisks denote a zonal average and the deviation, respectively. The overbar and prime denote a time average (e.g., monthly mean) and deviation. The meridional heat flux vT then becomes

$$vT = ([\overline{v}] + \overline{v}^* + v')([\overline{T}] + \overline{T}^* + T') \tag{3.3}$$

Averaged zonally and over time (e.g., a month or a season), the flux for a given latitude can be written as the sum of the contributions by the zonal-mean circulation, the quasi-stationary waves, and transient eddies (Fig. 3.7). Note that the separation between stationary and transient eddies is statistical and depends on the averaging period; we use 1 month in this book for convenience because many fields are only available as monthly means. It is important to note that the flux by the mean circulation can be in opposite direction in the lower and upper troposphere and thus needs to be integrated vertically. In a nutshell, this is how the machinery exchanges energy meridionally.

3.1.3 Stratospheric Circulation

Climatic changes since 1700 cannot be understood without considering the stratosphere. It is common knowledge that the stratosphere hosts the earth's ozone layer, which shields the surface from harmful UV radiation. But the stratosphere is much more than that. It is able to affect climate near the ground radiatively and dynamically and thus provides pathways through which external forcing factors such as volcanic eruptions or changes in solar activity (both of which are known to affect the stratosphere) can alter climate. It also plays a role in modulating the internal variability of weather and climate (see Polvani (2010) and Kidson et al. (2015) for an overview).

When Brückner wrote his book in 1890, the stratosphere was not yet discovered. Scientists had just started to explore the vertical dimension of the atmosphere (Labitzke and van Loon (1999) offer a historical review of stratospheric science). Soon, free-flying registering balloons reached altitudes of 20 km. In 1902, Léon Teisserenc de Bort (1902)[7] and Richard Assmann (1907)[8] independently published their first careful measurements of an upper-level isothermal layer or inversion—the tropopause was discovered (Hoinka 1997). A few years later, Arthur Berson[9] observed this layer in the tropics, but at a much higher altitude (Süring 1910). The existence of a stratosphere was established, and soon it was found that the stratosphere was the location of the ozone layer, which was discovered at about the same time from observations of solar spectra (Fabry and Buisson 1913).

In the 1920s, G. M. B. Dobson[10] started measuring the total column ozone in the atmosphere with a global network (Dobson and Harrison 1926). This contributed to the understanding of ozone and the dynamics of the stratosphere, which could not be observed directly back then (see Brönnimann et al. (2007a), Brasseur (2008), Müller (2009), and Bojkov (2012) for historical overviews).

After the development of the radiosonde to operational standards in the 1940s and particularly with the IGY, radiosonde measurements became abundant enough to compile the first climatologies of the stratosphere and to study stratospheric dynamics (Scherhag 1948). In 1952, Richard Scherhag[11] discovered the phenomenon of sudden warmings of the polar stratosphere (Scherhag 1958). At about the same time, the Brewer-Dobson circulation was described (Dobson 1956) and in 1961 the Quasi-Biennial Oscillation was discovered (Reed et al. 1961; Veryard and Ebdon 1961).

Around 1970, stratospheric chemistry (e.g., Crutzen 1970) became the focus of debates on possible side effects of nuclear explosions, fleets of supersonic transport, or the Space Shuttle (Goldsmith et al. 1973; Johnston et al. 1973). Ozone depletion by anthropogenic chlorofluorocarbons was projected at that time (Cicerone et al. 1974)–and later turned into reality in a much more severe form: the heterogeneous loss of ozone in the polar lower stratosphere known as the ozone hole (see Farman et al. (1985), Solomon et al. (1986), and Solomon (1999) for a review), which will be discussed in Sect. 4.5.2.

---

[7]Léon-Philippe Teisserenc de Bort, 1855–1913, was a French meteorologist and pioneer of aerology. He founded the observatory in Trappes.

[8]Richard Assmann, 1845–1918, was a German meteorologist and leading aerologist. From 1905 to 1914, Assmann was director of the aerological observatory in Lindenberg (see Emeis (2012)).

[9]Arthur Berson, 1859–1942, was a German meteorologist and aeronaut and the assistant of Assmann in Lindenberg. For Berson's work in East Africa, see Brönnimann and Stickler (2013).

[10]Gordon Miller Bourne (G. M. B.) Dobson, 1889–1976, was a British physician and meteorologist. He is well known for his decade-long work on ozone observations (Walshaw 1989).

[11]Richard Scherhag, 1907–1970, was a German meteorologist. Scherhag made important contributions to upper-level meteorology, including the Berlin weather charts and its stratospheric maps.

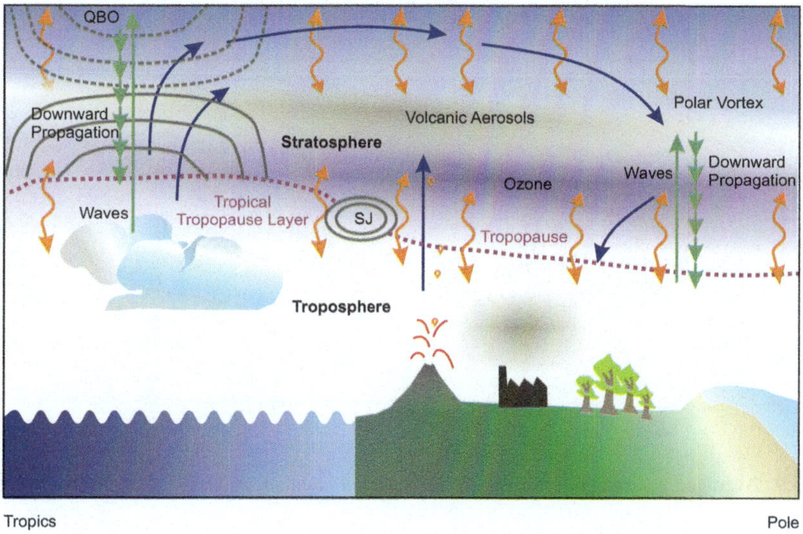

Fig. 3.8 Schematic of stratospheric processes and their role in climate. *Orange*, *blue*, and *green arrows* denote fluxes of radiation, mass, and momentum, respectively. *Grey lines* indicate zonal winds (*QBO* and *SJ* denote Quasi-Biennial Oscillation and subtropical jet, respectively)

3.1.3.1 Demarcation and Zonal-Mean Circulation

So, what is the stratosphere and how does it relate to climate? Figure 3.8 schematically depicts the main components and characteristics of the stratosphere. It extends from the tropopause at ca. 16 km (tropics) or 8 km (polar regions) altitude to the stratopause at ca. 50 km altitude. It is very dry, hosting the ozone layer and occasionally volcanic aerosols, and is characterised by a temperature increase with altitude. The lower boundary, the tropopause, can be defined based on the temperature, moisture, or ozone profiles or based on potential vorticity. In any case, the tropopause is not a well-defined surface. This is particularly relevant in the tropics as this region controls the entry of water vapour and other trace gases into the stratosphere (Fueglistaler et al. 2009). The term tropical tropopause layer is often used to address the transition zone.

The stratosphere is coupled to the troposphere through fluxes of energy (mostly in the form of radiation, with ozone playing a dominant role), mass, and momentum. Momentum fluxes are largely exerted through waves and their interaction with the mean flow, and they are important for understanding circulation. Circulation, in turn, is important to understand trace gas distribution.

The main circulation features of the stratosphere are the polar vortices in the winter hemisphere, the Quasi-Biennial Oscillation (QBO) of zonal-mean zonal wind in the equatorial regions, and the slow meridional circulation known as Brewer–Dobson circulation (see Holton et al. 1995). Mean geopotential height fields in

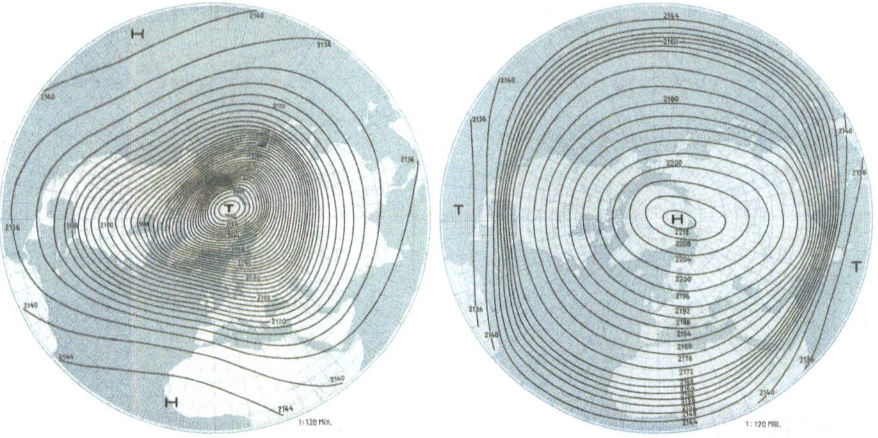

Fig. 3.9 Averaged geopotential height at 41 hPa in (*left*) January and (*right*) July (From Scherhag 1948)

the lower to midstratosphere (Fig. 3.9) show a strong low pressure system centred over the North Pole in January–the polar vortex. This vortex is not strictly annular, but has a wavenumber 3 structure. Farther up, the Aleutian high over the North Pacific displaces the vortex. In midwinter, the polar vortex encompasses the entire extratropical stratosphere, with a westerly wind maximum called "polar night jet". In summer, the polar low pressure system is replaced by a high pressure system above around 20 km and westerly winds by easterlies. Because the sunlit part of the stratosphere is heated more strongly than the other, these seasonal changes can be understood from the thermal wind relation (Eq. 3.1).

3.1.3.2 The Quasi-Biennial Oscillation (QBO)

In the equatorial region, stratospheric zonal winds are dominated by the Quasi-Biennial Oscillation—a peculiar alternation of easterly and westerly winds (see the review by Baldwin and Dunkerton (2001)). Today, the QBO is understood to be caused by a broad spectrum of vertically propagating waves—gravity waves, inertia gravity waves, and equatorial Rossby gravity waves (see Table 3.1)—that break in the stratosphere and interact with the mean flow. The contributing waves have different propagation characteristics, but how high they can propagate depends on the zonal flow and its shear, which thus acts as a filter. In simple terms, only eastward-propagating waves can penetrate in easterly winds and vice versa for westerly winds. Let us assume that the wind is easterly throughout the stratosphere. Eastward-propagating waves penetrate to the upper stratosphere where they break and dissipate. Thereby they deposit their eastward momentum and change the mean flow, which (in the absence of the Coriolis force at the equator) will eventually become westerly. This change in wind direction and shear changes the propagation characteristics for subsequent waves, which have to break at a lower level. The boundary between easterlies and westerlies thus propagates downwards. In the course of about a year, the boundary reaches the tropopause and is eroded. Now, the entire stratosphere has westerly winds, and westward-propagating waves can penetrate up to the upper stratosphere, where they break and eventually lead to easterly winds propagating downwards.

The QBO is a very regular phenomenon, with a periodicity of about 28 months (the easterly phase lasts slightly longer and is slightly more pronounced). It has been stable over the past six decades, for which radiosonde data are available, but recent work by Kawatani and Hamilton (2013) suggests a trend towards a weakening of QBO easterlies. Using historical upper-air data and the QBO imprint in other variables, we have recently reconstructed the QBO back to Brückner's time (Fig. 3.10).

The QBO is known to influence the extratropical stratosphere. An easterly QBO at 40 hPa favours a weak polar vortex (Holton and Tan 1980); a signal that can propagate down to the surface and affect weather in the northern extratropics (this mechanism is discussed in the next paragraph; see also Baldwin and Dunkerton (2001)). Perhaps this explains the unusual differences in Scandinavian snow cover between even and odd years that were noted during Brückner's time (Woeikof 1895).

The QBO might also directly affect the tropical troposphere through interaction with high-reaching tropical convection (e.g., Giorgetta et al. 1999). It has been suggested that the QBO modulates hurricane frequency (Gray 1984), the Indian monsoon, and the Pacific Walker circulation (see also Huang et al. 2012). Comparing the equatorial zonal-mean zonal wind in the lower stratosphere from our QBO reconstruction with corresponding circulation indices over the last 100 years (Brönnimann et al. 2009b) shows positive correlations with the Pacific Walker circulation—westerly QBO enhances the Walker circulation—and negative correlations with an Indian monsoon index (see Table 3.2 for definitions), but correlations are relatively low.

3.1.3.3 Meridional Circulation and Wave–Mean Flow Interaction

Winds in the stratosphere are essentially zonal. However, there is also a slow meridional circulation, which is particularly important with respect to trace constituents. Dobson's early ozone observations showed more ozone at midlatitudes than in the tropics and, in the extratropics, more ozone in late winter than in late summer (Fig. 3.11). This observation was puzzling as ozone is formed photochemically and thus expected to be higher in the tropics than in the extratropics and higher in summer than in winter. The peculiar distribution of ozone can be explained by a slow meridional circulation that moves ozone away from its tropical source regions towards the winter hemisphere, where its life time is longer. This meridional circulation is closed by upwelling in the tropics (ozone-poor tropospheric air enters

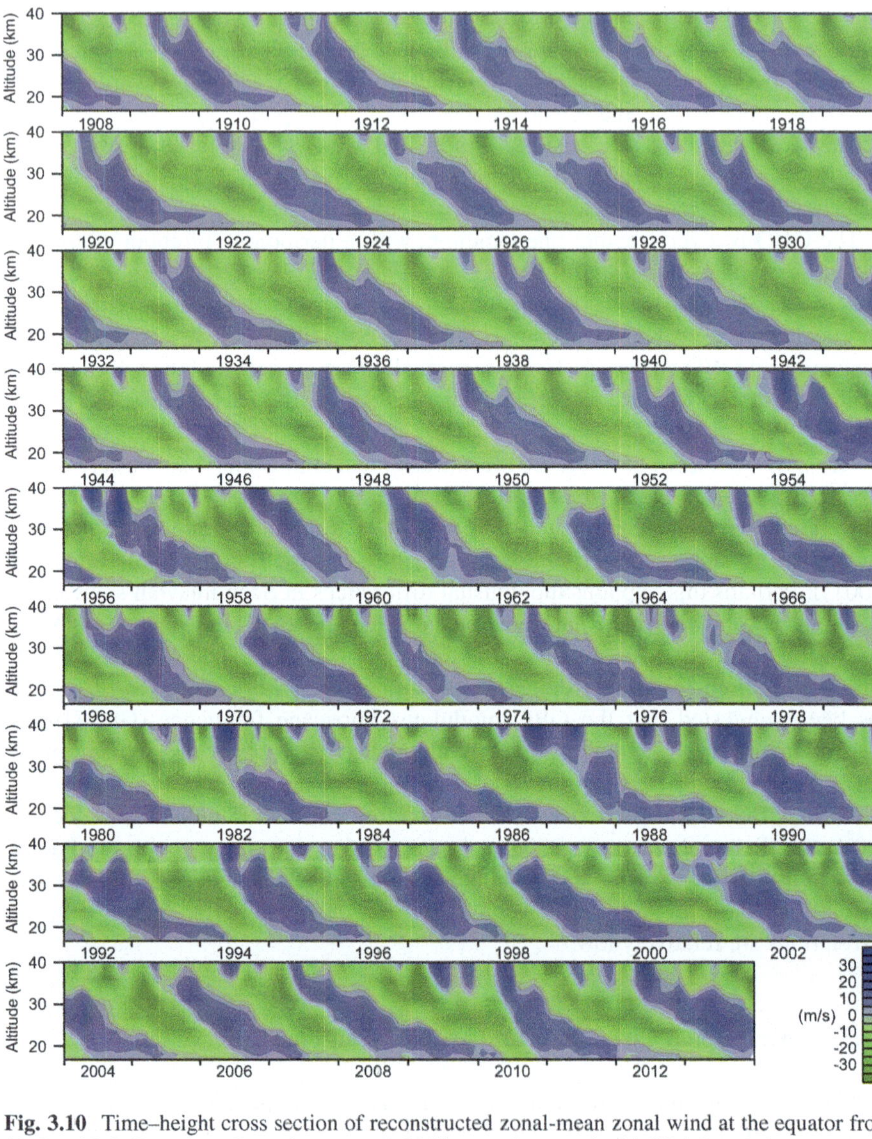

Fig. 3.10 Time–height cross section of reconstructed zonal-mean zonal wind at the equator from 1908 to 2013 (Data from Brönnimann et al. 2007a, supplemented with ERA-Interim)

the tropical stratosphere) and downwelling at subpolar and polar latitudes, where ozone eventually reaches the troposphere and is deposited (Holton et al. 1995).

This is called Brewer-Dobson circulation in the context of transport and mixing of trace constituents, or "residual circulation" in a dynamical context. It originates in vertically propagating long Rossby waves (mainly wavenumbers 1–3, see Table 3.1), which penetrate into the stratosphere and break. These waves require westerly winds

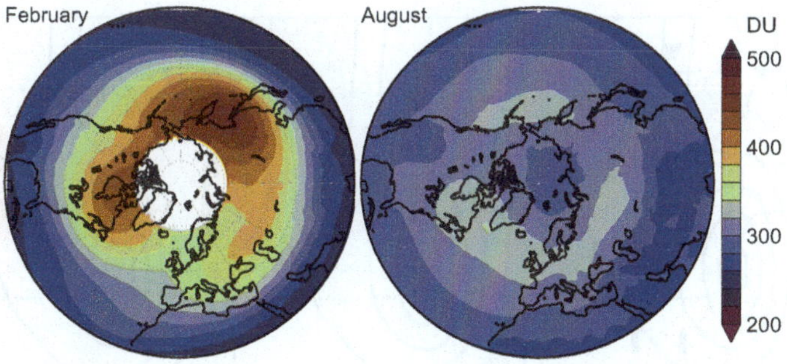

Fig. 3.11 Total column ozone for *(left)* February and *(right)* August from Total Ozone Mapping Spectrometer Version 8 data averaged over the 1979–2000 period (From Brönnimann 2002)

for vertical propagation, which they find in the winter season. Waves of a smaller length scale (wavenumbers 5 and higher, e.g., synoptic waves; see Table 3.1) cannot penetrate the stratosphere.

In the stratosphere, the waves eventually break and deposit momentum. The dissipation exerts a drag on the mean flow and leads to deceleration. The deceleration, via an adjustment of the geostrophic flow, leads to a small wind component towards the pole. Because the flow must satisfy the thermal wind relation (Eq. 3.1), the deceleration of westerlies must be accompanied by a cooling below that level on the equatorward side and a warming on the poleward side. These temperature anomalies are maintained through ascent (equatorward) and descent (poleward), resulting in a meridional circulation that is upwards in the tropics, poleward in the midstratosphere, and downwards in the polar and subpolar stratosphere. These processes overlay the diabatic descent within the polar vortex due to radiative cooling and the ascent in the tropics due to overshooting convection.

The processes driving the meridional circulation and its effects on stratospheric winds and temperatures are shown schematically as latitude–height cross sections in Fig. 3.12. The propagation of wave activity, as depicted by the Eliassen–Palm flux,[12] is shown as arrows in the figure. Convergence of this flux (dashed contours, also indicated by the shortening of the arrows) indicates wave breaking. Waves propagate upwards near 60°N, are refracted equatorward, and break at ca. 30 hPa. The induced residual circulation leads to cooling of the tropical lower stratosphere and to warming of the polar stratosphere (Fig. 3.12, right). Stratospheric westerly winds are weakened (Fig. 3.12, middle).

[12]The Eliassen–Palm flux is a vector quantity in the latitude–height plane. It is used as a measure of wave activity propagation (see Andrews et al. (1987) for details). Vertical and meridional components indicate the importance of the meridional eddy heat flux and the meridional flux of zonal momentum.

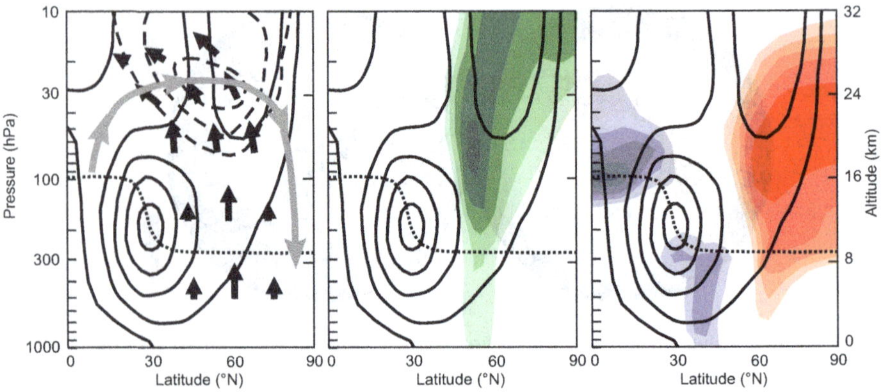

Fig. 3.12 Schematic latitude-height cross sections of the (*left*) effect of vertically propagating planetary waves on the mean flow and of the difference in (*middle*) zonal-mean zonal wind (*green* colours indicate weakening) and (*right*) zonal-mean temperature (*blue* and *red* colours indicate cooling and warming, respectively) in boreal winter for strong minus weak planetary wave activity. *Solid contours* indicate climatological zonal winds, *dotted lines* represent the tropopause. *Small arrows* and *dashed lines* at (*left*) indicate the Eliassen–Palm flux and its divergence, respectively, and the *long arrows* show the induced residual circulation (Adapted from Newman et al. 2001)

3.1.3.4 Sudden Stratospheric Warmings and Downward Propagation

Wave–mean flow interaction is not a continuous process. Large wave breaking events can have a particularly strong effect. If the wave activity flux is focused towards the pole, strong events may lead to a collapse of the wintertime polar vortex—a reversal of the flow from westerly to easterly accompanied by rapid warming of the polar stratosphere due to downwelling. These events occur almost exclusively in the Northern Hemisphere. The warming can reach an amplitude of 70 °C within a few days, as is shown in Fig. 3.13 for the case of January 2009. Events triggering a collapse of the vortex are called "Sudden Stratospheric Warmings". If they occur early enough in the season ("Major Midwinter Warmings"), the vortex can recover after 2 or 3 weeks (as in January 2009; see Fig. 3.13). If they occur later, the vortex may not recover and the event then marks the beginning of stratospheric spring ("early Final Warming").

Perturbations of zonal winds in the upper stratosphere can propagate down to the tropopause within a week or so, most likely by changing the propagation characteristic for subsequent waves, though the mechanisms are not fully understood. January 2009 is a good example for this process (Fig. 3.13). Once the perturbations have reached the tropopause, some vanish, while others affect the tropospheric circulation (see also Fig. 3.12) for 30–60 days. Following the January 2009 event (Fig. 3.13), easterly winds at the surface were more common than before the event. Downward propagation to the surface appears even more clearly when overlaying many such events (Baldwin and Dunkerton 2001). This makes the stratosphere interesting for seasonal prediction (Sigmond et al. 2013; Kidston et al. 2015).

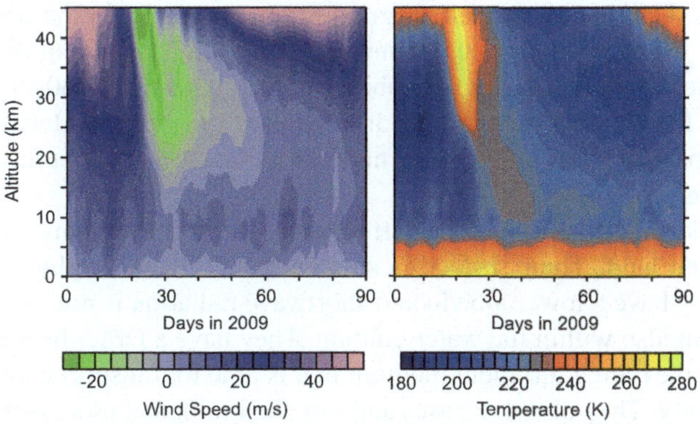

Fig. 3.13 Altitude–time cross sections of (*left*) zonal-mean zonal wind at 75°N and (*right*) temperature at 90°N from January to March 2009 (Data source is ERA-Interim)

Downward propagation is not only relevant for long-range weather forecasting, but also provides a mechanism of how climate forcings that primarily affect the stratosphere can have an influence on weather and climate near the ground. In relation to forcings, this is often termed a "top down" mechanism. Volcanic and solar activity as well as ozone depletion are examples of such forcings. The stratosphere may thus play an important role in some of the climatic changes addressed in this book—a major new perspective since the time of Brückner's writings.

3.1.4 The Ocean, Air–Sea and Land-Sea Interactions

Covering seventy percent of the earth's surface, oceans are an important component of the machinery. The atmosphere feels the ocean underneath and vice versa. Mechanisms of air–sea interaction are also essential for interannual-to-decadal climate variability and thus for understanding Chap. 4. The following pages give a superficial introduction to the ocean's effect on climate through air–sea interaction and land–sea mechanisms. The interested reader is referred to numerous textbooks (e.g., Sarmiento and Gruber 2006; Siedler et al. 2013; Wang 2004).

The influence of the ocean on mean climate has been known for a long time (e.g., Barrow 1819). For instance, since the mid-19th century, the relative warmth of Europe as compared to North America was explained by the Gulf Stream (the contribution of the atmosphere was not yet understood). James Croll[13] highlighted the role of the oceans for mean climate (e.g., Croll (1870); see also Kang et al. (2013)) and ascribed them a role in ice age mechanisms (Croll 1875). However, many climatologists at that time, including Brückner, did not devote much attention

[13]James Croll, 1821–1890, was a Scottish scientist. He developed one of the first physical ice age theories.

to the oceans. Despite substantial progress in oceanography in the first half of the 20th century, the climate science community only became aware of the importance of air–sea interaction for climate variability in the 1950s and 1960s (e.g., Bjerknes 1964, 1966). Even today, the ongoing discussion of the slowing global temperature rise over the past 15 years (Sect. 4.5.9) highlights that we do not yet fully understand the role of the oceans.

What makes oceans special, apart from their size? Oceans have different properties than land. Their surface is smoother (i.e., their roughness is smaller). They normally have a lower albedo and shortwave radiation is not only absorbed at the surface but also within the water column. They have a larger heat capacity than typical soils; they are a turbulent medium that is able to transport energy vertically and horizontally. They dissolve gases and link the biological processes in the ocean to the atmosphere. When ice covered, the properties of oceans area again entirely different. The differences in physical properties of land and ocean cause spatial gradients in climate variables which drive further processes.

3.1.4.1 Ocean–Atmosphere Interaction

Some important processes of air–sea interaction expressed in terms of the modulation of fluxes of heat, mass, and momentum are displayed schematically in Fig. 3.14. To the atmosphere, oceans are a large water reservoir, a heat reservoir, and (because they dissolve gases) a trace gas (such as CO_2) reservoir. Fluxes of heat, water, and trace gases between ocean and atmosphere operate in both directions and net fluxes are often the differences between two large quantities. Moreover, oceans are also a source of aerosols, which originate from evaporating water droplets (sea salt) or from sulphur compounds (dimethyl sulphide (DMS)) emitted by marine biota (see also Box 3.4, p. 161). The radiative properties of the ocean or ice surface, through modulating turbulent fluxes, also affect the atmosphere directly.

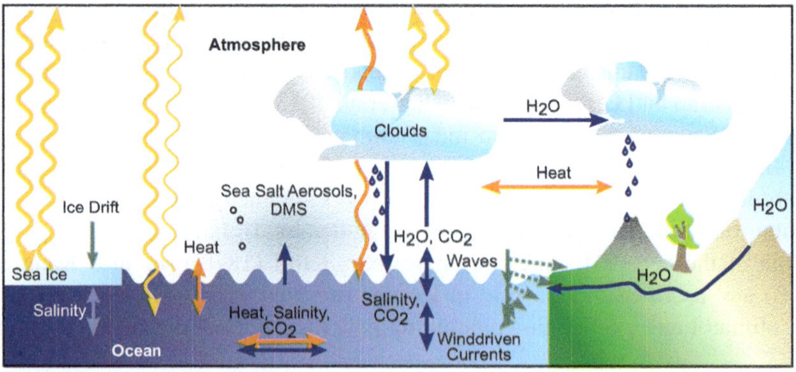

Fig. 3.14 Schematic displaying the processes at the air–sea boundary. *Yellow/orange*, *blue*, and *green arrows* denote fluxes of energy, mass, and momentum, respectively

To the ocean, the atmosphere is a source of momentum and freshwater, and a modulator of radiation. Heat fluxes from the atmosphere affect sea-surface temperatures. In terms of momentum fluxes, atmospheric wind stress sets ocean waters in motion and leads to ice drift. Due to the Coriolis force, the wind-induced motion of the water body is perpendicular to the wind stress, but with a vertical structure known as the Ekman spiral. The ocean system also reacts to fluxes of freshwater, heat, and CO_2 in ways ranging from thermohaline circulation (described in the next paragraph) to ocean acidification. Ocean salinity reflects freshwater fluxes between ocean, atmosphere, and land surface (precipitation, evaporation, and runoff).

Depending on the vertical structure of the atmosphere, ocean-atmosphere interactions affect different parts of the atmosphere. In the tropics, deep convection transfers the surface signatures to the entire tropospheric column. In the extratropics, a shallow marine boundary layer often restricts direct interaction to a small layer of the troposphere.

Similarly, within the ocean deep or shallow layers may participate in the interaction. For instance, thermal anomalies may be mixed into deeper layers in winter but then cut off from the atmosphere by a seasonal thermocline. When the column is mixed again in the next winter, the previous year's signal is mixed to the surface. In this way, the ocean may damp the atmospheric signal and return it back to the atmosphere, thus "reddening" the spectrum.

In regions of deep convection in the atmosphere, oceanic thermal anomalies affect the atmosphere through convective adjustment and subsequent dynamical effects. The response is important for ocean–atmosphere interactions related to El Niño–Southern Oscillation (ENSO) (see Sect. 3.2.6). Outside of the regions of deep convection, oceanic effects on the atmosphere (predominantly wind and clouds) are less well understood. Mechanisms involving moist convective processes also operate there (Czaja and Blunt 2011). Also near oceanic fronts, surface wind adjusts to sea-surface temperatures gradients through vertical momentum mixing (Xie and Carton 2004). Transient eddies in the ocean also affect the atmospheric flow by modifying turbulence (see Frenger et al. 2013).

Who rules who? In the tropics, ocean–atmosphere interaction is a close two-way coupling. In the extratropics, on short time scales (seasons), the ocean mainly responds to the atmosphere. On long time scales (decades), however, the ocean may influence the atmosphere (Bjerknes 1964; Gulev et al. 2013).

3.1.4.2 Ocean Currents

Surface currents of the oceans are mainly wind driven. The large-scale overturning circulation, in contrast, is density driven. Thus, both are strongly related to their interaction with the atmosphere through wind stress on the one hand, and freshwater (precipitation, runoff, and evaporation) and heat fluxes on the other hand. A map of global ocean surface currents is given in Fig. 3.15. The surface currents largely reflect wind patterns. In areas of anticyclones, Ekman transport to the centre (piling up of water) is balanced by the pressure gradient force, and an anticyclonic

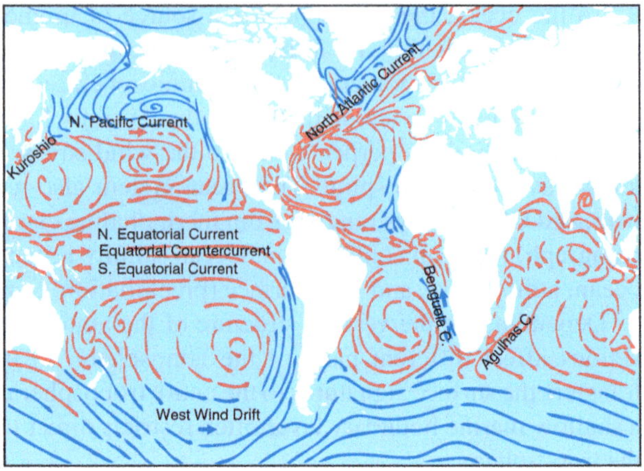

Fig. 3.15 Schematic of surface current (warm and cold). Based on U.S. Naval Oceanographic Office

geostrophic flow develops. The opposite occurs for cyclones. These circulations are called "gyres"; the subtropical gyres are particularly prominent. On their western sides, the flow is intensified due to the latitudinal change in the Coriolis force (western boundary currents). Due to the geometry of the ocean basins, a circumpolar current can be established around Antarctica but not in the Northern Hemisphere.

The slower, density-driven part of the global ocean circulation is the thermohaline circulation (e.g., Wunsch 2002). This is a global circulation driven by the formation of dense, cold waters in specific areas (such as in the North Atlantic), which subside to the deep ocean and continue to flow horizontally to other ocean basins. The Atlantic branch of the thermohaline circulation, known as the Atlantic meridional overturning circulation (AMOC), is of particular relevance for climatic changes since 1700. This circulation might exhibit variability on multidecadal time scales although observational evidence is scant (see Rahmstorf et al. 2015). The AMOC affects North Atlantic sea-surface temperatures, which in turn may affect regional and even global climate (see Sect. 3.2.8).

3.1.4.3 Land–Sea Interaction

Due to the different physical properties of land and ocean, spatial gradients develop that may induce specific circulation systems. Figure 3.16 shows atmospheric circulation over the land and ocean and associated phenomena. The most well-known effects are thermal wind systems (not to be confounded with the thermal wind relation introduced earlier) such as sea breeze. This forms during the day when the land heats up more strongly than the ocean, drawing an air flow towards the land. Land breeze at night is the reverse. On a larger (continental and seasonal) scale, monsoon systems develop, which are described in more detail below. Over the extratropical ocean, there is typically a shallow marine boundary layer. In tropical oceans, deep convection can develop. Near the coast, offshore winds or Ekman transport can drive the water away from the land. It will be replaced by upwelled

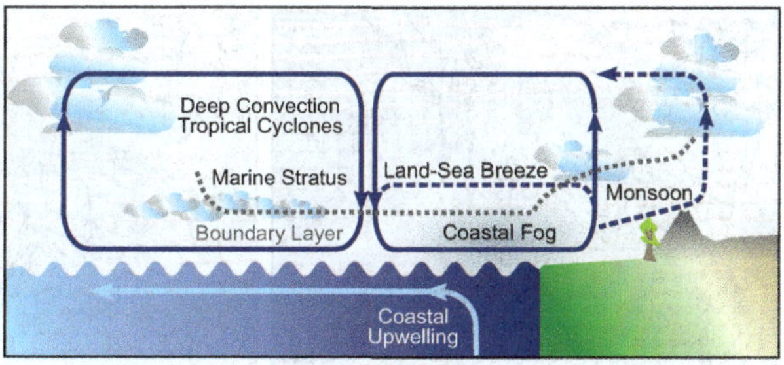

Fig. 3.16 Schematic of land–sea circulation systems and related phenomena on different space scales

cold deep water. Fog often forms near the coast when the overlying atmosphere is cooled to saturation.

Changes in either the sea-surface temperatures or the land surface may alter land–sea circulation systems such as monsoon. These mechanisms are discussed in the following chapter.

3.1.4.4 Monsoon

Monsoon circulations are large-scale, seasonally varying systems that are found on all continents. The large Eurasian land mass and Australian–Western Pacific region gives rise to the Australasian monsoon (Christensen et al. 2013), which consists of five monsoon systems: Indian, East Asian, Maritime Continent, Western North Pacific, and Australian monsoon. North and South America also have a monsoon system, and the West African monsoon is responsible for precipitation in the Sahel.

Arguably the most well-known monsoon circulation is the Indian monsoon (Fig. 3.17). Rains from the Indian summer monsoon feed one billion people. The Indian summer monsoon originates in the high pressure system off the coast of Madagascar (Fig. 3.16), from where air flows towards East Africa as a trade wind. Upon crossing the equator, the flow curves northeastward over coastal Africa and forms a low-level jet, known as the Somali jet. The flow then crosses the Arabian Sea and is drawn to the monsoon trough that stretches from northern India to the Bay of Bengal (see Fig. 3.6). Due to this low pressure area (pressure torque) and the effect of coastal mountain ranges (mountain torque), the monsoon flow keeps its westerly momentum. The moisture-laden air masses first rain out upon encountering the western Ghat Mountains in southern India. From there, the air flows eastward and curves north over the Gulf of Bengal towards the eastern Himalayas, from where it is deflected to the west. The onset of the monsoon follows a clear progression from early June, when monsoon rains start in southern India, to late July when the monsoon reaches the northwestern region of the Indian subcontinent. At upper levels (Fig. 3.17), westerlies are displaced north of the Himalayas and strong easterlies develop over India and the Indian Ocean, which can be thought of as a consequence of the thermal wind relation (Eq. 3.1) given the heated Asian land mass and comparably cold Indian Ocean.

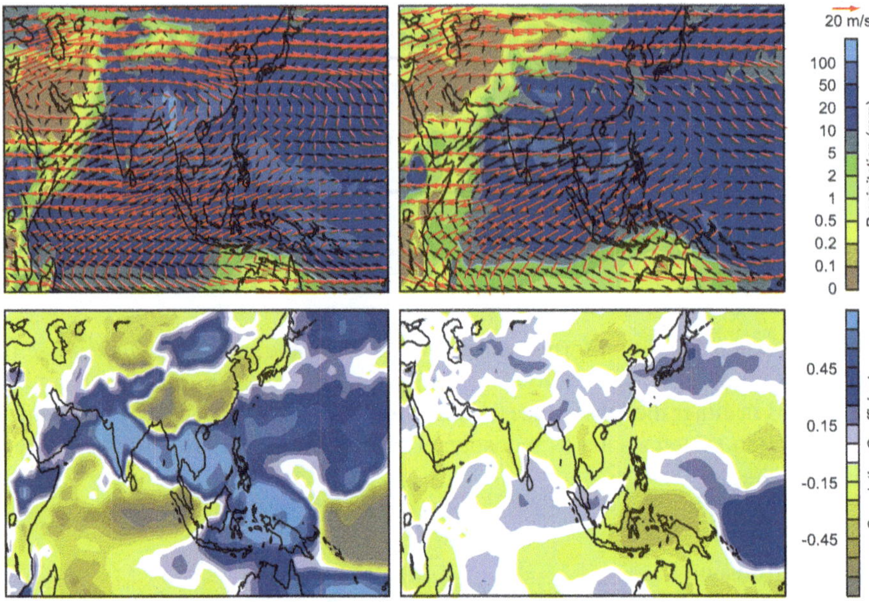

Fig. 3.17 Average precipitation (colour), 925 hPa wind (*black arrows*), and 200 hPa wind (*red arrow*) for the summer period (June-August) in (*top left*) ERA-Interim and (*top right*) CCC400. The *bottom figures* show the correlation of precipitation with (*left*) the DIMI index and (*right*) the NINO3.4 index in CCC400 in summer

What is driving the monsoon circulations? Traditionally it was held that the heating over the Tibetan Plateau is the main driving factor, establishing a cross-equatorial flow and drawing in the moisture-laden air from the Indian Ocean, similar to a large-scale sea breeze. Recently, this view has been modified. Boos and Kuang (2010) suggest that an important role is played by the Himalayas, which prevent the moist air from mixing with dry continental air ("barrier" effect, see also Chen et al. (2014)). Rather, the air rises and releases its latent heat, which further accelerates the monsoon. Rajagopalan and Molnar (2013) show that heating over the Tibetan Plateau is important for Indian summer monsoon rainfall during the early and late monsoon phases.

A special characteristic of the Indian monsoon is its sudden onset. The monsoon circulation can be established when the subtropical jet moves northward of the Himalayas in early summer (Fig. 3.17). An important role in monsoon onset seems to be played by interaction with extratropical eddies (Bordoni and Schneider 2008).

The West African monsoon is driven by the heating of the land mass; there is no plateau or barrier effect. The monsoon draws in moisture from the South Atlantic that rains out over the Sahel. In West Africa, above the monsoon westerlies at the surface, an easterly wind band forms known as African easterly jet (these easterlies, following the thermal wind relation, are a consequence of the thermal contrast between the heated Saharan desert and the cool Gulf of Guinea). Waves in this wind band can trigger tropical cyclones over the Atlantic.

3.1.4.5 Sea Ice and Snow Cover

Sea ice plays a particular role in the climate system (see Fig. 3.18) for several reasons. First, the albedo of ice is very high, making the surface energy balance completely different over ice and the neighbouring ocean. Second, sea ice temporarily separates the ocean from the atmosphere and interrupts heat and moisture fluxes. Third, sea ice influences the freshwater balance when precipitation falls on ice and when brine is rejected, which may affect the ocean circulation underneath. Many different feedback mechanisms that include sea ice have been suggested (see further discussion in Sect. 3.4.3). An overview on climate–sea ice links is given in DeWeaver et al. (2008).

Sea ice occurs both in the Arctic and the Antarctic. The Arctic is fully ice covered in winter; in Antarctica, where no land limits sea ice extension to the midlatitudes, seasonal changes are very large. Sea ice is formed when the upper ocean is cooled to the freezing level by the negative radiation balance during the polar winter and reaches its maximum area in spring. The rate of formation depends on wind speeds, oceanic heat flux, and other factors. A large fraction of the sea ice is seasonal (perennial), which means that the ocean freezes and thaws every winter and summer season. Multi-year ice forms particularly north of Greenland and the Canadian Archipelago.

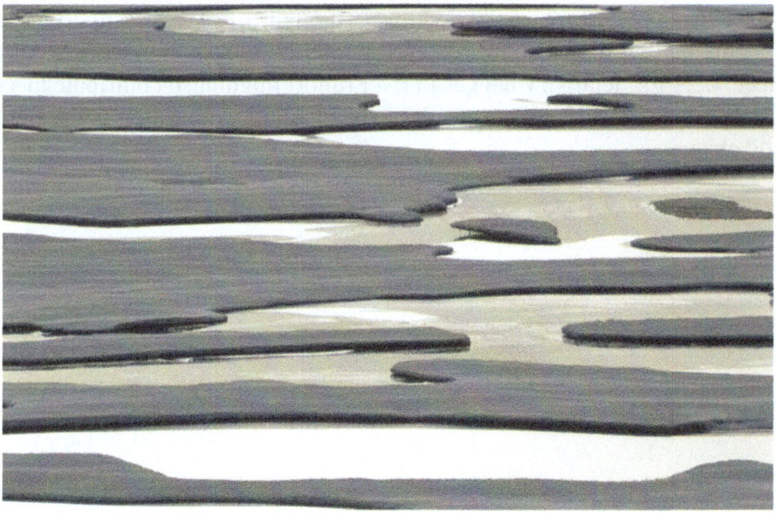

Fig. 3.18 Sea ice plays an important role in the climate system (NOAA Photo Library, Collection of Dr. Pablo Clemente-Colon, Chief Scientist U.S. National Ice Center)

Sea ice area and volume have been decreasing rapidly in the past three decades (see Sects. 4.3.2 and 4.5.8). Open ocean areas in polar regions are a local heat and moisture source that may affect static stability and promote the development and growth of cyclones (Jaiser et al. 2012). Changes in sea ice can have a major effect on regional climate and possibly even affect midlatitude climates. Petoukhov and Semenov (2010) find that reduced sea ice in the Barents and Kara Seas induces an anticyclonic circulation anomaly, which may promote advection of continental cold air to central Europe in winter. Declining Arctic sea ice has been suggested to cause more frequent weather extremes at midlatitudes (Francis and Vavrus 2012; Tang et al. 2013). The findings have however been debated (Wallace et al. 2014).

Another potential climate factor is snow cover. Changes in snow cover not only affect Arctic climates, but may also be relevant at midlatitudes and even affect tropical circulation. Effects of snow cover on the large-scale circulation were already studied during Brückner's time. In the 1880s, Henry F. Blanford, Chair of the India Meteorological Department, postulated a link between snow cover and the Indian Monsoon (Blanford 1884) that is still studied today though it appears to be rather weak (Fasullo 2004). Various authors have suggested that Siberian snow cover in fall, via changes in albedo and the surface energy balance, leads to increased Rossby wave activity flux in December, which, via the stratosphere and subsequent downward propagation, may affect the North Atlantic Oscillation in late winter (see Cohen et al. 2014, for a review).

Coupling mechanisms or feedbacks involving snow cover are of interest today because snow cover may provide a memory for atmospheric processes over longer time scales than the atmosphere itself, raising the hope for extended long-range predictability. The same holds for several other coupling processes with slower parts of the climate system such as sea ice, the ocean, or ice sheets.

Sea ice–ocean–atmosphere interaction is thought to play an important role in today's climate, but may also have played a role in the past via changes in salinity (see Sect. 4.2.1 for details). Changes in river runoff might also contribute to salinity anomalies. Hence, sea ice needs to be kept in mind even when addressing past climatic changes.

3.2 Expressions and Mechanisms of Climate Variability

Mean circulation and its governing processes are the basic mode of operation and working principles of the machinery. However, the machinery does not always operate at the same pace. Sometimes it slows down or speeds up, or it shifts gears. This chapter discusses temporal variability in atmospheric circulation and climate.

3.2.1 Statistical and Physical Perspectives of Climate Variability

The term "machinery" implies a focus on processes, which I will consider in this chapter. However, as introduced in Chap. 1, an alternative approach to climatic changes is statistical. Establishing scientific knowledge requires that the two perspectives are taken in concert. Empirical analyses need an explanation and, conversely, explanations need empirical evidence.

Historically, the empirical view has dominated climate sciences. For a long time, including Brückner's period, climatology was a descriptive science serving other disciplines (see Box 2.2, p. 22) and far from theory. The statistical approach was the only way forward. Before turning to the processes and in order to introduce this chapter, I would like to resume the statistical perspective from Chap. 1—the example of space and time dependence of temperature trends (Fig. 1.3)—and analyse the spatial and temporal structure of changes in more detail.

At a given location, precipitation and temperature vary in a different fashion. Large precipitation changes may occur from hour to hour or even from minute to minute; time scales over which temperature varies only smoothly. Conversely, temperature usually varies more strongly over a diurnal cycle. The annual cycle may be dominated by temperature changes or precipitation regimes, depending on the climatic zone. How about interannual to multidecadal variability—the time scales analysed in this book?

Maps of the spectral colour (the ratio between low- and high-frequency variability) of the two variables are shown in Fig. 3.19 (see also Fig. 2.29 with respect to proxies). The spectral colour differences indicate that temperature varies more strongly on decadal time scales (red) while precipitation varies more strongly on interannual time scales (blue) or shows no preference (white). This basic difference is also reproduced by the Twentieth Century Reanalysis and by climate models. The map also shows spatial differences in spectral colours. For instance, low latitudes and in particular oceanic regions (as compared to midlatitude land regions) seem to favour low-frequency variability of temperature and to some extent precipitation (note, however, that there are differences between climate models and the reanalysis). This different spectral behaviour is the expression of many different mechanisms of climate variability, which are addressed below.

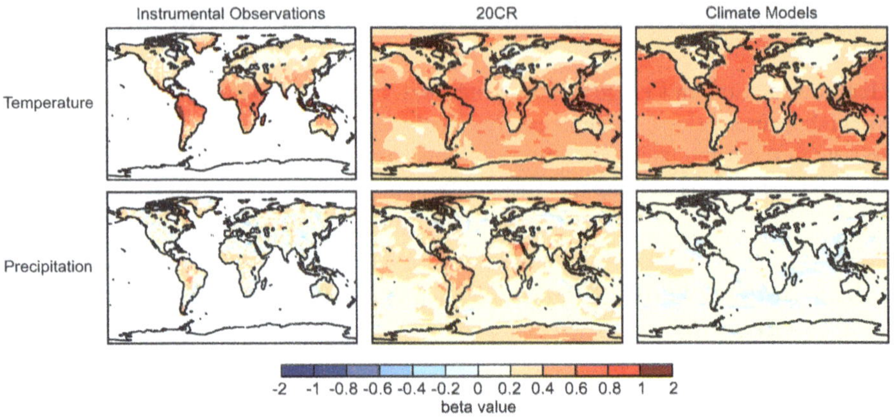

Fig. 3.19 Spectral colour of temperature and precipitation expressed as *beta*-values, calculated for periods between 2 months and 100 years. Maps are plotted for (*left*) instrumental data (data set CRUTS3 from the Climatic Research Unit, University of East Anglia), (*middle*) 20CR, and (*right*) climate models for the past millennium (adapted from Franke et al. 2013). Negative values denote *blue* spectra (high-frequency variability dominates); positive values denote *red* spectra

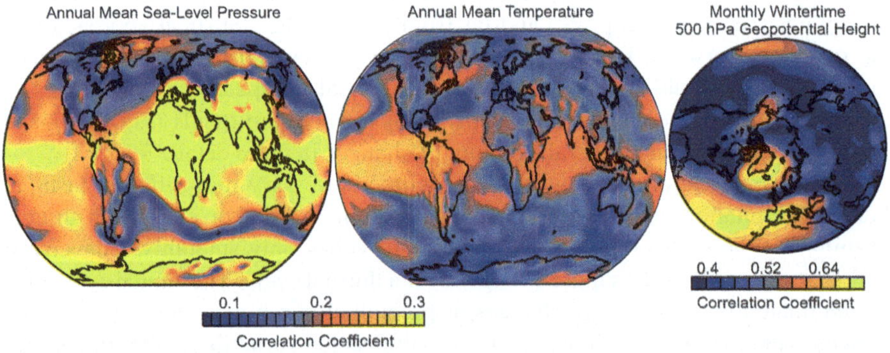

Fig. 3.20 Maps of the area-weighted average of the absolute value of the Pearson correlation coefficient between one location and all other locations for annual means of global sea-level pressure and linearly detrended temperature. (*right*) Map of the most negative correlation for each grid point for monthly boreal winter (December–February) northern extratropical 500 hPa geopotential height (data source is NCEP/NCAR Reanalysis, 1948–2013)

The temporal variability of climate variables has a spatial structure. How much variability does sea level pressure at one location share with the variability at all other locations? We can measure this with the area-weighted average of the absolute value of all correlation coefficients. For annual averages of sea level pressure (Fig. 3.20, left), the largest values are found in the tropics (the African–Australasian region) and in the southern high latitudes. These regions show the most coherent covariability with the rest of the globe. The planetary frontal zones, in contrast, show the least coherent variations. For temperature, the region with most coherent variability is the tropical Pacific.

This covariability equally weights positive and negative correlations. By plotting only the most negative correlation for each grid point, termed a teleconnectivity map, we can focus on the see-saws (Wallace and Blackmon 1983). For monthly values of the northern extratropical 500 hPa geopotential height in boreal winter (right), several see-saws appear over the Arctic, North Atlantic, and North Pacific. The centres with the highest negative correlations are the centres of action, whose variability we will now discuss in some more detail.

3.2.2 *Overview of Circulation Variability Modes*

Recall that the machinery balances differences in the energy budget and that energy conversion involves a cascade of mechanisms on different spatial scales. As the climate system combines atmosphere, ocean, land surface, and sea ice, each with very different typical temporal and spatial scales (see Fig. 1.6), a large range of climatic variations and phenomena can occur. Some of this variability is largely random; we consider it noise or dissipation of energy within our machinery. Other variations are not random but are related to preferred modes of operation of the machinery. One expression of this, in the atmosphere, are so-called circulation variability modes (see Stephenson et al. 2004, for an overview). These are large-scale modes of coherent variability, which have an imprint on different climate variables. Often, they are related to changes in the quasi-stationary waves (expressed most strongly in winter). As a consequence, statistical relations between climate variables at distant locations may appear, as seen in Fig. 3.20 (see Box 3.1, p. 104). These statistical relations are called teleconnections, a term coined in the 1930s (Ångström 1935).

Brückner's contemporaries (Hann 1890; Hildebrandsson 1897) described variability modes, but a statistical, large-scale, and multi-variable view was only introduced by the work of Gilbert Walker (see Box 4.1, p. 224). Analysing simultaneous and lag correlations among pressure, temperature, and precipitation in a global network of stations, Walker (1923, 1924) identified the Southern Oscillation (which we now know to be related to the El Niño phenomenon in the Pacific), North Atlantic Oscillation, and North Pacific Oscillation, variants of which are still the most widely used variability modes. Others have been added in recent years. In fact, dozens of variability modes have been suggested. Table 3.2, adapted from Hartmann et al. (2013), gives an overview of the most frequently used variability modes as well as other indices used later in this book. Some indices can be defined on a day-to-day scale, but they are typically used on a monthly or seasonal scale. These modes are described in the subsequent chapters, starting with the North Atlantic Oscillation.

Climate variability modes are not only indicative of the preferred internal variability of the climate system, but the underlying mechanisms may provide pathways of how climate responds to external forcing. Climate variability modes are therefore discussed in some more detail as they are crucial for understanding climatic changes since 1700 as outlined in this book.

Box 3.1 Statistical and physical interpretation of circulation variability modes

What is a circulation variability mode? From a statistical perspective, this describes spatially coherent variations in climate that may be expressed as correlations in climate variables over long distances. From a physical perspective, the term describes a mode of operation of the machinery. However, the two perspectives are often difficult to separate and a source of confusion.

For instance, defining a variability mode by the centres of action (e.g. Hurrell 1995; Jones et al. 1997; Rogers and van Loon 1979), which are known to affect weather, might appear physically rooted. However, teleconnectivity maps such as that shown in Fig. 3.20 identify the same centres using statistics. Therefore, how physically rooted is a centre of action index? Conversely, calculating the zonal-average circulation implies a statistical view, but there may also be physical reasons for defining variability modes via the strength of the zonal-mean zonal circulation as put forth by Namias (1950) and others.

The indices resulting from a statistical or physical definition may be similar or almost identical, but the perspective is different. Statistical definitions of variability modes often aim at expressing a maximum amount of variability in one variable; physical modes aim to capture a mechanism. The statistical perspective is helpful for impact research (e.g., by comparing indices–without asking for the mechanisms behind–with impacts) or for palaeoclimatological research (allowing us to describe a large amount of variability with few proxy time series). The physical perspective aims at the underlying mechanism, but often requires more specific diagnostics. After all, climate variability modes might well be a combination of several mechanisms.

3.2.3 The North Atlantic Oscillation (NAO)

Winters in central Europe are very variable. They may be cold and dry, such as in 1963, when many lakes in central Europe froze over for weeks (Fig. 3.21), or warm such as in 1990. Why were the two winters so different? Figure 3.22 shows (top) 500 hPa geopotential height, and anomalies of sea level pressure, temperature, and precipitation for January 1990 and January 1963. In January 1990, the 500 hPa geopotential height field shows a zonal structure with only smooth waves. In contrast, in January 1963, the flow at 500 hPa was much more meridional, particularly over the Atlantic. An isolated high pressure system over the Atlantic was "blocking" the zonal flow. Increased meridional flow is also found over the eastern North Pacific.

Table 3.2 Overview of some widely used variability modes (Adapted from Hartmann et al. 2013); for correlation maps see Figs. 3.23 and 3.24. Variables Z, ω, u, and ψ (followed by a number indicating the pressure level; hPa) denote geopotential height, vertical velocity, zonal wind, and the zonal-mean meridional mass streamfunction, respectively. Subscripts "a", "as", and "asc" denote anomalies, standardised anomalies, and anomalies standardised by calendar month, respectively. Brackets denote averages. The indices that are used in Chap. 4 are in italics

Phenomenon	Index	Definition	Reference
El Niño–Southern Oscillation	NINO1+2	SST_a (10°S–0°, 90°–80°W)	Rasmusson and Wallace (1983)
	NINO3	SST_a (5°S–5°N, 150°W–90°)	
	NINO3.4	SST_a (5°S–5°N, 150°E–150°W)	Trenberth (1997)
	NINO4	SST_a (5°S–5°N, 170°–120°W)	
	Southern Oscillation index (SOI)	SLP_{as} Tahiti minus Darwin	Trenberth (1984)
	Troup SOI	10 × SLP_{asc} Tahiti minus Darwin	Troup (1965)
	Equatorial SOI	SLP_{as} (5°S–5°N, 130°–80°W) minus (5°S–5°N, 90°–140°E)	
	Pacific Walker Circulation (PWC)	ω_{500} (10°S–10°N, 180°–100°W) minus (10°S–10°N, 100–150°E)	Oort and Yienger (1996)
Northern Annular Mode (NAM)	*North Atlantic Oscillation (NAO)*	SLP_{as} Azores minus Iceland SLP_{asc} Gibraltar minus Iceland Rotated PC of SLP_a (20°–90°N)	Hurrell (1995) Jones et al. (1997) NOAA
	Arctic Oscillation	Leading PC of SLP_a (> 20°N)	Thompson and Wallace (1998)
	Polar Vortex Index (PVI)	Z_{100} (75°N–90°) minus (40°N–55°)	Brönnimann et al. (2009b)
North Pacific Modes	*Pacific–North American pattern (PNA)*	$\frac{1}{4}(20°N, 160°W) - (45°N, 165°W) + (55°N, 115°W) - (30°N, 85°W)$ in Z_{500as}	Wallace and Gutzler (1981)
		Rotated PC of SLP_a (20°N–90°)	Barnston and Livezey (1987)
	North Pacific Oscillation	SLP_a (30°–65°N, 160°E–140°W)	Trenberth and Hurrell (1994)
Southern Annular Mode (SAM)	Antarctic Oscillation (AO)	Leading PC of Z_{850a} (> 20°S)	Thompson and Wallace (2000)
	Gridded Southern Annular Mode	SLP_a (40°S) minus (70°S)	Marshall (2003)
Pacific–South American pattern	*Pacific-South American pattern (PSA)*	2nd PC of SH seasonal Z_{500a} Z_{500a} (−(35°S, 150°W) + (60°S, 120°W)-(45°S, 60°W))	Mo and Paegle (2001) and Karoly (1989)
Hadley cell and subtropical jet	*Northern Hadley cell (HC)*	$(0°–30°N)_{max}$ of ψ_{500}	Brönnimann et al. (2009b)

(continued)

Table 3.2 (continued)

Phenomenon	Index	Definition	Reference
	Northern subtropical jet (SJ)	$(0°–50°\text{N})_{max}$ of zonal mean u_{200}	
Indian Monsoon	*Dynamic Indian Monsoon Index (DIMI)*	u_{850} (5°S–15°N, 40°–80°E) minus (20°S–30°N, 70°–90°E)	Wang et al. (2001)
Intertropical convergence zone	Intertropical convergence zone index	Lat. of min. of zonal mean ω_{500}	–

Fig. 3.21 Harbour of Lindau, Lake Constance (Photo: W. Stuhler). January 1963 was extremely cold in central Europe and led to widespread freezing of lakes

Interestingly, the sea level pressure anomalies from the long-term mean were almost in the same place in the two cases—they correspond to the main centres of action—but of opposite sign. In January 1990, the Icelandic low and Azores high were both strong (this was typical for the 1990s; see Fig. 3.6 and Sect. 4.5.4). This pattern is known as the positive mode of the North Atlantic Oscillation (Hurrell et al. 2003; Wanner et al. 2001). It is accompanied by mild winters over northern and central Europe (but cold winters in the eastern United States) and an increase in precipitation in western Europe. If both the Icelandic low and Azores high are weak, we call this a negative state of the NAO. Northern and central Europe are cold and dry, while winters are warm in the eastern United States. A statistical view is given in Fig. 3.23.

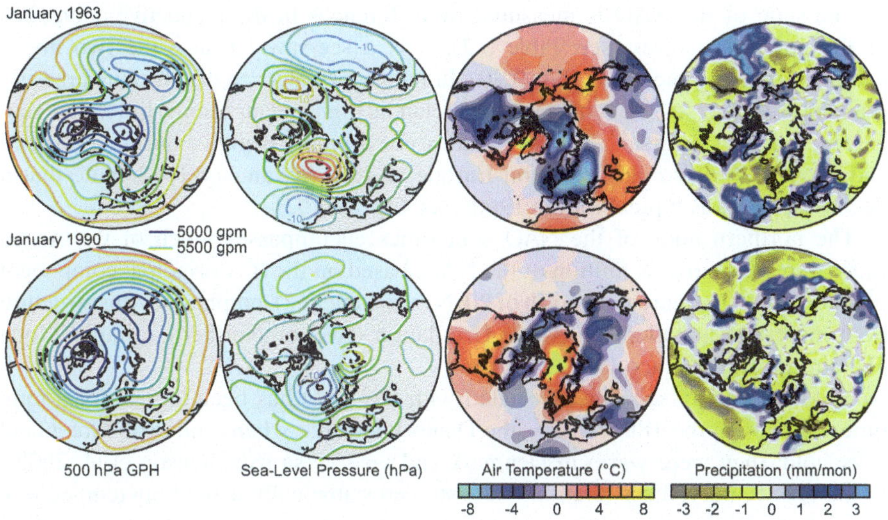

Fig. 3.22 500 hPa geopotential height and anomalies (with respect to 1961–1990) of sea-level pressure, temperature, and precipitation for (*top*) January 1963 and (*bottom*) January 1990 (Data source is NCEP/NCAR Reanalysis)

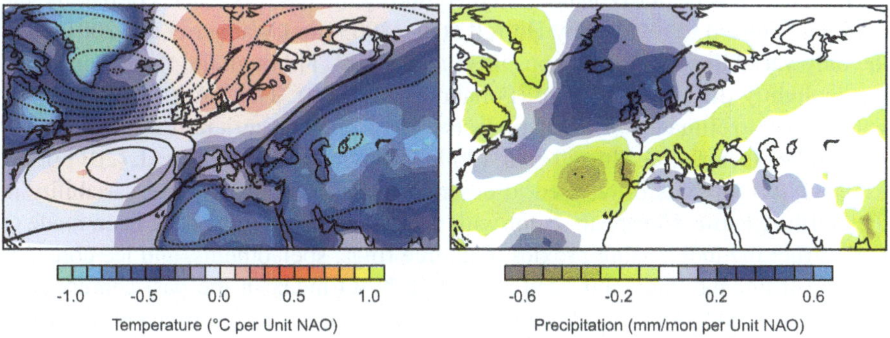

Fig. 3.23 Regression coefficients of (*left*) temperature and 500 hPa geopotential height (contours) and (*right*) precipitation on a boreal winter (December–February)-averaged NAO index (Table 3.2) (Based on 20CR data from 1900 to 2010)

During positive NAO phases, the North Atlantic storm track is tilted northeastward, the jet stream is more pronounced, and storms are more frequent. During negative NAO phases, the storm track is weakened and shifted to the south, increasing precipitation in the Mediterranean region. The NAO is also a mode of meridional heat flux partitioning. During a negative NAO, the stationary wave contribution to the heat flux (Fig. 3.7) increases and the transient eddy contribution decreases.

The state of the NAO is measured by differences in the respective anomalies of the pressure centres (see Table 3.2). It is a see-saw because the two nodes exhibit high teleconnectivity (Fig. 3.20). In January 1963, even the absolute pressure gradient was reversed (i.e., there was an Icelandic high and an Azores low). This rare case is called a "reversal". The NAO index is extremely negative during a reversal and it can be bitter cold in central Europe as seen in January 1963 (note that a negative NAO was typical for the 1960s; see Fig. 3.6).

The northern node of the NAO sometimes encompasses much of the Arctic Basin. An alternative definition of the NAO, based on the first principal component of the sea level pressure field north of 20°N (see Table 3.2), emphasises this structure even more and is called the Arctic Oscillation (AO) or Northern Annular Mode (Thompson and Wallace 1998, 2000).

The temperature see-saw associated with the NAO has been known for a long time. Already in the 18th century, the Danes knew that when winters in Greenland were cold, they were warm in Denmark and vice versa (Stephenson et al. 2003). In the 19th century, these relations were studied statistically using long temperature time series. With air pressure measurements becoming more abundant, scientists were able to attribute the temperature see-saw to systematic variations in pressure (Hann 1890). Brückner knew this work and studied similar relations, but on a decadal time scale.

The NAO has been extensively studied during the past 20 years. For details, the reader is referred to reviews by Hurrell et al. (2003) and Wanner et al. (2001). It is the only variability mode that appears in all seasons, although the "Summer NAO" has a slightly different structure (Folland et al. 2009). The NAO is also invoked with respect to climate variability impacts on hydropower, agriculture, fisheries, and many other sectors (see Hurrell et al. (2003) and Vicente-Serrano et al. (2011)).

The NAO index is often used to describe the interannual climate variability in the North Atlantic–European sector with one number. Attempts have been made to reconstruct the index using proxies (e.g., tree rings, speleothems, and ice cores) or early instrumental data calibrated against an NAO index in the past century (e.g. Cook et al. 2002; Jones et al. 1997; Luterbacher et al. 2002; Ortega et al. 2015). Chapter 4 will present variations in these indices.

3.2.4 Variability Modes in the Pacific and the Southern Hemisphere

3.2.4.1 The Pacific-North American Pattern

The North American climate also exhibits teleconnections. Centres with increased teleconnectivity (Fig. 3.20) are found over the North Pacific. When winters are warm and wet along the West Coast of the United States, they are often cold in the southeastern states. The quasi-stationary wave over the Pacific–North American sector (North Pacific trough–Rocky Mountains ridge–Atlantic trough) is pronounced.

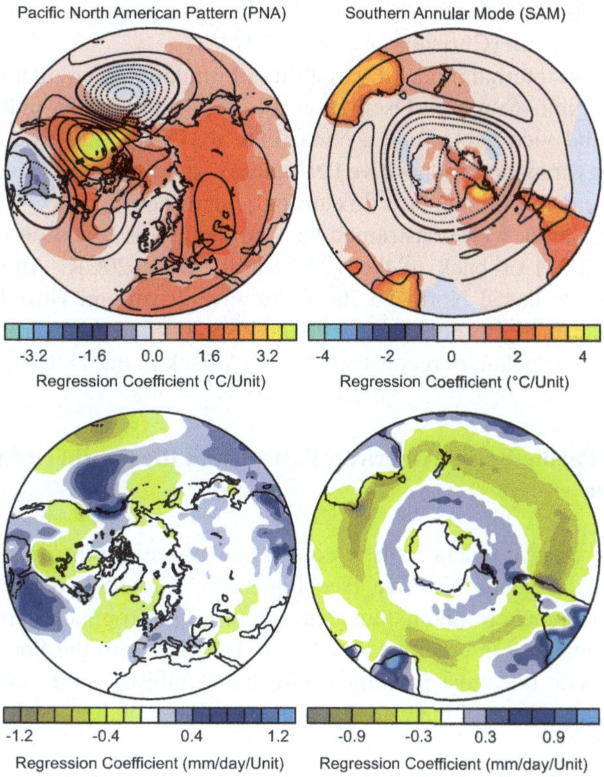

Fig. 3.24 Regression coefficients of (*top*) temperature and 500 hPa geopotential height (contour spacing is 10 gpm/Unit for the *left plot* and 50 gpm/Unit for the *right*) and (*bottom*) precipitation on a boreal winter (December–February) PNA index (*left*) and on an austral winter (June–August) SAM index (Table 3.2) (Based on 20CR data from 1900–2010)

Opposite climate anomalies are found when the wave amplitude is small (e.g., in January 1990; see Fig. 3.22). This circulation variability mode is called the Pacific–North American pattern. Its correlations are shown in Fig. 3.24.

The PNA is usually defined based on 500 hPa geopotential heights at four grid points over the southern North Pacific, the Aleutian region, the Rocky Mountains, and the southeastern United States (Wallace and Gutzler 1981). Together, the four points capture the quasi-stationary wave. The PNA is a winter pattern; summertime variability in the Pacific–North American sector is related to other flow structures (see Sect. 4.2.3).

A related variability mode over the North Pacific is the North Pacific Oscillation (see Table 3.2), which can be traced back to the work of Walker (1924). It represents a north–south dipole in sea level pressure over the North Pacific (Linkin and Nigam 2008; Rogers 1981) and north–south shifts of the Asian–Pacific jet stream and the

Pacific storm track. It affects the North Pacific gyre circulation and thus ocean-atmosphere interaction (Christensen et al. 2013).[14]

The PNA is an internal mode of variability of midlatitude circulation, but is also affected by El Niño–Southern Oscillation. El Niño events often induce a positive PNA pattern and vice versa for La Niña. Like the NAO, the PNA is a mode of poleward energy transport. In positive PNA phases, the stationary wave contribution to the zonal-mean meridional heat flux strongly increases.

The PNA index is well constrained back to the 1920s based on upper-air data (Ewen et al. 2008) or back to the 19th century from 20CR. Alternatively, the sea level pressure–based North Pacific index (Trenberth and Hurrell (1994); see Table 3.2), which is highly correlated with the PNA, is often used. Trouet and Taylor (2010) used tree ring data to reconstruct the PNA back to the 18th century.

3.2.4.2 The Pacific-South American Pattern and the Southern Annular Mode

The Pacific-South American pattern consists of a wave train across the Pacific from Australia to South America (see Fig. 3.24). The PSA is understood as a Rossby wave response to anomalous convection in the tropical Pacific. It exhibits both intraseasonal and interannual variability. Together with the Southern Annular Mode, the PSA is the most prominent climate variability mode of the Southern Hemisphere and affects climate in the South Pacific–South American sector and Antarctica.

The Southern Annular Mode is the dominant mode of variability of the southern extratropics and is clearly indicated on the global correlation map (Fig. 3.20, top left). Because the circulation over the southern high latitudes is more zonally symmetric, this mode is more "annular" than its northern equivalent, the Arctic Oscillation (which is also called the "Northern Annular Mode"). It is characterised by anomalies in geopotential height of opposite sign at southern mid- and high latitudes and extends through the depth of the troposphere (Christensen et al. 2013). In the positive SAM phase, circumpolar winds are stronger and jets and storm tracks move poleward. The SAM is relevant for climate in the extratropical Southern Hemisphere (see Fig. 3.24). It also affects ocean–atmosphere interaction in the Southern Ocean, which is important for the global carbon cycle (Thompson et al. 2011).

With very few exceptions, observations from Antarctica, which are key to diagnosing the state of the SAM, reach only back to the International Geophysical Year in the 1950s. Several attempts have been made to reconstruct the SAM and related atmospheric circulation patterns based on station sea level pressure data

[14]There is also a see-saw between the North Pacific and North Atlantic. The strengths of the Aleutian and Icelandic lows are anticorrelated (Honda et al. 2001) and a weak negative correlation is also found between NAO and PNA indices (Ewen et al. 2008).

(Jones et al. 2009b), tree rings (Villalba et al. 2012), lake sediments (Saunders et al. 2015), and ice cores (Russell et al. 2006). These reconstructions will be revisited in Chap. 4.

3.2.5 Extratropical Circulation Variability Modes and the Stratosphere

The stratospheric meridional circulation is driven by waves that originate in the troposphere (see Table 3.1)—waves whose statistics change in accordance with variability modes. Conversely, through the mechanism of downward propagation, the stratosphere affects the tropospheric circulation and thus the variability modes. In this chapter, we will take a closer look at this two-way coupling between the stratosphere and circulation variability modes.

Let us return to January 1963 and the large, meridionally elongated waves in the monthly mean 500 hPa field (Fig. 3.22; repeated below in Fig. 3.25). Can we trace the wave structure vertically? At 200 hPa near the tropopause, the wave pattern is similar. However, some of the smaller-scale structures (e.g., the narrow ridge to the west of California) have weakened. At 50 hPa in the stratosphere, the smaller-scale waves have disappeared and only a pronounced wavenumber-2 pattern remains (note that these are monthly mean fields).

The Eliassen–Palm flux allows us to track the vertical and meridional flux of wave activity (Fig. 3.26, top left). Most of the wave activity is confined to the troposphere. However, near 60°N, waves can propagate to the stratosphere. This upward flux was higher in January 1963 (see the difference in Fig. 3.26, bottom left) than in January 1990 (Fig. 3.26, top right). Increased convergence of the Eliassen–Palm flux (blue colours) indicates more wave breaking and thus a strengthened meridional circulation. In fact, the difference in the late winter (January–April) ozone fields of the two years (Fig. 3.26, bottom right) shows more ozone in the polar stratosphere and less in the tropical stratosphere, a result of altered transport. Note, however, that ozone depletion between 1963 and 1990 also affects the difference field.

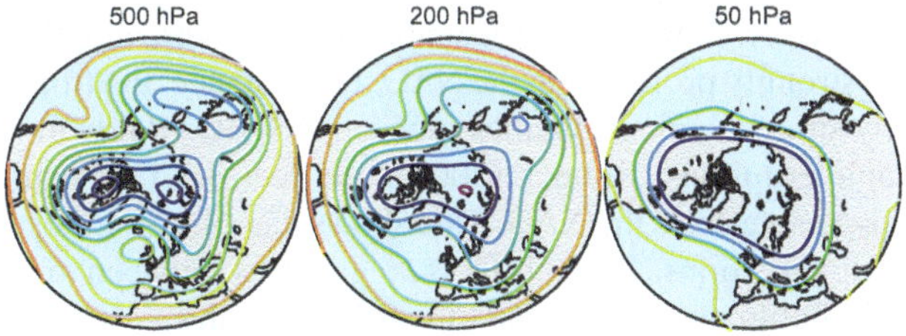

Fig. 3.25 Geopotential height at 500, 200, and 50 hPa for January 1963 (Data source is NCEP/NCAR Reanalysis)

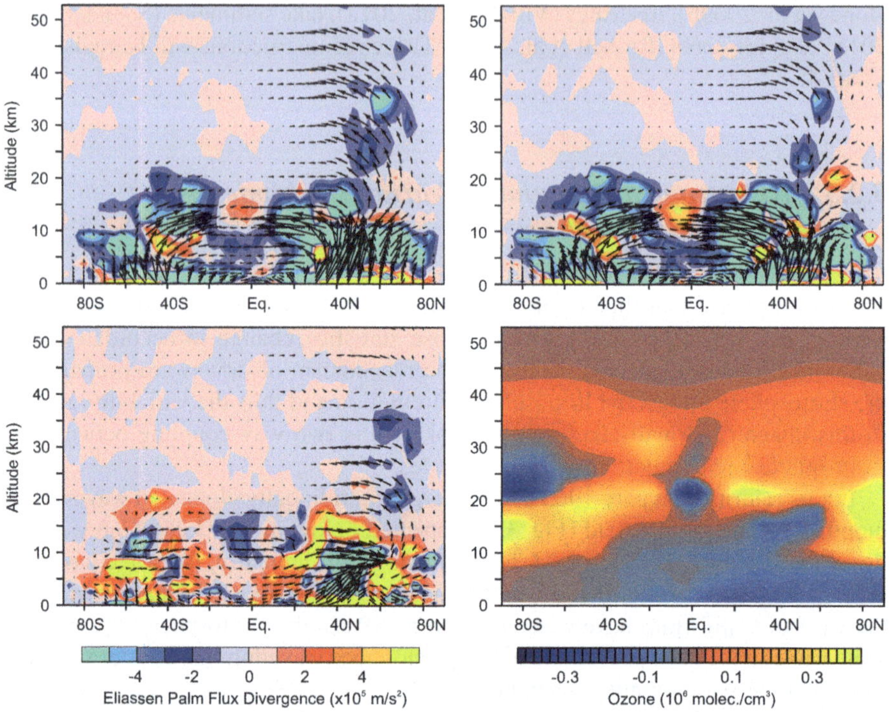

Fig. 3.26 Vectors of the Eliassen–Palm fluxes (for display purposes, fluxes were weighted with the square root of pressure and the vertical component was multiplied by 100) and its divergence for (*top left*) January 1963 and (*top right*) January 1990. (*bottom left*) The difference between 1963 and 1990 (Data source is ERA-40). (*bottom right*) The zonal-mean ozone difference for January–April (Data source is the historical ozone data set HISTOZ)

The amount of wave activity reaching the stratosphere at 60°N (see also Fig. 3.12) changes according to the large-scale flow pattern in the troposphere. In fact, January 1990 shows the typical situation during positive NAO winters and January 1963 is typical for negative NAO winters, as found in correlation analyses (Hartmann et al. 2000). During a negative NAO or AO, more wave activity reaches the stratosphere and converges. Hence, the NAO or AO affects the stratospheric circulation. Conversely, downward-propagating stratospheric circulation anomalies (see Sect. 3.1.3) strongly project onto the NAO. For instance, the tropospheric signature of primarily stratospheric anomalies such as the QBO (Baldwin and Dunkerton 2001) or volcanic eruptions (Robock (2000); see Sect. 3.3.1) resembles the NAO. Thus, the stratosphere affects the NAO. This shows that NAO–stratosphere coupling operates in both directions. The NAO is a gateway between tropospheric and stratospheric variability on weather and climate time scales.

3.2.6 Tropical Variability Modes: El Niño–Southern Oscillation (ENSO)

3.2.6.1 The ENSO Phenomenon

Everyone knows (or believes to know) "El Niño". Ever since the strong 1983 event, El Niño events have triggered worldwide media attention—for a good reason. It is the dominating mode of climate variability at the global scale.

More technically, "El Niño–Southern Oscillation" is a coupled mode of variability in the tropical Pacific. It emerges from the interaction of oceanic and atmospheric processes. The atmospheric part of ENSO is the "Southern Oscillation", which is a mode of variability of the Pacific Walker circulation (see Sect. 3.1.2). It is defined by the pressure difference between Tahiti and Darwin, Australia (see Table 3.2). If the Southern Oscillation is positive, convection over the western tropical Pacific and trade winds are strong. If it is negative, trade winds are weak or reversed. The connection of the Southern Oscillation to the oceanic El Niño phenomenon was discovered by Bjerknes (1966). A negative Southern Oscillation concurs with a warming of the eastern tropical Pacific, "El Niño". The opposite phase is called "La Niña". Often, the expressions "ENSO warm phase" and "ENSO cold phase" are also used.

Four examples of the state of tropical Pacific sea-surface temperatures are shown in Fig. 3.27. During El Niño (left), local temperature anomalies may reach as high as 5 °C.

ENSO is the dominating mode of interannual variability in the global climate system. It is a truly global mode with teleconnections to many parts of the globe, and it is a truly coupled mode because both oceanic and atmospheric processes act together to maintain the mode. ENSO variability is important for understanding climatic changes since 1700 and hence is dealt with in some more details below.

3.2.6.2 ENSO Mechanisms

Today, the mechanisms behind ENSO are relatively well understood (schematically shown in Fig. 3.28). Consider the atmospheric circulation over the Pacific, the Walker circulation (see also Sect. 3.1.2), which consists of trade winds at the surface, upwelling in the western tropical Pacific, upper-level westerly flow along the equator, and subsidence in the eastern tropical Pacific. How does this flow affect sea-surface temperatures? Under ENSO-neutral conditions, the trade winds are moderately strong and transport water along the equator to the western tropical Pacific. These waters have been in contact with the atmosphere and have warmed. In fact, the western tropical Pacific is one of the warmest regions of the globe. Conversely, waters are cold off the coast of South America, where temperatures are controlled by coastal upwelling rather than solar heating. This "cold tongue" stretches well into the central Pacific. Equatorial upwelling due to divergent Ekman

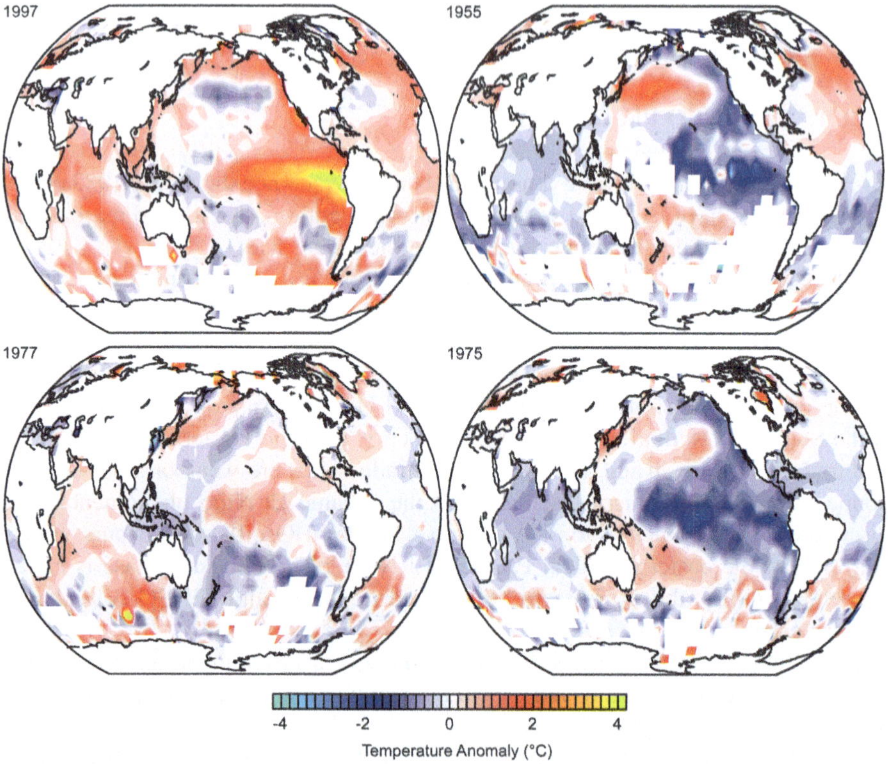

Fig. 3.27 Sea-surface temperature anomalies from July to January for (*left*) two El Niño and (*right*) two La Niña events. East Pacific events are shown at (*top*) and central Pacific events are shown at (*bottom*) (Data source is the Hadley Centre sea-surface temperature data set HadSST3, Kennedy et al. 2011)

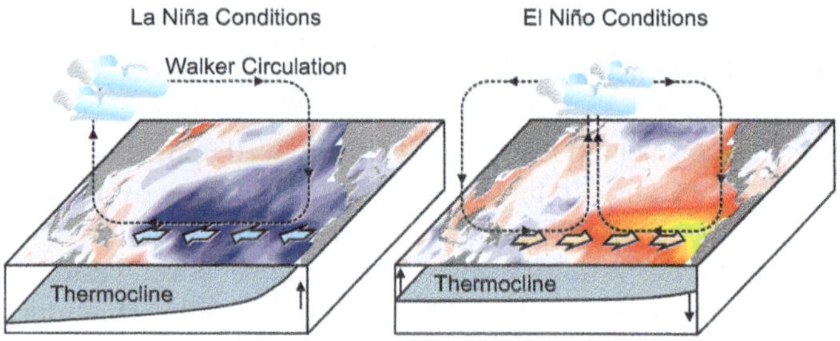

Fig. 3.28 Schematic of tropical Pacific sea-surface temperature anomalies, thermocline, surface currents, and atmospheric circulation during La Niña and El Niño phases

transport (perpendicular to the trade winds north and south of the equator, thus pointing away from the equator) enhances cooling in the central Pacific. The west–east difference in sea-surface temperatures affects the sea level pressure field and promotes convection in the west and subsidence in the east. In turn, the altered sea level pressure enhances the trade winds. Stronger trade winds further enhance the wind stress and thus the upwelling. A positive feedback develops, known as "Bjerknes feedback".

The thermocline is shallow in the eastern tropical Pacific and near the equator, but deep in the western tropical Pacific. The atmospheric circulation transports water towards the western tropical Pacific and leads to a significant change in sea level. The cross-basin difference amounts to several decimetres.

The ocean has its own characteristic response to short atmospheric disturbances. If at a given longitude the trade wind is temporarily interrupted and becomes eastward, equatorial Ekman transport reverses locally and leads to an area with increased sea level at the equator and two areas with decreased sea level off the equator. From the former, a downwelling wave develops and propagates eastward. From the latter, upwelling waves develop and propagate westward.[15] Within a few months, the waves cross the Pacific. In the eastern tropical Pacific, the wave may interrupt the upwelling and warm surface waters. The positive feedback between wind stress and sea-surface temperatures sets in and changes the atmospheric circulation. Convection shifts away from the western tropical Pacific to the central or eastern tropical Pacific—an El Niño event.

ENSO is a coupled ocean–atmosphere phenomenon. The atmosphere provides a feedback that, in the absence of the ocean, tends to stabilise either "El Niño" or "La Niña". The ocean, in the absence of the atmosphere, develops an oscillation through Kelvin and Rossby waves that are reflected off the basin ends. In reality, the two systems are coupled. The shallow thermocline in regions of tropical upwelling facilitates communication of oceanic signals to the atmosphere. As a consequence of the coupling, an oscillatory mode develops.

In the 1980s, several models were developed to explain the coupled behaviour of the system, which are still used today (e.g., Zebiak and Cane 1987). In the "delayed oscillator" concept (Suarez and Schopf 1988), temperature changes in the eastern tropical Pacific are assumed to depend on temperature itself (the Bjerknes feedback), temperatures at a previous time (the oceanic oscillation), and a nonlinear damping term. This simple model is able to explain many of the properties of the ENSO system. A slightly different approach is the "recharge oscillator" model (Burgers et al. 2005), which in essence is a damped oscillator depending on the recharge time of the warm pool and a time delay between the east and west Pacific.

The question of the trigger and possible forcing mechanisms remains open. Atmospheric triggers of ENSO events may develop in the tropical Pacific, but they may also originate in other regions. It has recently been suggested that the Indian Ocean (Izumo et al. 2010) or the tropical North Atlantic may affect the ENSO system (Ham et al. 2013). ENSO affects the global energy budget. The tropical Pacific acts as a gateway of energy from the atmosphere into the upper ocean and vice versa.

[15]The downwelling wave is an equatorial Kelvin wave, the upwelling waves are slow Rossby waves. Both waves have very large wavelengths.

3.2.6.3 ENSO Variability and "Flavours" of ENSO

ENSO is a quasi-periodic phenomenon that occurs every 3–8 years and lasts for 1–2 years. This can be explained by the delayed oscillator. Another characteristic of ENSO variability that can be explained by a delayed or a recharge oscillator is the phase locking of ENSO to the seasonal cycle. ENSO often starts in early boreal summer and matures in December. It decays in boreal spring.

Apart from 3–8 years, the tropical Pacific also exhibits variability on longer time scales. This imprints on the "redness" of the spectrum (Fig. 3.19). To what extent this variability differs from ENSO is a matter of debate. Section 3.2.8 will further explore Pacific decadal variability modes.

Not every El Niño has the same appearance. Some events exhibit their strongest temperature anomalies off the coast of Peru. They are often called "eastern Pacific El Niños". An example of such an event is given in Fig. 3.27 (top left). Other events have their strongest sea-surface temperature signal in the central Pacific (Larkin and Harrison 2005) and are called "dateline El Niño events" (or "El Niño Modoki"; Fig. 3.27, bottom left). Recently, it has been argued that the two types of El Niño need to be distinguished because they are associated with different teleconnections and might change differently in a future climate (e.g., Ashok et al. 2007). Also, the vertical thermal structure of the atmosphere of the two types of events is different (Trenberth and Smith 2006, see also Zubiaurre and Calvo (2012)). Whether different mechanisms are responsible is an active area of research.

Because of its importance for the global climate, ENSO has been the target of many climate reconstructions. These will be further discussed in Chap. 4. Reconstructions are based on tree rings (e.g., Li et al. 2013), corals (Cobb et al. 2013), documentary data (Garcia-Herrera et al. 2008), or multiple archives (Emile-Geay et al. 2013). Because few proxies are available close to the central tropical Pacific, many reconstructions exploit the strong ENSO teleconnections that are presumed to be stationary. Next, ENSO teleconnections will be discussed.

3.2.6.4 ENSO Teleconnections

ENSO affects weather around the world, but the most prominent responses are in the tropics. When the Pacific Walker circulation weakens, the Atlantic Walker circulation weakens and the Indian Ocean Walker circulation changes its structure. With a delay of several months, these changes lead to a warming of both the Indian Ocean and the tropical Atlantic.

Over the Indian Ocean, the ENSO-induced circulation anomalies interact with the monsoon circulation. The South Asian and Indian monsoons (Gilbert Walker's actual target) tend to be weaker during El Niño events, although the link is stronger during dateline than east Pacific El Niños (Kumar et al. 2006).

ENSO's teleconnections also reach the extratropics. They result from the eastward shift in atmospheric convection. During El Niños, the Hadley circulation weakens over the western tropical Pacific but strengthens over the central tropical

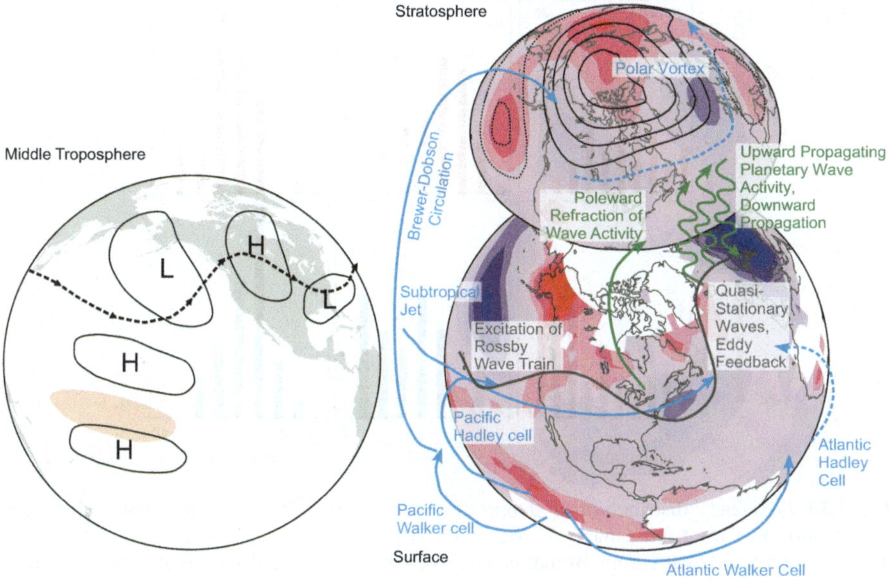

Fig. 3.29 Schematic of the effect of El Niños on the northern extratropics and the stratosphere. (*left*, adapted from Horel and Wallace (1981)) Initiation of a wave pattern over the Pacific due to anomalous upper-level divergence (colour). (*right*, adapted from Brönnimann (2007)) Possible downstream effects to Europe and the Northern Hemisphere's stratosphere. *Solid* and *dashed arrows* denote strengthening and weakening circulations, *colours* indicate temperature anomalies

Pacific. The associated changes in upper-level divergence—increased divergence over the central tropical Pacific and decreased divergence over the western tropical Pacific—may trigger wave trains (Trenberth et al. 1998). This can be seen in the sea level pressure shown in Fig. 3.29. Depending on the flow conditions, these waves may reach the extratropics and interact with the mean flow or be trapped in the subtropical jet. Again, the teleconnections are different for dateline and east Pacific El Niños (Larkin and Harrison 2005).

The extratropical effects of El Niños include projections onto some of the variability modes described in the previous chapter and are well expressed in the respective winter seasons (Fig. 3.29). Over the North Pacific, the Aleutian low is strengthened and a positive mode of the PNA can be initiated, leading to warm and wet winters along the west coast of North America and cold winters in the southeastern United States. Cooling is also found over northern Eurasia, whereas warming occurs over the Middle East and central Asia. In the Southern Hemisphere, El Niños promote: a low pressure centre to the east of New Zealand, a high pressure centre in the eastern South Pacific, and a low pressure centre over South America (Karoly 1989).

Over Europe, the impact of ENSO is small but discernible. In a previous study, we analysed long (ca. the past 300 years) climate records and found significant differences between the frequency distributions of winter temperatures in Sweden,

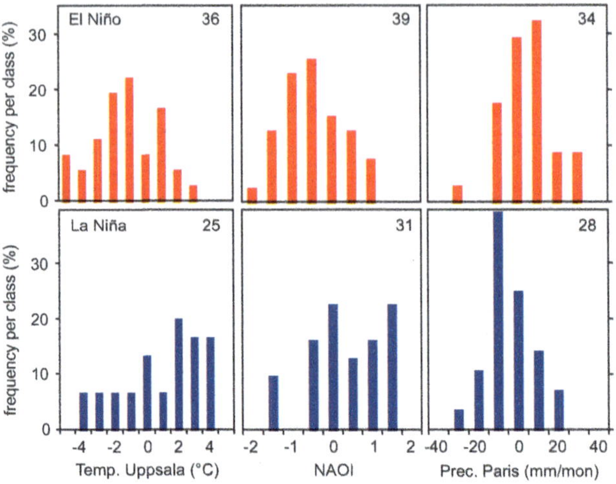

Fig. 3.30 Frequency distribution of temperature anomalies in Uppsala, the NAO index, and precipitation in Paris from January to March for strong El Niño and La Niña events based on data from 1706 to 2000 (From Brönnimann et al. 2007b, with kind permission from Springer Science+Business Media)

precipitation in Paris, and the NAO index during strong El Niño and strong La Niña events (Fig. 3.30; Brönnimann et al. 2007b). Climate models reproduce these teleconnections. However, the relation is weak and mainly occurs for "dateline" events (Graf and Zanchettin 2012).

ENSO also affects the stratosphere. Over the Pacific, the response consists of an east–west dipole in lower stratospheric temperature and column ozone, reflecting the shift in convection. In addition, ENSO imprints on the high-latitude stratosphere. During ENSO events, the northern polar vortex is weakened (Labitzke and van Loon 1999; Manzini et al. 2006; Sassi et al. 2004). Moreover, column ozone is decreased in the tropics and increased at high latitudes due to a strengthened Brewer–Dobson circulation.

The effect on the high-latitude stratosphere is presumably related to an increase of planetary wave activity, which propagates from the troposphere to the stratosphere and accelerates the meridional circulation. A schematic of this mechanism is given in Fig. 3.29. This increase in planetary wave activity is an ENSO teleconnection to the extratropics. Garfinkel and Hartmann (2008) have suggested that the strengthening of the Aleutian low is an important contributor.

Through the mechanism of downward propagation (Sect. 3.1.3), the weakened polar vortex during El Niño might induce a negative NAO pattern in late winter. In fact, this might be one cause for the statistical ENSO signal in the Atlantic–European region in late winter (Fig. 3.29; Ineson and Scaife 2009). Alternatively, the European signal might also be a downstream effect from the North Pacific towards Europe (see discussion in Brönnimann 2007). Graf and Zanchettin (2012) suggest that central Pacific events may activate the subtropical jet as a wave guide and excite a wave disturbance that propagates to the Atlantic and weakens the Azores high.

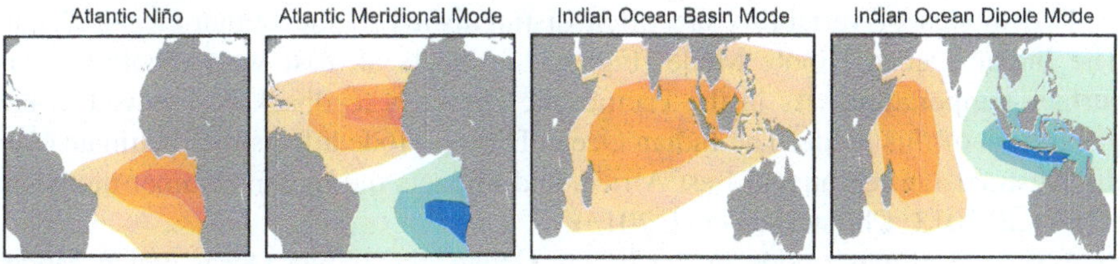

Fig. 3.31 Schematic depiction of Atlantic and Indian Ocean modes of variability in terms of sea-surface temperature anomalies

3.2.7 Variability Modes in the Atlantic and Indian Ocean

What is the role of other oceans in climate variability? Their effects on climate may not be as far reaching as those related to the tropical Pacific, but they are important on regional-to-continental scales. In particular, the tropical Atlantic and Indian Ocean have several modes of variability. Their sea-surface temperature patterns are shown in Fig. 3.31.

The Bjerknes feedback also works in the tropical Atlantic Ocean and in a similar way as in the Pacific. However, the Atlantic Ocean has a smaller east–west extension than the Pacific. Therefore, a strong coupled mode such as ENSO cannot grow in the same manner. The corresponding Atlantic mode is called the Atlantic Equatorial Mode or Atlantic Niño (Fig. 3.31). It is an interannual, internal mode of the Atlantic, but it can be excited by Pacific ENSO events (e.g., Keenlyside and Latif 2007) or advection of warm Atlantic waters from north of the equator (Richter et al. 2013). The Atlantic Equatorial Mode affects the West African monsoon and rainfall in coastal regions.

In addition to this interannual mode, the tropical Atlantic exhibits a decadal mode known as the Atlantic Meridional Mode (e.g., Xie and Carton 2004). This consists of an interhemispheric north–south gradient in sea-surface temperatures (Fig. 3.31; e.g., Chang et al. 1997; Nobre and Shukla 1996) and is understood to be related to ENSO, the NAO, and radiative forcings such as aerosols (see Christensen et al. 2013).

Finally, an important feature of tropical North Atlantic sea-surface temperatures is the formation of a warm pool in the western tropical Atlantic, Caribbean Sea, and Gulf of Mexico. The Atlantic Warm Pool varies in size from year to year. This affects, among other things, precipitation in the Caribbean region and the southern United States (Wang et al. 2007a).

The tropical North Atlantic is the genesis region of tropical cyclones. These can travel westward, develop into North Atlantic hurricanes, and then turn north where they can impact North America. Tropical Atlantic variability modes impact the genesis and path of hurricanes through changes in sea-surface temperatures as well as wind. An important role in this variability is also played by the Azores high, which affects the paths of the hurricanes (Colbert and Soden 2012).

Two modes of variability are often distinguished in the Indian Ocean Basin. The Indian Ocean Dipole Mode (Fig. 3.31) reflects an east–west dipole in sea-surface temperatures. As in the Pacific, the Bjerknes feedback is involved in the generation of this mode. The Indian Ocean Dipole Mode affects hydroclimate over the Australasian region and East Africa, and impacts the Indian summer monsoon (Cai et al. 2013; Christensen et al. 2013).

A second climate mode in the Indian Ocean, the Indian Ocean Basin Mode, reflects a basinwide warming or cooling (Fig. 3.31; see Xie et al. 2009) that affects precipitation in the Indian and western Pacific regions. This mode can be understood as the response of the Indian Ocean Walker cell to Pacific ENSO events, as outlined in the previous chapter (Cai et al. 2013).

Do we have information on the past behaviour of these modes? Sea-surface temperatures have been reconstructed based on multiple proxies by Mann et al. (2009a) and were used to drive our model simulations shown in this book. These reconstructions allow us to explore Atlantic and Indian Ocean variability modes to an extent. The Indian Ocean Dipole has been reconstructed using corals (Abram et al. 2008). Chapter 4 will further explore the past behaviour of oceanic variability modes.

3.2.8 Decadal Climate Variability Modes

Is the climate stable? Is the year-to-year variability found in the Atlantic, Indian, and Pacific random around a stable mean state? Today, there is consensus that climate varies on all time scales, including from decadal to multidecadal. During Brückner's time that consensus did not exist. The predominant belief was that climate was essentially stable. Against this belief, Brückner proposed multidecadal changes in his book "Climatic Changes since 1700" (his view is summarised in Sect. 4.1.2). Though his claim of a 35-year cycle of global climate may not be correct, his focus on multidecadal scales is noteworthy.

Today, variability on decadal scales is considered relevant for decadal predictions and climate diagnostics (such as the current "global warming hiatus"). Many decadal-to-multidecadal variability modes have been suggested. With respect to underlying mechanisms, three basic families of explanations exist: mechanisms inside the machinery that operate on decadal-to-multidecadal scales, quasi-periodic external forcings, or stochastic forcing of a system with a very large inertia (in other words, the slow response of the machinery). Many modes are believed to be related to oceanic processes, although external forcings (solar, volcanic, and aerosol forcings) also have multidecadal components. Because most observational data only cover the past 150 years, quasi-oscillatory behaviour on multidecadal time scales is hard to detect. Therefore, palaeodata and climate models are often used to acquire more information.

In this chapter, I mention two prominent modes of decadal-to-multidecadal variability: the Atlantic Multidecadal Oscillation (AMO) and the Pacific Decadal

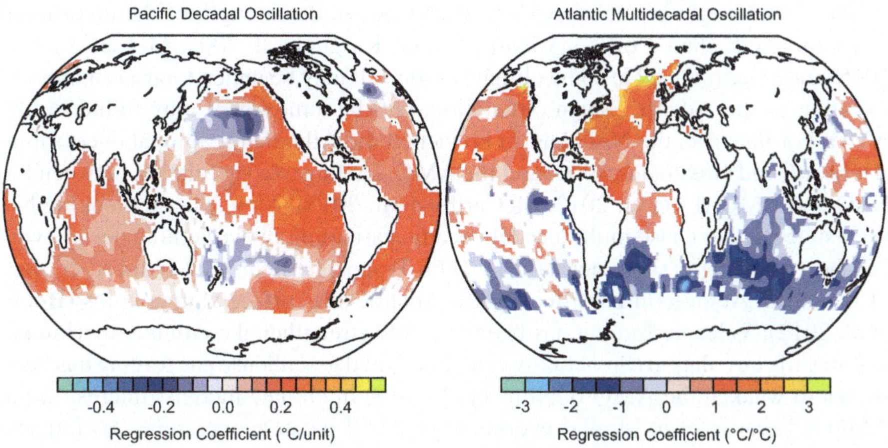

Fig. 3.32 Regression coefficients of annual mean sea-surface temperatures on indices of (*left*) the Pacific Decadal Oscillation (Mantua et al. 1997) and (*right*) the Atlantic Multidecadal Oscillation (Trenberth and Shea 2006) (Based on HadSST3 data)

Oscillation (PDO). In addition, multidecadal changes have been suggested for the Arctic Ocean. Figure 3.32 shows the main decadal-to-multidecadal variability modes. Most of the definitions used for multidecadal modes involve either a detrending or a subtraction of global means, that is, the modes are independent of long-term climate trends.

The Pacific Decadal Oscillation (Mantua et al. 1997), which is very similar to the Interdecadal Pacific Oscillation (IPO; Folland et al. 1999), describes a mode of variability of extratropical North Pacific (PDO) or global (IPO) sea-surface temperature anomalies (see Christensen et al. 2013). Both modes are similar and describe a dipole between sea-surface temperature anomalies in the eastern tropical Pacific and along the west coast of North America on the one hand, and the western and central North Pacific on the other hand (Fig. 3.32).

The PDO and AMO can be captured with indices based on sea-surface temperatures; the time series of these indices for the past 400 years will be discussed in Chap. 4. The PDO index reached its last high phase in the 1980s and 1990s and has become increasingly negative since the 2000s. The PDO is linked to changes in the Aleutian low as captured by Walker's North Pacific Oscillation or the North Pacific index (Trenberth and Hurrell 1994). It is considered to be an internal, unforced mode of variability. However, the mechanisms underlying the PDO remain unknown.

In the North Atlantic, prominent variability on scales of around 70 years appears in basinwide sea-surface temperatures (e.g., Delworth and Mann 2000; Enfield et al. 2001). The North Atlantic was warm in the 1870s, cold in the 1910s, warm in the 1940s, cold in the 1970s, and again warm in the 2000s. The AMO has been linked to various changes in the summer climate of North America (Seager et al. 2008) and Europe (Sutton and Hodson 2005), changes in the West African monsoon, and changes in hurricane frequency (see Christensen et al. 2013; Knight et al. 2006).

The AMO might be an expression of changes in the strength of the meridional overturning circulation of the Atlantic Ocean (Knight et al. 2005; McCarthy et al. 2015) and is reproduced in some climate models. It may be understood as an internal variation in the Atlantic ocean circulation, but the atmosphere may influence the AMO. For instance, the NAO has an impact on the Atlantic meridional overturning (Delworth and Greatbatch 2000) and the AMO, leading to a lag relation between the NAO and AMO (Li et al. 2013; McCarthy et al. 2015). Atmospheric blocking has been suggested to cause multidecadal oceanic variability by weakening ocean gyres and heat exchange (Häkkinen et al. 2011). Tropospheric aerosol-induced cooling of the North Atlantic might affect North Atlantic sea-surface temperatures (Booth et al. 2012). Volcanic forcing has been found to strengthen the Atlantic meridional overturning circulation (Stenchikov et al. 2009) and greenhouse gas forcing has been shown to weaken the AMO (Collins et al. 2013) in climate models. Interest in the AMO is because of its similar evolution as global-mean temperatures, its impacts (including on hurricane frequency), and its potential for decadal predictability.

AMO and PDO indices have been reconstructed by various groups using corals and tree rings (Biondi et al. 2001; Gray et al. 2004; MacDonald and Case 2005). They can also be derived from the global sea-surface temperature reconstructions of Mann et al. (2009a). These reconstructions will be further discussed in Chap. 4.

3.3 Forced Climatic Changes

The previous chapters outlined the working principles of the machinery. Due to scale interactions and interactions of subsystems, the machinery produces both random and coherent variations in macroclimate. This section addresses how the machinery responds to external influences, that is, forcing factors. Understanding the response mechanisms is crucial for assessing the role of external forcings for climatic changes since 1700. The section covers the most relevant forcings on the time scales considered in this book. I do not consider orbital or geogenic forcings, which are of little relevance on a 300 year time scale. The main radiative forcings of the climate system during the past 300 years, according to current understanding, were exerted by volcanic aerosols, solar irradiance changes, greenhouse gases, tropospheric aerosols, and land surface changes. These forcing factors are considered in this section. Feedback mechanisms in the climate system, which may enhance these forcings, are discussed in Sect. 3.4.

3.3.1 Volcanic Effects

Anyone who eye witnesses a large volcanic eruption with its impressive plume has no doubt that volcanic eruptions are able to affect climate. Imagery of eruption plumes, such as the one produced by Mount Pinatubo (Fig. 3.33), demonstrate the irresistible force with which material from the earth's interior penetrates the atmosphere. Volcanic eruptions are the main external factor affecting global climate on an interannual time scale. However, it is not the visible plume, consisting of volcanic ash and water, which affects climate. Volcanic ash only has a short residence time in the atmosphere. It is quickly washed out or settles gravitationally. Most of the climatic effects of volcanic eruptions are related to sulphur compounds that are emitted in gaseous form. Once in the atmosphere, these are oxidised to sulphuric acid, which then forms aerosols.

Volcanic effects on the atmosphere are shown in Fig. 3.34. Here, the volcano is tropical and it is assumed that a large amount of sulphur reaches the stratosphere (later, more detail regarding eruption processes will be provided). Within weeks, stratospheric sulphur gases react with the hydroxyl radical (OH) or oxygen to form sulphuric acid (H_2SO_4). The sulphuric acid condenses and forms aerosols. Both gases and aerosols quickly spread in the longitudinal direction and travel around the globe, similar to the ash cloud of the Krakatau eruption (see Box 3.2, p. 124). However, the meridional transport through the stratospheric meridional circulation (see Sect. 3.1.3) is much slower. This is illustrated by satellite-borne measurements of optical depth (a measure of extinction through aerosols) before, during, and after the Pinatubo eruption in Fig. 3.35. Before the eruption (top left), very low levels of aerosol extinction were observed. This state is often called the "stratospheric background". During the eruption (top right), optical depth increased over the

Fig. 3.33 Volcanic plume of the Mount Pinatubo eruption (Photograph was taken on 12 June 1991 from the Clark Air Base by Richard P. Hoblitt (United States Geological Survey/Cascades Volcano Observatory))

tropics but a large fraction of sulphur compounds was still in gaseous form. Three months after the eruption (bottom left), an aerosol layer had formed in the tropics. Although optical depth also already started to increase in the extratropics, there was a very clear meridional gradient.

Box 3.2 Learning from volcanic eruptions
In 1783, a strange haze was observed in various places in Europe and the following winter was very harsh. Benjamin Franklin[16] (Franklin 1784)

(continued)

[16]Benjamin Franklin, 1706–90, was an American scientist, inventor, and statesman. Despite never being a president, Franklin was one of the founding fathers of the United States.

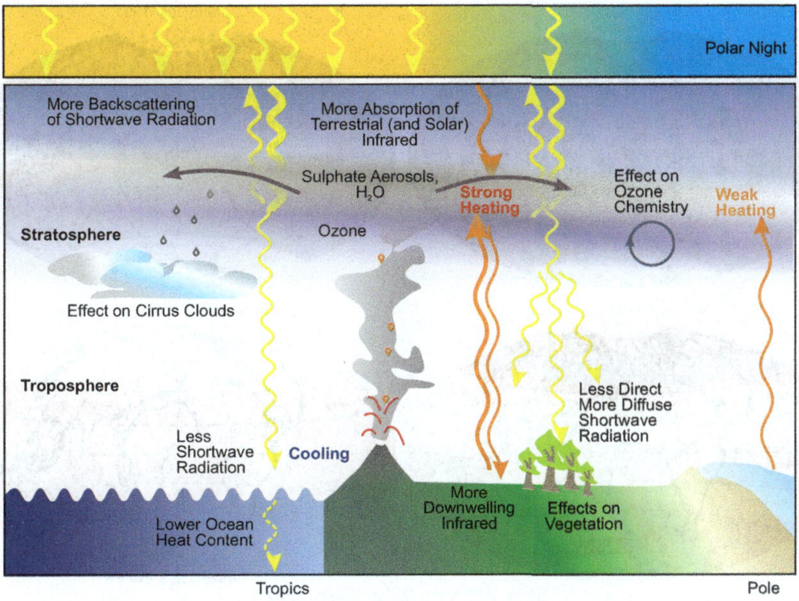

Fig. 3.34 Schematic of the atmospheric effects of a tropical volcanic eruption

Box 3.2 (continued)
speculated that both could be a consequence of the Laki eruption on Iceland. This fissure eruption injected large amounts of sulphur into the troposphere over a period of 8 months. Franklin's paper is often quoted as the first scientific study on the effects of volcanic eruptions on climate.

Volcanic eruptions can be considered "experiments of nature". They provide an opportunity to learn, depending on the research questions of interest (see also Box 2.2, p. 22). The Krakatau eruption of 1883, a few years before Brückner's book appeared, triggered not only immediate media interest, but also "global science" (see Dörries 2003). The report commissioned by the Royal Society (Symons 1888) made lasting contributions to atmospheric sciences (although climatic effects were not the focus). For instance, the report noted that the volcanic cloud circled the globe in little over 2 weeks, which points to strong easterly winds at high altitudes. These so-called "Krakatau easterlies" (Hamilton 2012) are now understood as a phase of the Quasi-Biennial Oscillation in the stratosphere (see Sect. 3.1.3).

(continued)

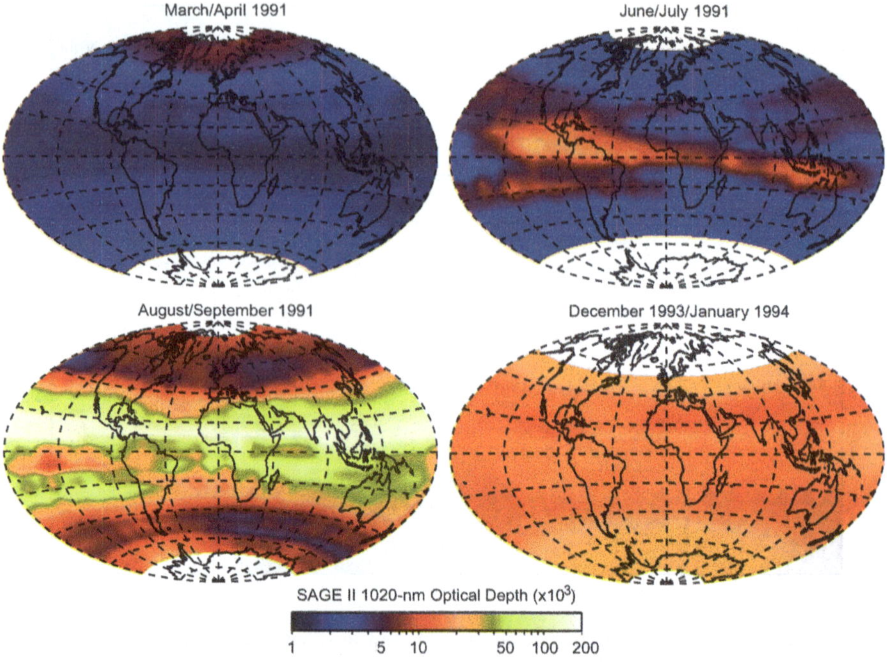

Fig. 3.35 Aerosol optical depth at 1020 nm before, during, and after the eruption (Data source is Stratospheric Aerosol and Gas Experiment II. Figure is provided by Larry Thomason)

Box 3.2 (continued)

Large eruptions in the early 20th century, namely Santa Maria in 1902 and Katmai in 1912, triggered further interest in volcanic effects. Humphreys (1913) and others discussed the cooling effect of large eruptions in the light of ice age theories—then a topic of interest. The first radiation measurement series were available, allowing observational studies. During each subsequent eruption, science was at a different stage, posed different questions, and got different answers.

Using historical data, science can also revisit previous eruptions and test new research questions using old "experiments" (Bodenmann et al. 2011). For instance, Humphreys (1913) analysed the climatic effects of the Tambora eruption of 1815. Since then, this historical eruption has been revisited in many different contexts (see also Box 4.3, p. 280). In fact, as shown by this book, we can still learn from this eruption that occurred almost 200 years ago.

(continued)

Box 3.2 (continued)

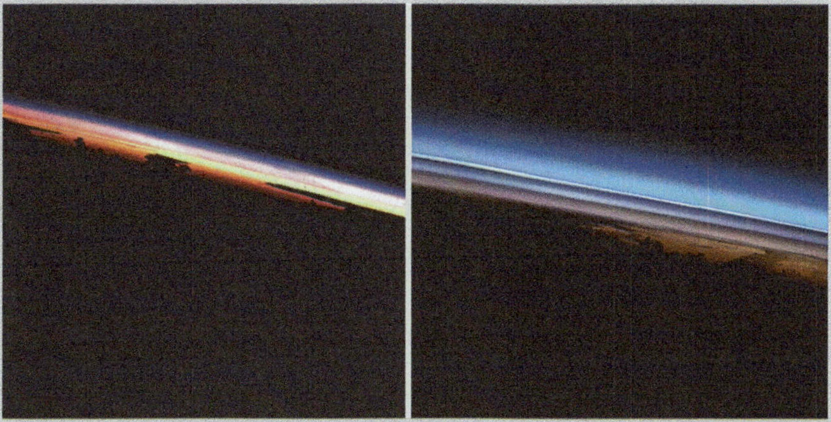

Fig. 3.36 Photographs of the earth's atmosphere taken from Space Shuttle missions. (*left*) Before the Pinatubo eruption. (*right*) After the eruption. The aerosol layer is clearly visible (NASA images STS41D-32-14 and STS043-22-23)

Today, the main effects of volcanic eruptions on the atmosphere are relatively well understood. However, some puzzles remain. Our current knowledge is based, to a considerable extent, on a single eruption—Pinatubo in 1991 (Fig. 3.36). This was the biggest tropical eruption of the century and it occurred when a satellite-based observing system was in place and modelling capabilities were developed. This allowed a plethora of fruitful studies.

The transport towards the poles is much slower than the zonal transport and takes 1–3 years. Figure 3.35 (bottom right), representing the situation 2.5 years after the eruption, shows a globally increased optical depth, although the level is not as high as 2 years earlier in the tropics. Interestingly, optical depths are higher in the extratropics than in the tropics, where aerosol-laden air had already been largely replaced by cleaner air after the eruption.

3.3.1.1 Direct (Radiative) Effects

Sulphate aerosols are nearly spherical and both absorb and scatter light of different wavelengths (the ratio of scattering to the sum of absorption and scattering is called single-scattering albedo). The scattering is mainly relevant for short wavelengths. Because part of the solar radiation is scattered back to space, less solar radiation (a high fraction of which is diffuse radiation) reaches the ground. This direct volcanic

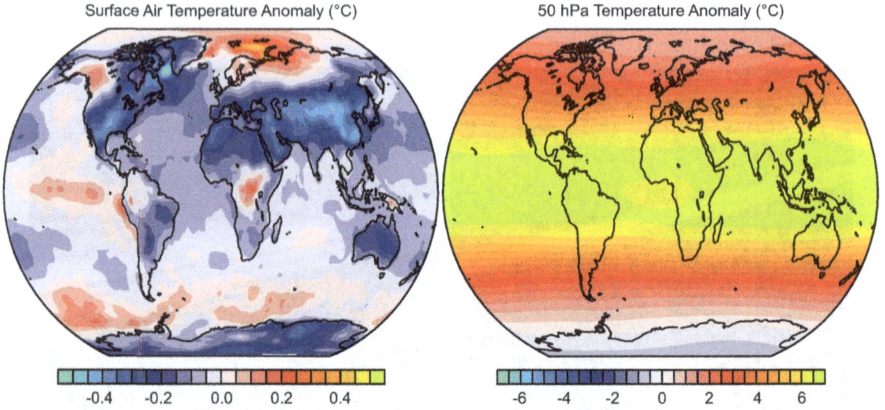

Fig. 3.37 Temperature at the (*left*) earth's surface and (*right*) stratosphere at 50 hPa in months 4–15 following 14 tropical volcanic eruptions in the CCC400 model simulation since 1600. Shown is the average of 30 ensemble members relative to a reference period surrounding the date of eruption (see Wegmann et al. 2014)

effect must lead to a cooling of the earth's surface, which is largest in areas and seasons with high incoming surface shortwave radiation. In the following, I will elaborate on volcanic effects on the atmosphere on the basis of our CCC400 model simulations.

In the annual mean, tropical and midlatitude continents show the strongest cooling (Fig. 3.37, left). Because of their larger heat capacity and strong thermal inertia, the oceans respond slower to the change in radiation. The initial change in sea-surface temperature is damped, but the signal persists longer and multiple eruptions within a short time can have large effects.

The absorption is mainly relevant for (solar and terrestrial) near-infrared and (terrestrial) infrared radiation. The absorption of radiation heats the aerosol layer in the stratosphere considerably (Fig. 3.37, right). The warming (which is arguably overestimated in the model) is strongest in the tropics. Here, surface temperature and hence upwelling longwave radiation fluxes are higher. Further radiative effects of volcanic eruptions are due to the amount of water vapour entering the stratosphere and the interaction with ozone chemistry. Aerosol particles provide a surface for heterogeneous reactions that reduce nitrogen oxides (NO_x) and enhance chlorine oxide radicals (ClO_x), In the presence of anthropogenic halogens, this can lead to ozone depletion but to an ozone increase prior to the era of chlorofluorocarbons (CFCs).

3.3.1.2 Indirect Dynamical Effects

In addition to the direct radiative effects, indirect dynamical effects of volcanic eruptions develop. Many of the climatic effects of volcanic eruptions owe their origin to

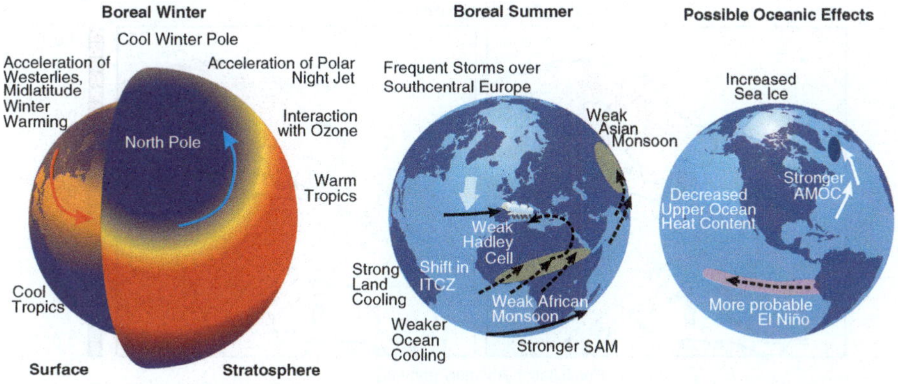

Fig. 3.38 Schematic of the volcanic influence on (*left*) boreal winter circulation, (*middle*) boreal summer circulation and (*right*) the ocean

these indirect effects. On the interannual time scale, arguably the most prominently studied are dynamical effects of the changes in stratospheric circulation, the land–sea thermal contrast, and the sea-surface temperature gradient across the tropical Pacific. On decadal and longer time scales, effects that operate via changes in ocean circulation may become important. These effects are discussed in the following.

Due to the unequal latitudinal distribution of aerosols as well as the unequal latitudinal distribution of infrared (and solar) radiation, particularly in the winter hemisphere, stratospheric heating is not globally uniform (depicted in Fig. 3.38). This causes a substantial anomaly in the equator-to-pole temperature gradient in the lower stratosphere. This change in the gradient alters stratospheric circulation, leading to a strengthening of westerly winds in the winter hemisphere (i.e., a strengthening of the polar vortex). Through downward propagation (Baldwin and Dunkerton 2001, see Section 3.1.3), anomalies in stratospheric zonal winds may affect tropospheric circulation through wave–mean flow interaction.

This mechanism is believed to cause the counter-intuitive winter warming that is often found over the northern continents in winters following major tropical volcanic eruptions (Robock 2000). This warming also appears clearly in European temperature reconstructions reaching back to 1500 when analysing post-eruption winters (Fischer et al. 2007a, see Fig 3.39). The warming is related to circulation anomalies; specifically, increased pressure over the Mediterranean Sea and the subtropical North Atlantic (i.e., an eastward-extended Azores high) and a strengthened Icelandic low. This situation, which is similar to a positive mode of the NAO (Wanner et al. 2001), favours advection of mild Atlantic air masses to Eurasia.

The winter warming used to be well reproduced in model simulations (Shindell et al. 2004). However, more recent generations of coupled climate models (Driscoll et al. 2012) have greater difficulties representing the warming. Our atmospheric model simulations of the past 400 years reproduce both the winter warming (Fig. 3.39) and the summer cooling over Europe (Wegmann et al. 2014). A corre-

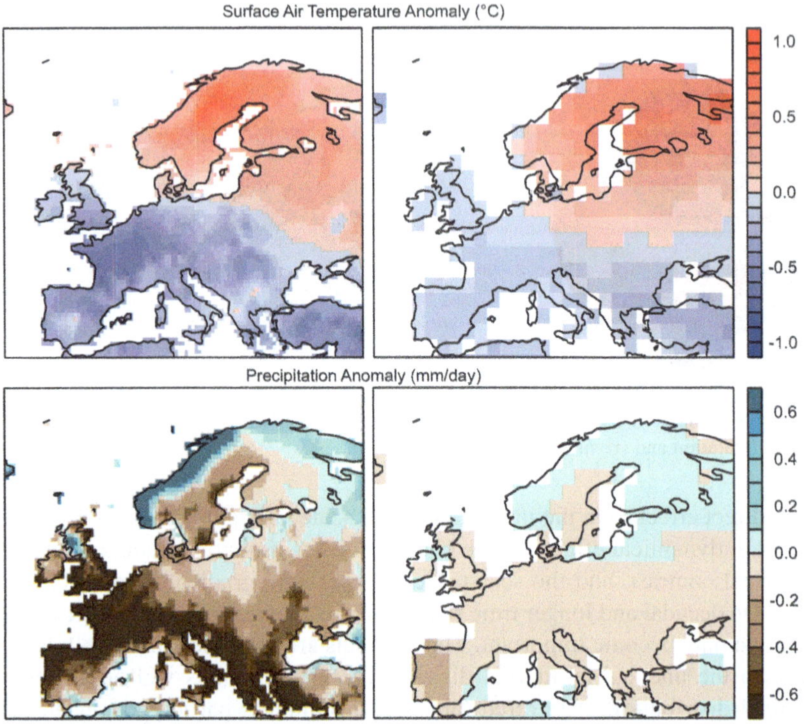

Fig. 3.39 Boreal winter (December-February) (*top*) temperature and (*bottom*) precipitation anomalies averaged across 14 tropical volcanic eruptions. Difference from statistical climate reconstructions (*left*) and ensemble-mean difference in the CCC400 simulations (*right*; composite of 420 simulated eruptions) are given. Differences are relative to a reference period around each eruption and are averaged across all 14 eruptions (Wegmann et al. 2014, © copyright 2014 AMS)

sponding SAM response in the Southern Hemisphere is found in model simulations for strong eruptions, but empirical evidence is lacking (Krüger and Toohey 2015).

The direct radiative effect due to reduced shortwave radiation, that is, the cooling of the surface, is not globally uniform (see Fig. 3.37). Due to their large thermal inertia, oceans cool slower than the land surface, which leads to a weakening of the land-sea temperature contrast in summer. As a consequence, the summer monsoon circulation systems and monsoon rainfall are weakened after volcanic eruptions (Fig. 3.40). This is not only seen in climate models, where the West African, Indian, and Asian summer monsoons systems are affected (Joseph and Zeng 2011; Wegmann et al. 2014), but also in tree ring–based summer drought reconstructions in Asia (Anchukaitis et al. 2010, see Fig. 3.40).

The weakening of monsoon rainfall following tropical eruptions deserves particular interest in the light of the debate on climate engineering through sulphate injection (Crutzen 2006). Monsoon rainfall is vital to a billion people. Hence, any

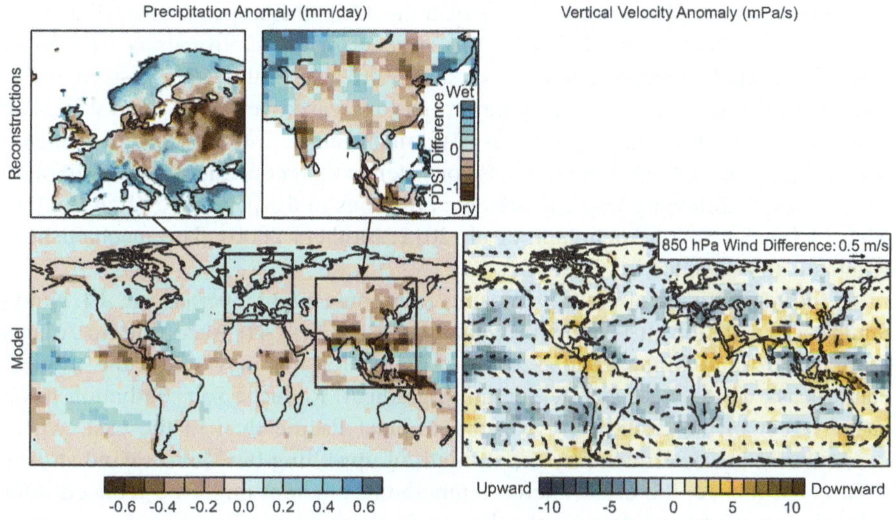

Fig. 3.40 Climate anomalies during boreal summers following tropical volcanic eruptions. (*Top*) Reconstructed anomalies of precipitation over Europe and of the Palmer Drought Severity Index over East Asia. (*Bottom left*) Precipitation anomalies in CCC400 simulations (a composite of 420 simulated eruptions). (*Bottom right*) Anomalies of vertical velocity at 500 hPa and wind at 850 hPa. All fields represent an average over 14 eruptions since 1600. Differences are relative to a reference period around each eruption, then averaged across all eruptions (Wegmann et al. 2014, © copyright 2014 AMS)

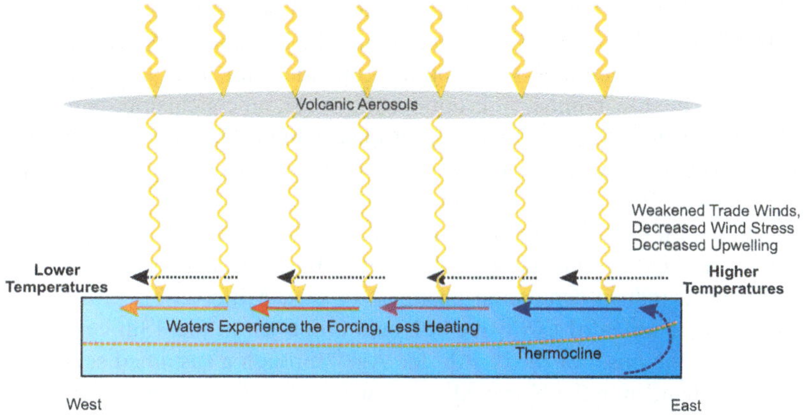

Fig. 3.41 Schematic of the influence of volcanic forcing on tropical Pacific longitudinal sea-surface temperature gradients (note that this simple schematic does not consider effect on clouds)

intervention that carries the danger of changing the monsoon system needs to be carefully evaluated (see also Haywood et al. 2013).

Our simulations suggest that the weakened Asian and West African summer monsoons also decrease convection over the Sahel-Sudanese region and thus weaken the northern Hadley circulation over the African-European sector (Fig. 3.40, right; see also Gaetani et al. 2011). The descent in the subtropics is weakened, which promotes convection and allows a southward shift of the Atlantic storm track. As a consequence, summer precipitation may increase over south-central Europe following tropical volcanic eruptions (Wegmann et al. 2014). This mechanism, which appears in our simulations and is consistent with reconstructions, may explain why summers following tropical volcanic eruptions in the past were often rainy in parts of Europe (e.g., Auchmann et al. 2012; Stothers 2000). The mechanism is schematically depicted in Fig. 3.38.

A further indirect effect of volcanic eruptions might operate via El Niño–Southern Oscillation. It is shown schematically in Fig. 3.41. According to Adams et al. (2003) and Mann et al. (2005), volcanic forcing may affect the tropical Pacific east–west sea-surface temperature gradient. Reduced solar radiation has a larger effect in the central and western tropical Pacific than in the east, where temperatures are determined by the strength of upwelling (and hence wind stress). As a consequence, the cross-Pacific temperature gradient might be reduced after volcanic eruptions, weakening the Pacific Walker circulation. Via the Bjerknes feedback (see Sect. 3.2.6), this weakens wind stress and upwelling over the eastern tropical Pacific and hence increases temperature. In this situation, El Niños have a higher chance of occurrence and an ENSO cycle may be initiated, as was found in empirical analyses and simple model simulations (e.g., Mann et al. 2005). However, the mechanism is still a matter of debate (McGregor and Timmerman 2011; Maher et al. 2015).

A brief analysis of the timing of the 134 strong ENSO events listed in Brönnimann et al. (2007b) relative to 16 strong tropical volcanic eruptions since 1500 shows that 7 out of 73 El Niños, but only 1 out of 61 La Niñas, occurred in winters following volcanic eruptions (8 eruptions coincide with neutral or weak ENSO years). Due to the low number of eruptions, however, no conclusions can be made.

Indirect climatic effects of volcanic eruptions might also exist on longer time scales. The winter warming discussed above might have a decadal component (Zanchettin et al. 2013). It has also been suggested that multiple eruptions occurring in a short interval might strengthen the Atlantic meridional overturning circulation (e.g. Iwi et al. 2012; Stenchikov et al. 2009). Thus, volcanic eruptions may have the potential to induce centennial changes. In fact, Miller et al. (2012) recently suggested that a series of explosive volcanic eruptions in the 13th century might have triggered the onset of the "Little Ice Age" through a sustained sea ice–ocean feedback mechanism in the North Atlantic (see further discussion in Sect. 4.2.1.).

3.3.1.3 Aerosol Size

What determines whether a volcano has a strong or a weak effect? Arguably, the most important factor is the amount of sulphur that reaches the stratosphere, which again depends on various factors. Different magmas have a different composition and the sulphur content can change; a "mass factor". Very important is the eruption itself. The sulphur must reach the stratosphere in order to be effective; an "injection factor". Global climatic effects almost exclusively come from explosive eruptions, but only a minor part of this is due to the initial kinetic energy of the eruption. The rise of volcanic plumes is mainly buoyancy driven. Often it is not the initial eruption column (called a "Plinian plume") that is too dense to rise high, which injects SO_2 into the stratosphere. Rather, the collapse of this initial column may create a buoyant plume (called a "co-ignimbrite plume") which may rise into the stratosphere (Herzog and Graf 2010). Entrainment of air and water vapour during the rise also plays an important role. How high will the plume rise? A good measure of the plume height is the so-called "neutral buoyancy height", which denotes the level where buoyancy vanishes, although some overshooting will occur initially. The neutral buoyancy height depends on, among other things, the temperature of magma and the temperature profile of the atmosphere. The neutral buoyancy height is typically in the 20–25 km range (Arfeuille et al. 2013).

In the stratosphere, the meridional transport plays a role: a "transport factor". The residence time of volcanic aerosols is longest if the eruption occurs in the tropics. Then, the aerosols are transported with the stratospheric meridional circulation until the air parcels are mixed again into the troposphere at mid- to high latitudes (although other transport and mixing routes may also play a role).

Because the Brewer–Dobson circulation (see Sect. 3.1.3) is directed towards the winter hemisphere, the transport factor has a distinct seasonality. As a consequence, the interhemispheric distribution depends on the eruption season. Tropical eruptions in boreal summer and early fall are more likely to affect the Northern Hemisphere (Fig. 3.42), whereas eruptions in boreal late winter and early spring might affect the Southern Hemisphere more strongly.

Once the sulphur has reached the stratosphere, there is another factor that matters, the "aerosol size factor". The optical properties of aerosols depend on their geometry and size. Scattering depends on the number of aerosols, whereas absorption depends on the aerosol mass. Small particles efficiently scatter solar shortwave radiation. At this stage, the mass and aerosol size factors may cause a "concentration factor".

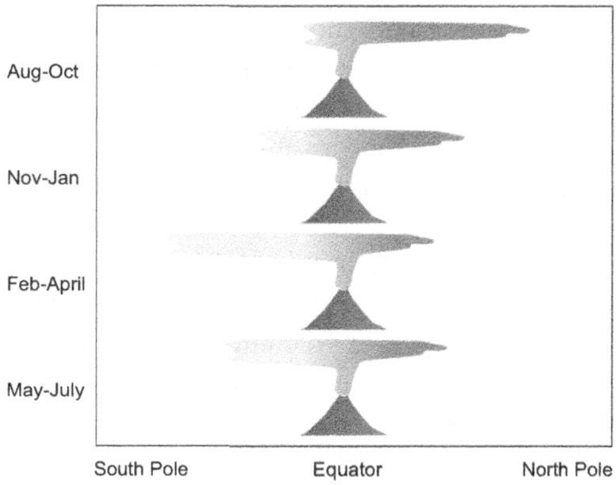

Fig. 3.42 Schematic of the effect of the eruption season on the spreading of the stratospheric aerosol cloud

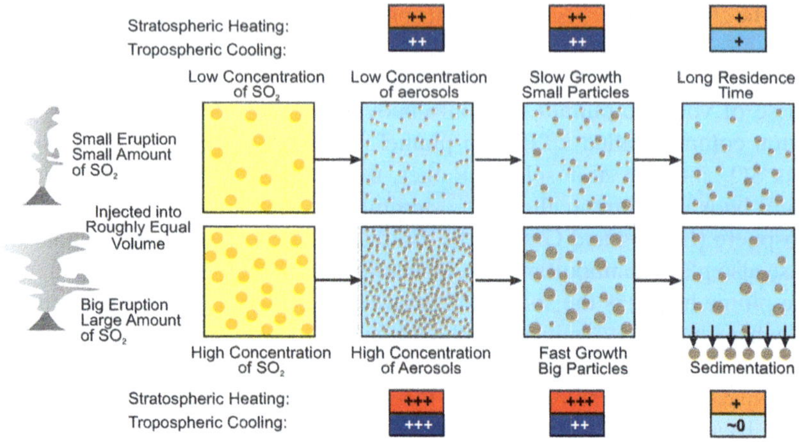

Fig. 3.43 Schematic depiction of the difference between high and low initial concentrations of sulphur dioxide in the stratosphere

Let us assume, for simplicity, that the mass of sulphur injected into the stratosphere depends on the eruption, whereas the volume into which it is injected depends on the thermal structure of the stratosphere. In the case of a big eruption, a larger mass of sulphur would be injected into the stratosphere than for a small eruption. However, the mass would be injected into a similar volume of air. As a consequence, the concentration of sulphur gases would be higher. Initially, this would lead to more small particles and hence more scattering and absorption. However, due to the higher concentrations, the particles coagulate more quickly. As a consequence, growth is faster. Large particles are less efficient in scattering than small particles, but are still efficient absorbers. Thus, after a big eruption, the stratosphere is heated very strongly, but scattering does not necessarily scale with the eruption size (Timmreck 2012). Figure 3.43 shows this situation.

After further growth, sedimentation will remove the large particles so that after 2 or 3 years, the aerosol mass may actually be smaller after a big eruption than a small eruption—as should the climate effect (e.g., Timmreck et al. 2009). In fact, the concentration may be one of the factors behind a question that is still open in palaeoclimate research and that has caused intense debate (see Anchukaitis et al. 2012). Why is the presumed big eruption of 1257 (now identified as Samalas, Indonesia; see Lavigne et al. (2013)) and others such as Kuwae 1458, the unknown eruption of 1808 or 1809, and Tambora in 1815 prominent in all model simulations but less so in some proxies? Was the climate effect weaker than estimated because of larger particle sizes or do we not correctly understand the proxies or models?

Climate effects of volcanic eruptions are a fascinating research topic because they involve many different components of the climate system (from the stratosphere to the oceans) and incorporate microphysical, chemical, radiative, and dynamical effects. It is an opportunity to test our knowledge and to learn continuously.

3.3.1.4 Quantification of Past Volcanic Forcing

This brings us to the question of how the strength of volcanic eruptions is measured today and for the past. Today, stratospheric aerosols can be measured from satellites and ground-based instruments. However, we still have difficulties quantifying aerosol properties (Arfeuille et al. 2013). Uncertainties increase farther back in time. Based on solar radiation measurements, stellar extinction, and lunar eclipses, aerosol optical depth can be computed back to the late 19th century. However, there are large uncertainties due to errors in observations and sparse coverage (Sato et al. 1993; Stothers 2001).

Another way of constraining the forcing is to analyse the composition of polar ice cores. Volcanic aerosols, after entering the troposphere, are washed out and may end up in polar ice cores. Using ice cores from Greenland and Antarctica and making assumptions concerning the stratospheric meridional transport, several authors have tried to estimate the amount of sulphur injected into the stratosphere during an eruption (Gao et al. 2008; Sigl et al. 2013, 2015). From this, aerosol optical depth can be estimated (Crowley et al. 2008). Instead of parameterisations, the aerosol forcing can be calculated from an aerosol microphysics model that is fed with the assumed sulphur emissions from ice core estimates; an approach that we have used in our group (Arfeuille et al. 2014). A comparison of two recent reconstructions of aerosol optical depth is shown in Fig. 3.44. Differences between different estimates are still considerable—even more so when addressing hemispheric partitioning (not shown).

Concerning the period analysed in this book, there are approximately two or three big eruptions per century plus another two or three smaller (but stratospheric) eruptions. However, the eruptions are not equally distributed over time.

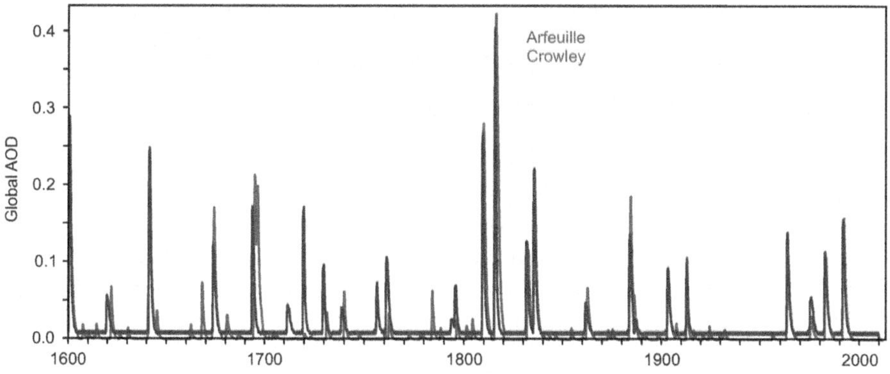

Fig. 3.44 Aerosol optical thickness since 1600 from two recent estimates (Arfeuille et al. 2013; Crowley et al. 2008). Note that most of the eruptions are known, which reduces the error in the attribution of sulphate spikes in the ice cores records, on which both estimates are based

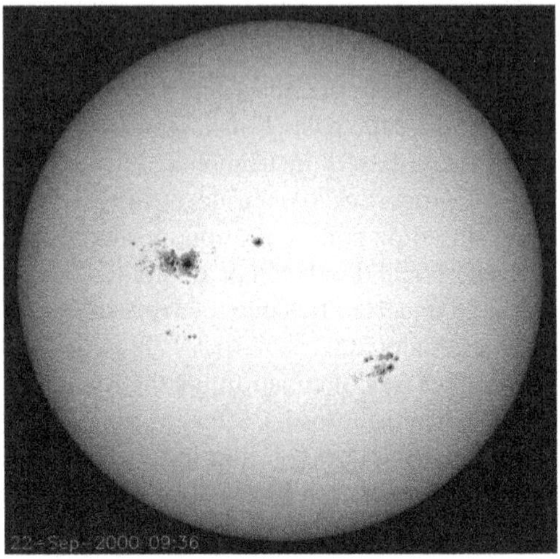

Fig. 3.45 Large sunspot group in September 2000 (Picture is from the MDI instrument onboard the Solar and Heliospheric Observatory (SOHO) space mission of ESA and NASA)

3.3.2 Solar Influences

The sun is the main energy source of earth's climate system. The wind systems, storms, jets, and anticyclones are ultimately driven by the sun. The diurnal and seasonal cycles, both of which are solar, are the most prominent variations in the climate system. Therefore, it is natural to think that changes in the energy the earth receives from the sun, even just slight changes, might have a strong effect on climate.

Sun–climate relations are an old topic of research. Systematic sunspots observations (Fig. 3.45) with telescopes started in 1609 and demonstrated the sun's variability. At least since the late 18th century, there have been speculations linking climatic anomalies to sunspots. Herschel (1801) compared sunspot data with wheat price data (a climate proxy) and sunspots were blamed for causing good or bad weather by learned (or less learned) citizens throughout the 19th century. The discovery of the 11-year sunspot cycle by Samuel Heinrich Schwabe, further elaborated by Rudolf Wolf,[17] in the 1840s further fuelled speculation because periodicity implies predictability. Weather was seen by many as an overlay of cycles and the sunspot cycle was certainly a promising one. Imprints of the sunspot cycle were not only looked for in European weather, but Meldrum (1872) and others claimed an effect of solar variability on Indian Ocean cyclones. Brückner (1890) tried to summarise the already overwhelming literature. Although he denied that sunspots could have caused the multidecadal changes he was investigating, he

[17]Samuel Heinrich Schwabe, 1789–1875, was a German pharmacist, astronomer, and botanist. Schwabe is known for his work on sunspots. Rudolf Wolf, 1816–93, was a Swiss astronomer and mathematician. Wolf was professor of mathematics and astronomy in Bern and in 1855 became director of the Swiss astronomical observatory. Wolf's method of counting sunspots is still in use.

nevertheless saw changes in the sun as the only possible cause for decadal climatic changes (see Sect. 4.2.1 for a discussion of Brückner's view).

Sunspots provide evidence for changes in solar activity, at least in its visual appearance, but does irradiance vary over a sunspot cycle? In 1902, Charles Abbot and Fredrick Fowle[18] began a programme with the Smithsonian Institution to measure the solar constant (see Hoyt 1979). They measured radiation in the visible and near infrared at many dry, high-altitude sites. Measurements continued until 1962, but remained inconclusive with respect to variations in solar irradiance. The solar constant is extremely difficult to measure from the earth due to variable atmospheric effects. Even from space, the absolute measurement of solar irradiance is difficult because of the rapid degradation of radiometers in space. Although all satellite missions provide very precise measurements, most are inaccurate (see Fig. 2.13 for a definition of these terms). A very recent space mission succeeded and present total solar irradiance is now estimated in the range of 1361–1362 Wm^2 rather than 1367–1369 Wm^2 as held earlier (Kopp and Lean 2011).

In addition to changes in the sun's output of energy (or changes in the earth–sun geometry, which are not discussed here), there are additional ways of how the sun might affect earth's climate. Examples include effects of changes in UV radiation, energetic particles, or galactic cosmic rays. Sun–earth connections have become a hotly debated topic in climate science and the public. In this section, I briefly follow the main lines of this discussion (for more details, see recent reviews by Gray et al. (2010) and Lockwood (2012)).

3.3.2.1 Changes in Solar Activity

The most prominent features of activity on the sun are sunspots (dark regions appearing in the midlatitudes of the sun during the early phase of the sunspot cycle and near the equator towards the end of a cycle) and faculae (bright regions occurring concomitantly with the sunspots). Both are expressions of magnetic activity. The features change in frequency and size with an 11-yr periodicity—the sunspot cycle.

Sunspots are cool regions with decreased irradiance; the opposite is true of faculae. The latter dominate the solar radiation response. Hence, when more sunspots are visible there is more total solar irradiance (TSI) reaching the earth. Variations in TSI are relatively well quantified (although the absolute value is not) for the past 30 years through space-borne instruments. Over a solar cycle, TSI varies by approximately 1 Wm^2 or 0.1 %—a relatively small amount. Ultraviolet radiation, which originates from different regions of the solar atmosphere, changes much more than TSI (by 1–10 %; Ermolli et al. 2013).

[18]Charles Abbot, 1872–1973, and Frederick Fowle, 1869–1940, were American astrophysicists at the Smithsonian Institution. Abbot was the director of the Smithsonian Observatory and later the secretary of the Smithsonian Institution.

Changes in magnetic activity and the solar magnetic field have other relevant effects. During certain phases of the solar cycle, the emission of energetic particles by the sun is enhanced. Another effect is that electrons in the earth's outer atmosphere can be accelerated towards the earth by the interaction of the sun and earth's magnetic fields. Finally, the solar magnetic field also modulates the flux of galactic cosmic rays to the earth. The flux is higher during periods of weak solar activity (Gray et al. 2010).

Much is still unknown about the changes outlined above, despite our observationally well-constrained satellite period. For instance, recent work found much stronger UV changes than expected, but changes of the opposite sign in the visible range over the declining phase of the last sunspot cycle (see Haigh et al. 2010). The uncertainty increases when trying to quantify changes back in time. There is also uncertainty about the mechanisms of how these changes (some of which only affect the middle atmosphere) are able to change climate on the ground. Before I turn to mechanisms and reconstructions of past solar activity, I briefly summarise the (limited) well-established observed effects of solar variability on climate. More detailed reviews can be found in Gray et al. (2010) and Lockwood (2012).

3.3.2.2 Observed Solar Effects

The solar imprint in the earth's atmosphere on various time scales is still neither well documented, nor well understood. Among the most established effects are the imprints of the 11-yr solar cycle on temperature and on ozone in the upper stratosphere (Fig. 3.46). During a solar maximum, ozone and temperature increase near an altitude of 40 km. Both also show an increase (although patchy) in the tropical lower stratosphere and in the polar regions.

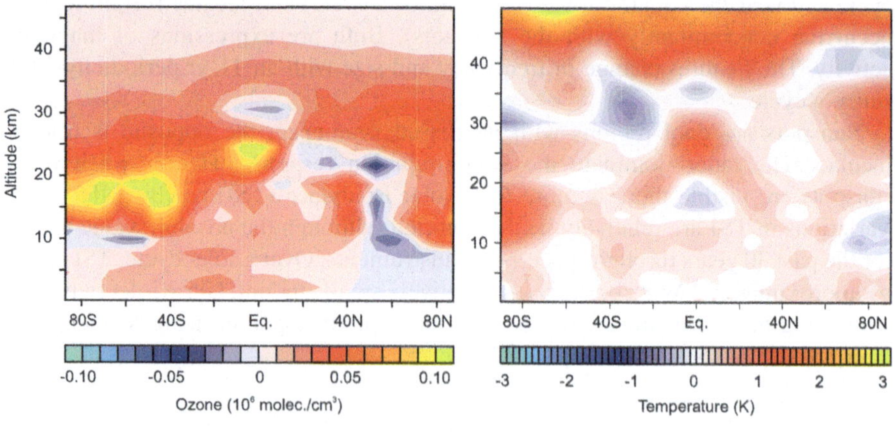

Fig. 3.46 Latitude–height cross section of the difference between solar maxima and minima (defined as the 24-month averages with the highest and lowest sunspot number per cycle) (*left*) ozone (HISTOZ 1927–2007, see Brönnimann et al. (2013a) for details) and (*right*) temperature (ERA-Interim, 1979–2013 covering 4 maxima and 3 minima)

Effects on stratospheric winds depend on the season. The northern polar stratospheric vortex is weaker in midwinter and stronger in late winter during solar minima compared to maxima. However, the effects of solar activity and the QBO (see Sect. 3.1.3) modulate each other, complicating the analysis (Labitzke et al. 2006).

At the earth's surface, the solar cycle imprints on the global-mean temperature, but the effect is not very strong—only about 0.1 °C over a cycle (Lean and Rind 2009). Regionally, the effects are more pronounced. This indicates that it is not only the total energy change, but the redistribution of energy through an altered circulation that is relevant. In zonal-average temperature and geopotential height fields, positive correlations with solar activity are found at midlatitudes, indicative of a strengthening and poleward shift of the Ferrell cell (Gray et al. 2010; Haigh 2003). Over the Atlantic–European sector, circulation in winter tends to be more meridional during solar minima, with increased cold-air advection to central and eastern Europe (Fig. 3.47; see Brugnara et al. 2013). Several authors report a signature of the 11-yr solar cycle in sea level pressure. For solar minima, this corresponds to a pressure increase between Iceland and the British Isles (see Fig. 3.47, left), in line with cold winters in central and eastern Europe (e.g., Ineson et al. 2011; Woollings et al. 2010). Effects over the Pacific were found by van Loon and Meehl (2008). However, the detection of an influence of the 11-yr solar cycle at the earth's surface is difficult and depends on the statistical methods, the incorporation of other influencing factors, and the incorporation of a lag (Gray et al. 2013).

3.3.2.3 Possible Mechanisms

How can changes in solar activity affect weather and climate? Figure 3.48 shows the main mechanisms responsible for solar imprints in climate. The imprints of the 11-yr solar cycle on temperature and on ozone in the upper stratosphere (Fig. 3.46) are understood to be caused by changes in the UV part of the spectrum (e.g., Sitnov

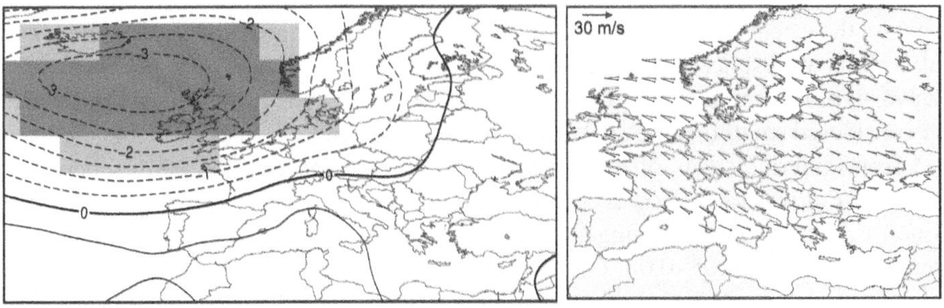

Fig. 3.47 Solar cycle imprint on (*left*) sea level pressure from 1749 to 2002 (solar maxima minus solar minima) and (*right*) wind vectors at 300 hPa from 1927 to 2002 (*red*: solar maxima, *blue*: solar minima) during January–March (Brugnara et al. 2013)

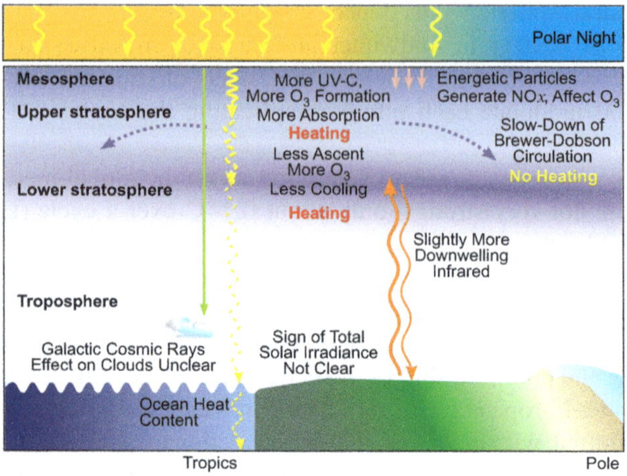

Fig. 3.48 Schematic of the effects of solar variability on climate (see also Gray et al. 2010)

2009). More UV radiation is absorbed in the process of ozone formation and as more ozone is produced, more energy is absorbed. The overall effect is a heating of the upper stratosphere.

The response in the lower stratosphere cannot be explained as a direct effect. This is because the UV radiation necessary for ozone formation does not penetrate this far down. Rather, this can be explained through an indirect effect. Because the solar effect in the upper stratosphere is stronger in the tropics than in the polar night region, the meridional temperature gradient is strengthened. The polar night jet strengthens and the meridional circulation weakens (Kodera and Kuroda 2002). The reduced upwelling in the tropics leads to higher temperatures and more ozone.

Energetic electrons mainly affect the thermosphere, mesosphere, and upper stratosphere. Here, they interact with N_2 and O_2. Through subsequent changes in nitrogen oxides (NO_x) and hydrogen oxide radicals (HO_x), energetic particles can deplete ozone (e.g., Päivärinta et al. 2013). From the mesosphere or upper stratosphere, the effect may propagate downwards to the mid- and lower stratosphere and also affect atmospheric circulation. Strong particle events (e.g., solar proton events) leave significant imprints in the stratosphere, but their effect on longer time scales is considered small (Funke et al. 2011, but see also Andersson et al. 2014).

Tropospheric effects of solar variability may originate in the stratosphere and propagate downward. However, are there also direct effects of changes in TSI? Energy balance models suggest that solar cycles only cause small changes in temperatures (ca. 0.05–0.08 °C). The changes are even smaller and lagged over the oceans (North et al. 2004). Note that the change in TSI during a sunspot cycle of about 1 Wm^2 translates to a radiative forcing of only 0.18 Wm^2 due to the ratio between earth's cross section and surface area (factor of 4) and albedo. To put this number into context, this is one-sixth of the estimated change in anthropogenic forcing from 1980 to 2011. Nevertheless, the direct forcing is not negligible, particularly because the amplitude of the forcing on longer time scales remains unknown.

It is possible that the solar effect on temperature is larger than would be estimated from radiative forcings alone due to mechanisms other than direct radiation and feedback effects within the machinery. A candidate for the former are galactic cosmic rays, which may lead to ionization and the formation of cloud condensation nuclei. It has been suggested (Svensmark and Friis-Christensen 1997) that through such a mechanism, solar activity (which modulates galactic cosmic rays) could change cloud cover on earth. This work has bred controversy for over a decade. Other observational studies have not confirmed the results (Calogovic et al. 2010). Recent experiments at the Conseil Européen pour la Recherche Nucléaire (CERN) research facility (Kirkby et al. 2011) showed that cosmic rays affect aerosol formation in the presence of certain specific precursor gases. However, the climatic relevance could not be established.

Amplification of the global signal or regional feedbacks might operate through water vapour (see Sect. 3.4.1) or the enhancement of natural modes of variability such as the NAO or ENSO. The NAO might be affected from the stratosphere. In fact, the imprint of the 11-yr sunspot cycle on circulation over the North Atlantic is sometimes interpreted as a signature of stratosphere–troposphere coupling (e.g., Ineson et al. 2011; Woollings et al. 2010). This is plausible because the 11-yr solar cycle is known to affect the stratosphere and downward propagation is empirically established (Baldwin and Dunkerton 2001). However, the mechanism is still unclear.

What about ENSO? According to the "thermostat" mechanism (Fig. 3.49, top; Mann et al. 2005), an increase in TSI contributes to stronger heating of waters on their path from the eastern to the western tropical Pacific. This leads to warmer waters in the western Pacific, but not in the east, where temperatures are dominated by upwelling. A stronger temperature gradient emerges, which, through the Bjerknes feedback, might change the wind stress and upwelling and lead to cooler waters in the east. This situation would damp the probability of El Niño and favour La Niña.

Meehl et al. (2009) suggest a different mechanism that involves a cloud feedback (see Fig. 3.49, bottom): The effects of changes in solar irradiance on the surface are strongest over the cloud-free areas such as the subtropics. Here, evaporation increases, and more latent heat is transported into the ITCZ, where it is released and contributes to a strengthened Hadley circulation. In the downwelling branch of the Hadley circulation, the strengthened subsidence leads to more cloud-free areas, and thus more solar radiation.

3.3.2.4 Beyond the Sunspot Cycle

The sun also varies on longer time scales than the 11-yr cycle. Cycles of 90 years (Gleissberg cycle), 205 years (De Vries cycle), and longer have been suggested. During the past 1200 years, several "grand maxima" and "grand minima" of solar

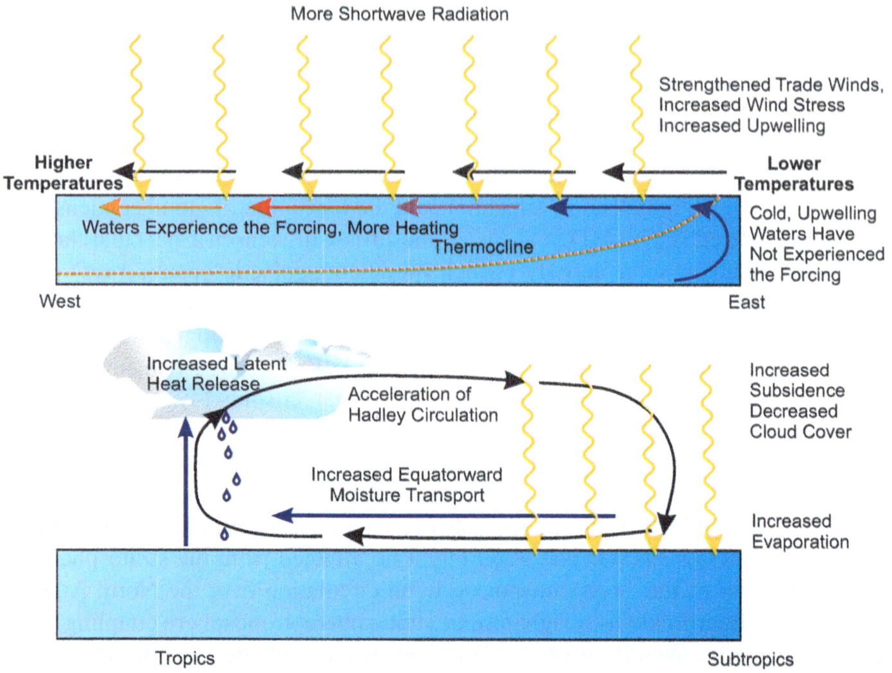

Fig. 3.49 Schematic of the possible effects of solar irradiance changes on (*top*) ENSO and (*bottom*) Hadley circulation (see also Meehl et al. 2009)

activity occurred. These include the Wolf, Oort, Spörer, Maunder, and Dalton minima (e.g., Steinhilber et al. 2012). Presently, we are experiencing a grand maximum or perhaps the end of it. The weak last solar cycle suggests that solar activity may be dropping into a new minimum.

Periods of solar maxima coincide with warmer temperatures in the Northern Hemisphere and vice versa. It is generally assumed that the sun played a role in forcing climatic events such as the "Little Ice Age". Chapter 4 will go into more detail with respect to some of these events. However, a fundamental uncertainty in statistically linking these climatic changes to the sun is that during the past millennium low solar activity often coincided with high volcanic activity (Masson-Delmotte et al. (2013); see also Sect. 3.3.1). In our work, separating volcanic and solar signatures from very long temperature-sensitive tree ring chronologies (Breitenmoser et al. 2012), we fitted a simple empirical model of the volcanic effect:

$$T_t = aF + T_{t-1}e^{-b\Delta t} \tag{3.4}$$

In this model, the anomaly of temperature (or tree ring width) T at time t depends on the instantaneous volcanic forcing F (in our case negative global aerosol optical depth) and a decay term that depends on the previous year's temperature anomaly

T_{t-1} ($\Delta t = 1$ year). I will use this simple, illustrative approach again in Chap. 4 to address the role of forcings. After subtracting T_t from the tree ring series, the residual series showed significant variability near the time scale of the solar De Vries cycle (ca. 205 years), confirming the existence of a centennial-scale solar imprint in tree ring climate proxies.

Most studies on longer-term solar effects use climate proxies and derived reconstructions, which allow limited interpretation with respect to atmospheric circulation changes. Documentary sources might be more informative. For instance, Sirocko et al. (2012) show a relation between solar activity and the freezing of the Rhine that is related to an altered frequency of atmospheric blockings. Moffa-Sánchez et al. (2014) find connections between solar activity and North Atlantic hydrography at centennial scales. However, some studies have considerable methodological uncertainties (van Oldenborgh et al. 2013) and further studies must confirm these links. Therefore, a clear relation between weather-scale events (such as blocking) and longer-term changes of the sun is not yet established.

3.3.2.5 Time Series of Solar Forcing

Solar variability encompasses changes in many different properties. Consequently, different measures are used to describe changes in solar activity. Quantifying these changes is even difficult today, in the satellite era. In fact, 30-yr time series of TSI produced by three different groups by merging satellite data provide conflicting results for whether TSI during the recent solar minimum was lower or higher than during previous minima (Myhre et al. 2013). Reconstructing TSI or other measures back in time is even more difficult and current estimates of past solar activity differ widely.

Solar radiation measurements from the earth's surface do not help because atmospheric effects dwarf solar changes. The 10.7-cm radio flux, observed since the 1940s, is not perturbed by the atmosphere. Therefore, it is often used as a measure of solar activity once calibrated against TSI.

In addition to irradiance, solar magnetic activity can be measured from the earth's surface by observing changes in the earth's magnetic field that occur due to its interaction with the solar magnetic field. The so-called *aa* index of geomagnetic activity, for which measurements date back to the 19th century, contains information about changes in solar activity.

Observations of the solar disk from the ground also reveal changes of the sun's magnetic activity. Photographs that reach back to the early 1900s and that are taken in certain wavelengths allow us to explore changes in activity. By far, the most widely used measure of solar magnetic activity is the number and area of sunspots. Time series of observations reach back to the 17th century and most reconstructions of solar activity covering the last 400 years use sunspots. Also the sunspot data have uncertainties and a substantial revision and switch to a new sunspot number has recently taken place (http://www.stce.be/press/01/welcome.html).

Because the flux of galactic cosmic rays is modulated by solar activity, reconstructions of past solar activity may be acquired through measuring the imprint of galactic cosmic rays. Data from neutron monitors or cloud chambers reach back to the 1950s or 1930s, respectively, and are a direct measure of the interaction of galactic cosmic rays with the atmosphere. Further back in time, climate scientists use cosmogenic nuclides (^{10}Be and ^{14}C), which are formed through spallation processes in the stratosphere, as a measure of galactic cosmic rays. They are recorded in natural archives such as ices cores (^{10}Be) and tree rings (^{14}C). Reconstructions need to consider many different system effects. First, the earth's magnetic field affects the production and distribution of cosmogenic nuclides and variations in space and time of earth's magnetic field need to be considered. Second, the two nuclides take very different routes until they enter the archive. ^{10}Be attaches to aerosols, which are then quickly deposited (a process that depends on atmospheric circulation). ^{14}C is globally distributed and eventually taken up by organisms including trees.

After accounting for all system effects, cosmogenic nuclides provide a reconstruction of the so-called solar modulation potential (see Steinhilber et al. 2012), which is calibrated against TSI or used for more complex, physical–empirical approaches (Shapiro et al. 2011). Statistical calibration is limited by the availability of observations (which only cover recent decades and have large uncertainties even in the satellite age), while other approaches are limited by our understanding of the sun.

Unfortunately, results from different approaches differ widely, even for TSI alone. Reconstructed time series going back to 1700 are shown in Fig. 3.50 (Myhre et al. 2013). While they are qualitatively consistent, they differ in the amplitude of

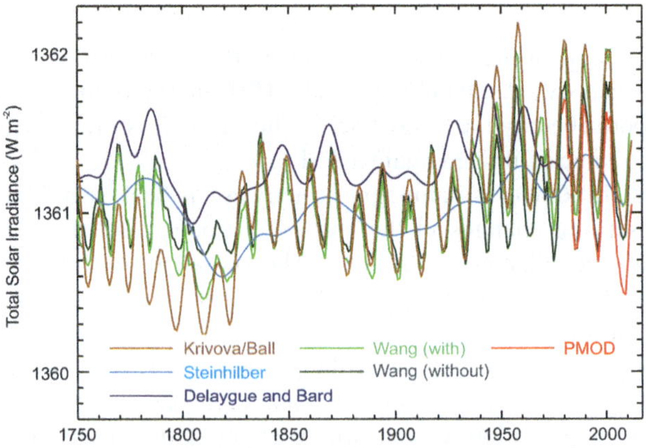

Fig. 3.50 Reconstructions of total solar irradiance since 1745; annual resolution from Wang et al. (2005), with and without an independent change in the background level of irradiance, Krivova et al. (2010) combined with Ball et al. (2012) and 5-year resolution time series from Steinhilber et al. (2009) and Delaygue and Bard (2011). The series are standardized to the Physikalisch-Meteorologisches Observatorium Davos (PMOD) measurements (From Myhre et al. 2013)

the forcing. Hence, while we do know that total solar irradiance was lower during the Dalton minimum in the early 19th century than today, we do not know by how much. Furthermore, feedback mechanisms within the earth's climate system are not well known. This explains why, despite more than two centuries of research, our knowledge about the sun's influence on climate is still limited.

3.3.3 Well-Mixed Greenhouse Gases

During Brückner's time, the most prominent and thought-provoking concept of climate change was the ice age theory. In fact, it was a conglomerate of many theories—some of which were peculiar. One of these theories was that of carbon dioxide's greenhouse effect by De Marchi (1895a) and Arrhenius (1896) (see Box 4.2, p. 256). Now that humans have profoundly altered the atmospheric composition, this mechanism has become the dominant driver of long-term climate change. It plays a major role for understanding climatic changes not only during the last 40 years, but also the last 400 years.

In the following, I will briefly introduce the effect of well-mixed greenhouse gases. This is shorter than the text on solar and volcanic forcings because the greenhouse effect is well covered in other studies and assessments. Specifically, I do not attempt to cover the ability of models to reproduce future climate. Instead, the reader is referred to the technical literature and to the Fifth Assessment Report of the IPCC (IPCC 2013).

3.3.3.1 Mechanism

The greenhouse effect is exerted by gases with a dipole moment. They absorb infrared radiation to change their vibrational state. They also reemit radiation. Important greenhouse gases are water vapour, carbon dioxide, and methane. In the presence of these gases, the atmosphere is partly opaque for longwave radiation, but rather transparent for shortwave radiation.

The incoming shortwave radiation must be approximately balanced by outgoing longwave radiation at the top of the atmosphere (Sect. 3.1.2). However, due to water vapour and other greenhouse gases, a large fraction of the outgoing radiation is absorbed within a short distance and re-emitted both up- and downwards. How can the outgoing longwave radiation be maintained? Consider the atmosphere as a thin layer, opaque for longwave radiation but transparent for shortwave radiation (Fig. 3.51, left). In equilibrium it will emit the same amount of radiation to space as the earth receives as shortwave radiation, but the same amount of radiation will also be emitted from that layer downwards. To maintain this double flux, the layer must receive this energy from below. If this occurs in the form of longwave radiation, upward longwave radiation from below must be twice the amount of shortwave radiation. Assume that below there is a second layer. This layer must emit this

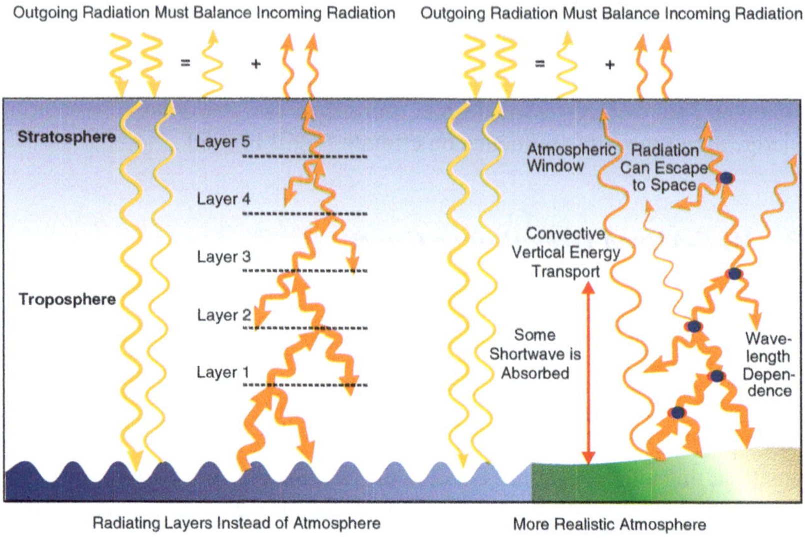

Fig. 3.51 Schematic of the greenhouse effect in (*left*) a hypothetical world in which the atmosphere consists of thin layers and (*right*) in a more realistic world

double flux upwards and the same amount downwards, and thus receive 4 times the amount of shortwave radiation. One-quarter of this amount comes from the layer above. The remaining three-quarters must come from below.

Expanding this thought to an atmosphere with many thin and opaque layers, we find that longwave fluxes must increase the further down in the atmosphere we go. At the surface, the up- and downwelling longwave radiation greatly exceeds the incoming solar shortwave radiation. In order to deliver such a large flux, the surface must heat up considerably.

In reality (Fig. 3.51, right), the atmosphere is not completely transparent for shortwave radiation and not completely opaque for longwave radiation. For some specific wavelengths, the atmosphere is even transparent (e.g., 8–12 μm atmospheric window) and wavelength dependencies are crucial. Furthermore, processes in addition to radiation play a role. Turbulent heat fluxes transport energy from the surface upwards through the troposphere. In the stratosphere, chemical processes contribute to heating. However, if the amount of greenhouse gases increases, the atmosphere becomes more opaque. In the example above, this is similar to adding another layer—the lower layers of the atmosphere must maintain higher longwave fluxes. The earth's surface, which is cooled by turbulent heat fluxes, will warm further. From the heated surface, the energy is transported upwards into the troposphere via turbulent fluxes (sensible and latent heat), and the ocean is warmed.

The stratosphere, in contrast, is not heated from the ground. Rather, it has its own heating source: the ozone layer. The stratosphere is heated by absorbing UV radiation. As vertical turbulent fluxes are suppressed in the stratosphere, energy is more efficiently transported by means of longwave radiation. From the stratosphere, longwave radiation can easily escape to space without being absorbed and re-emitted. Here, longwave radiation cools and additional greenhouse gases amplify this cooling.

Incremental amounts of greenhouse gases (the same holds for thin clouds) have a different net effect depending on their altitude. The largest net effect occurs in the cold tropopause region. Normally this region does not emit much radiation, and most of the radiation seen from space originates from warmer, lower layers. With more efficient absorption and radiation in the cold tropopause region, the radiation from the warmer, lower layers can less easily escape and downwelling radiation increases in the levels below.

The climate system responds to an increase in greenhouse gases by tropospheric warming and stratospheric cooling. This leads to a rising tropopause. Furthermore, an increase in greenhouse gases leads to warming of the upper ocean, melting of ice and permafrost, and acceleration of the hydrological cycle due to larger latent heat fluxes. Our understanding of the fundamental physics of the greenhouse effect is very good (Myhre et al. 2013), and climate models capture the main effects well (Flato et al. 2013). Uncertainties arise when it comes to feedbacks, effects on the weather scale, and effects on atmospheric circulation (Shepherd 2014).

3.3.3.2 Impacts on Future Climate

Greenhouse gases are the dominant global force in climate evolution in the twenty-first century. Figure 3.52 shows the projected warming for the period 2081–2100 assuming that the world follows a radiative forcing pathway of 4.5 W m^{-2}.[19] The figure is based on the median of multiple models (see also Collins et al. 2013). The salient features are that warming takes place everywhere, but is more pronounced over land than over ocean and more pronounced over polar latitudes, particularly in winter, than in the tropics. The vertical structure shows that in the tropics, the warming is higher in the upper troposphere (where the latent heat is released) than near the ground (Collins et al. 2013). In contrast, in the Arctic the heating is strongest near the ground because feedback processes involving sea ice operate near the ground and because strong inversions in winter inhibit the transport of heat to upper layers.

Precipitation is far more difficult to model than temperature and projected changes are less certain and less consistent among models. Projections show a

[19]The Fifth Assessment report of the IPCC uses scenarios with a given approximate total radiative forcing in the year 2100 relative to 1750. They are termed "Representative Concentration Pathways" (RCP). For instance, RCP2.6 means that the total radiative forcing in the year 2100 is 2.6 W m^{-2} higher than in 1750. RCPs are based on a combination of integrated assessment models, climate models, and atmospheric chemistry and global carbon cycle models (IPCC 2013). The RCPs encompass a very low forcing scenario (RCP2.6), two stabilization scenarios (RCP4.5 and RCP6), and a high greenhouse gas emission scenario (RCP8.5).

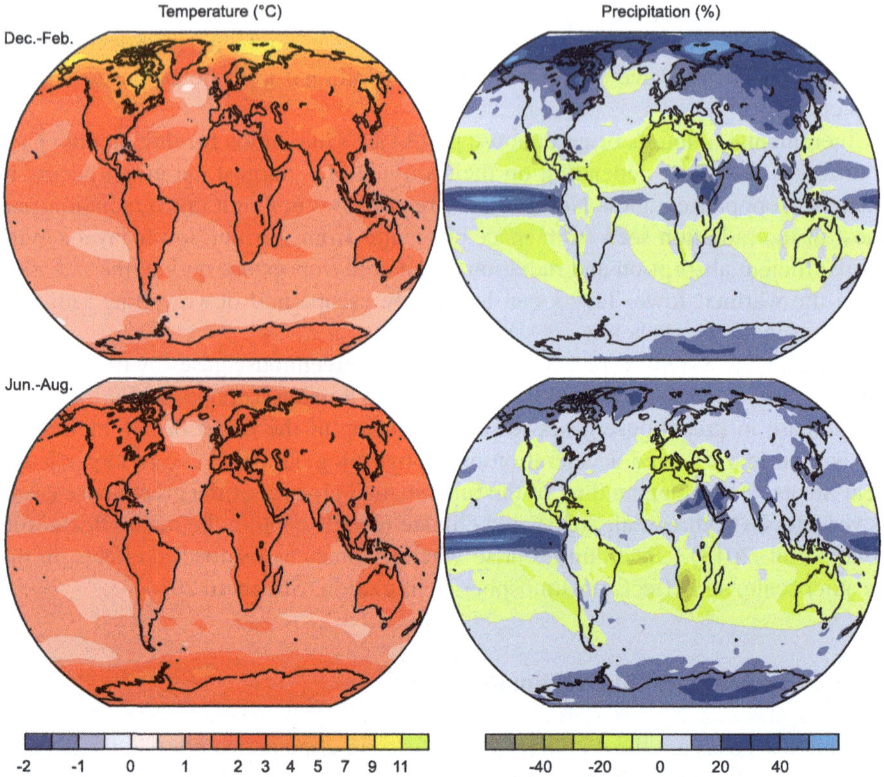

Fig. 3.52 Projections of (*left*) temperature for (*top*) December–February and (*bottom*) June–August, and (*right*) precipitation for (*top*) October–March and (*bottom*) April–September for the RCP4.5 emission scenario (multi-model median). Shown is the difference 2081–2100 minus 1986–2005 (Generated on the Climate Explorer, see IPCC (2013))

precipitation increase in the inner-tropical regions, the midlatitudes, and the polar regions, but a decrease in the subtropical regions, particularly in summer. This pattern is also called "the rich get richer, the poor get poorer".

3.3.3.3 Effects of Greenhouse Gases on Past Climate

How important were greenhouse gases in the past? Very important, at least for the past 50 years. The Fifth Assessment Report of the IPCC states (IPCC 2013):

> "It is extremely likely that human influence has been the dominant cause of the observed warming since the mid-20th century."

Statements linking observed climate trends to underlying causes are termed attributions (see Box 3.3, p. 149). Since the first attribution study was published in the 1990s (Hegerl et al. 1996), attribution of climate trends to greenhouse gases was successful for trends in surface air temperature (global and regional temperature, global temperature extremes), free-atmospheric temperature, tropopause height, atmospheric humidity, upper-ocean heat content, and Arctic sea ice and snow cover (Bindoff et al. 2013). Studies have also addressed the response of the machinery. For instance, anthropogenic signals were detected in sea level pressure (Gillett et al. 2003).

The same detection and attribution methodology has also been used to analyse past climatic changes (Hegerl et al. 2011; Schurer et al. 2014). More simple approaches include comparisons of observations with model simulations or multiple regression methods (Hegerl and Zwiers 2011). In this book, for illustrative purposes, I will use a simple regression approach. Chapter 4 will go into more detail as to the roles of greenhouse gases and other forcings prior to the 1950s.

Box 3.3 Detection and attribution of climate change

How can scientists conclude that humans contribute to climate change? During the last 20 years, statistical techniques and a methodological framework have been developed for this purpose. This framework is known as "Detection and Attribution", which highlights the two-step procedure. Detection of observed climate change requires demonstrating that the observed change is unlikely to have occurred by chance due to internal variability alone. However, detection does not provide a cause for that change (IPCC 2013). Internal variability is often estimated from model simulations, although observations or proxies can also be used.

Attribution then is the *"process of evaluating the relative contributions of multiple causal factors to a change or event with an assignment of statistical confidence"* (IPCC 2013). Attribution to a given factor, for example, greenhouse gases, requires demonstrating that (a) a trend has been detected, (b) the observed change cannot be explained by any combination of other factors, and (c) the observed change can be explained by a combination of factors that include greenhouse gases.

A statistical method for attributing climate change is the "optimal fingerprint method". Here, the expected spatiotemporal response patterns to individual forcings (solar, volcanic, greenhouse gases, tropospheric aerosols) is specified beforehand based on climate model simulations (Hegerl and Zwiers 2011). The method then aims at finding linear combinations of the patterns that best explain the observed trend. Recent work has addressed not just trends, but also interannual variability (Estrada et al. 2013). Attribution of individual extreme events to anthropogenic factors is also often attempted (Pall et al. 2011).

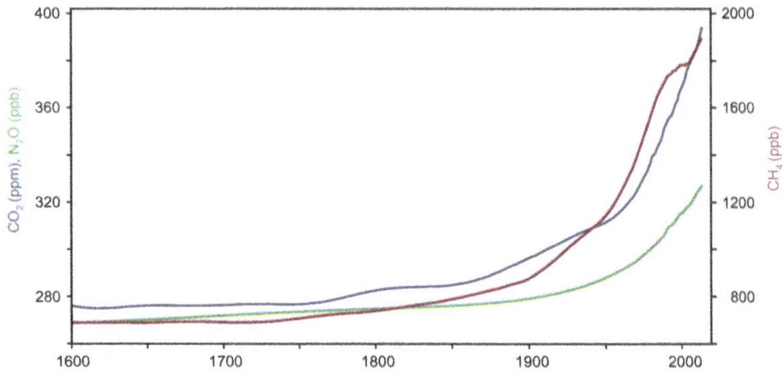

Fig. 3.53 Concentrations of CO_2, CH_4, and N_2O in the atmosphere since 1600 based on ice cores and measurements (Data from Yoshimori et al. 2005)

3.3.3.4 Changes in Greenhouse Gas Concentrations Since 1600

The greenhouse gas forcing changed considerably during the last part of the period considered in this book (see Hartmann et al. 2013; Myhre et al. 2013). Atmospheric concentrations of carbon dioxide, methane, and nitrous oxide are shown in Fig. 3.53 and are based on direct measurements for the past few decades and ice core data before (Yoshimori et al. 2005). From 1600 to 1850, CO_2 concentrations remained below 285 ppm, but since then have been increasing almost exponentially. Concentrations have now reached 400 ppm. Most of this increase is from fossil fuel combustion. Land use changes further contribute.

The second most important greenhouse gas is methane, which is emitted by wetlands and through agricultural (but also industrial) activities. Methane concentrations have more than doubled since preindustrial times, but with peculiar variations in the growth rate in the 1990s and early 2000s. The third most important gas is nitrous oxide, which is emitted by soils treated with synthetic or organic fertiliser. N_2O is not only relevant as a greenhouse gas, but also as an ozone-depleting substance (Ravishankara et al. 2009). Finally, ozone also acts as a greenhouse gas. With respect to climatic changes since 1700, Fig. 3.53 implies that most changes in greenhouse gases have occurred since the 1950s, but there were also changes in the first half of the twentieth century and earlier. In fact, Ruddiman (2003) argued that humans have been altering greenhouse gas radiative forcing since the early Holocene through methane emissions.

3.3.4 Tropospheric Aerosols and Reactive Trace Gases

> "The sky at the horizon up to 20 degrees towards the zenith is almost always blurred by a hazy mist, often preventing a good view. This is due to the smoke of the enormous grass fires, which take place in the entire tropical South African region every year during the dry

season. They produce an amount of smoke that can hardly be imagined and in comparison to which the smoke produced by all steam engines, moor fires etc. in Europe is insignificant." (von Danckelman 1884)

The effect of tropospheric aerosols on climate and, particularly, their interaction with clouds, remains one of the largest uncertainties of climate change. While effects of changing aerosols since the 1940s and recent changes in the spatial patterns are relevant for assessing current climate change, this quote from the 19th century suggests that aerosols were present before that time and may also play a role in our discussion of climatic changes since 1700.

3.3.4.1 Properties of Aerosols

Aerosols are suspensions of liquid or solid particles in the air. They encompass mineral particles from soils or volcanic ash, biogenic particles (e.g., pollen or spores), sea salt, combustion products (e.g., soot), sulphate, and secondary organic aerosols. Figure 3.54 shows a range of different particles. Their sizes span three orders of magnitude (ca. from 10 nm to 10 μm), and their properties vary widely.

Their importance for atmosphere and climate derives from their physical, radiative, and chemical properties. An overview is given in Fig. 3.55. Tropospheric aerosols scatter and absorb radiation, similar to sulphate aerosols in the stratosphere. Black carbon (soot) and mineral dust are efficient absorbers; for most others, scattering dominates. Aerosols also serve as cloud condensation nuclei. Thus, they have a large effect on the properties of the cloud that is forming. Some are efficient ice nuclei. The altered cloud properties affect the short- and longwave radiation balance and thus affect surface climate, as will be discussed below.

Aerosols also provide surfaces for heterogeneous reactions in the atmosphere which may affect, for example, ozone chemistry (see Solomon 1999). Finally, after being deposited, aerosols may change the albedo of snow or ice surfaces, changing

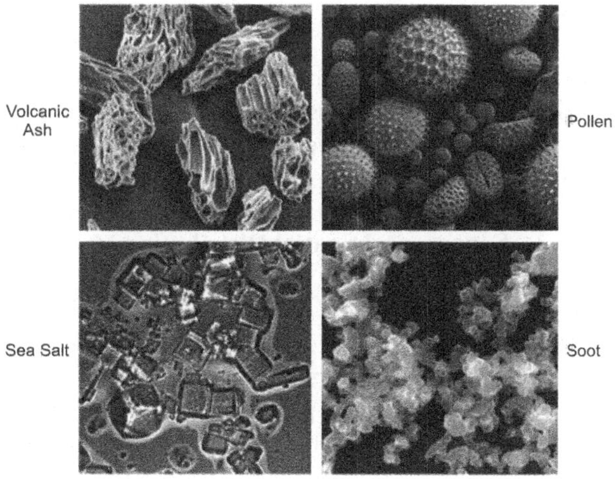

Fig. 3.54 Scanning electron microscope images (not at the same scale) of volcanic ash, pollen, sea salt, and soot. Micrographs courtesy USGS, UMBC (Chere Petty), and Arizona State University (Peter Buseck)

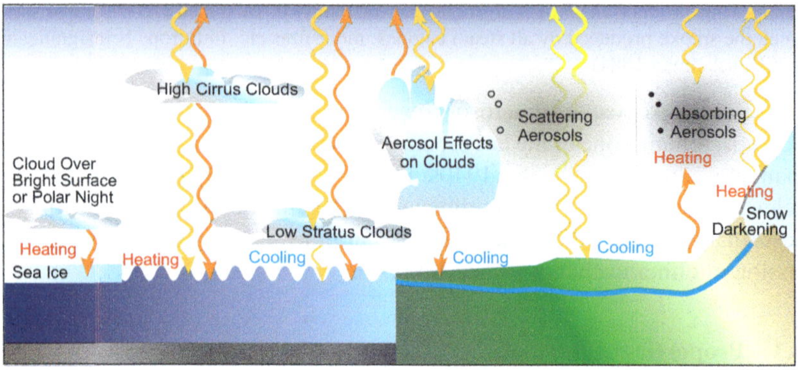

Fig. 3.55 Schematic depiction of the effects of aerosols on climate

Fig. 3.56 Sources of aerosols (*left*) biomass burning, (*middle*) mineral dust from a sandstorm transported to the North Atlantic on 26 February 2000, and (*right*) industrial combustion (Pictures by NASA, STS048-71-090, the SeaWiFS Project, NASA/GSFC and ORBIMAGE)

the radiation balance. Aerosols are quickly rained out or washed out or they settle gravitationally. Their atmospheric lifetime is therefore from days to weeks. Aerosols and their climate effects are thus spatially and temporally variable.

3.3.4.2 Sources and Spatial Distribution of Aerosols

Aerosols have a variety of sources; some of which are shown in Fig. 3.56. Natural aerosols, which dominate by mass, include mineral dust, sea salt, pollen, fungal spores, natural biomass burning aerosols (ash and black carbon) and biogenic sulphate particles. Anthropogenic aerosol sources include combustion processes such as domestic burning, biomass burning (black carbon), combustion processes in

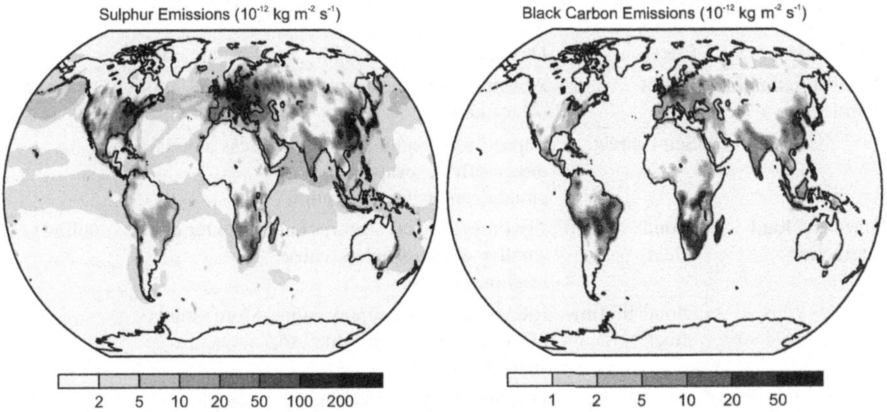

Fig. 3.57 Emissions of (*left*) sulphate and (*right*) black carbon in the 1990s (September) as used in our ECHAM5.5-HAM simulations (Sect. 2.9)

vehicles or industry (sulphate and black carbon), agricultural activities (ammonium nitrate), or land use change (mineral dust). Secondary organic aerosols are formed from anthropogenic or natural emissions of hydrocarbons.

These aerosol sources have different spatial emission patterns (Fig. 3.57). Because of their short lifetime, the emission patterns are well reflected in the aerosol spatial distributions. Domestic burning (cooking and heating) mirrors the population density, particularly in tropical and subtropical regions. Biomass burning occurs in the tropical savannah regions (Africa, Brazil, and Borneo). Aerosols from industrial activities primarily originate from North America, Europe, and East Asia. As to the mineral dust, a large fraction originates from a few localised desert sources such as the Bodélé depression in the Sahara.

3.3.4.3 Impacts on Climate

Von Danckelman described the Sun as a matte disk covered by biomass burning smoke. He also speculated that smoke might act as cloud condensation nuclei and affect precipitation type (von Danckelman 1884). These are today addressed as "direct" (aerosols directly interfering with radiation) and "indirect" effects (aerosol indirectly interfering with radiation through cloud modification). In addition to these effects in the atmospheric column, the spatial distribution of aerosol radiative effects may induce changes in atmospheric circulation and atmospheric stability, and further effects such as snow darkening need to be taken into account. This makes aerosols one of the most complex climate factors. There are still significant gaps in our knowledge.

Direct effects are comparably easy to understand. Scattering aerosols increase the planetary albedo and lead to shortwave cooling of the surface. Absorbing aerosols

Table 3.3 Characterisation of aerosol effects (Adapted from Boucher et al. 2013)

Mechanism	Effect	Description	Climate effect
Aerosol–radiation interaction	Direct	Absorption or scattering of solar radiation	Surface cooling (warming aloft for absorbing aerosols)
	Semi-direct	Lapse rate adjustment to direct effect, evaporation of cloud droplets by absorption	Less clouds
Aerosol–cloud interaction	Cloud albedo effect	"Twomey effect": more but smaller cloud droplets cause brighter clouds	Brighter clouds (cooling)
	Cloud lifetime effect	Life time effect: suppression of rain formation due to smaller droplets	More clouds
		Glaciation effect: ice condensation nuclei change precipitation efficiency	Complex
		Thermodynamic effect: smaller droplets require cooler temperatures to freeze	Complex

heat the layer in which they reside, but pass less radiation to the ground and thus decrease the surface solar radiation, leading to cooling.

The indirect effects of aerosol resulting from their interaction with clouds are more difficult to study. Some aerosols act as cloud condensation nuclei or ice nuclei and change cloud properties such as the number and size of droplets and the precipitation efficiency. This leads to different optical properties and lifetime of clouds and, indirectly, to changes in radiation. Table 3.3 gives a brief summary of the direct and indirect effects of aerosols.

Aerosols increase the number of cloud condensation nuclei. Therefore, the resulting clouds have more (although smaller) droplets. The optical properties of clouds depend on droplet size. "Dirty" clouds with smaller droplets are brighter (Fig. 3.58, top) and reflect more radiation (albedo effect). However, as a consequence of the small droplets, it takes longer for clouds to form rain droplets. Thus, the clouds last longer (lifetime effect). Clouds shield the surface from shortwave radiation, contributing to cooling, but also radiate longwave radiation, contributing to warming. Low, optically thick water clouds tend to cool more. High, thin cirrus clouds tend to warm. The resulting net effect of these mechanisms depends on the albedo and temperature of the surface and cloud tops and other factors. Over bright surfaces, shortwave cooling is less important. Overall, the climatic effects of clouds induced by aerosol–cloud interactions act to cool the globe (Boucher et al. 2013).

Another indirect effect of aerosols unrelated to clouds is snow darkening. In models, black carbon efficiently lowers the albedo of snow, which can have rather strong effects in the Arctic. This effect has been verified in the laboratory (Hadley and Kirchstetter 2012).

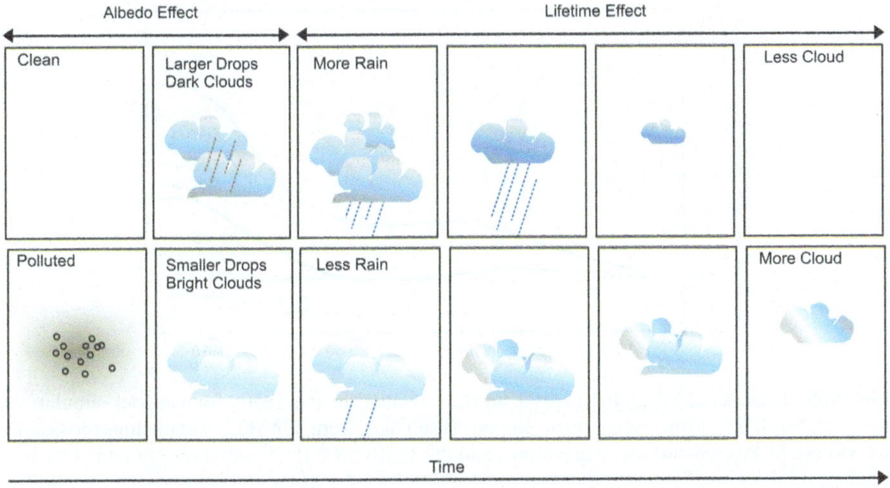

Fig. 3.58 Schematic depiction of the albedo effect and the lifetime effect of aerosols on clouds (Adapted from Stevens and Feingold 2009)

Finally, many recent studies suggest that the spatially heterogeneous imprint of aerosols on sea-surface temperature may induce anomalous circulation. For instance, aerosols have been suggested to affect North Atlantic hurricanes (Dunstone et al. 2013), the West African monsoon (Rotstayn and Lohmann 2002), the Indian monsoon (Wang et al. 2009a) and Southern Hemisphere circulation via ocean–atmosphere interactions (Cai and Cowan 2007). In Chap. 4, some of these mechanism will be discussed in more detail.

How has aerosol forcing changed over time? As the quote at the beginning of this chapter shows, aerosols were emitted from various sources—natural and anthropogenic—long before the onset of industrialisation. Although often intentionally ignited, substantial biomass burning can also occur naturally if ignition sources (e.g., lightning) are available. Historical biomass burning is a function of temperature and precipitation (affecting the amount of fuel), population density, and land use changes. Sedimentary charcoal suggests that biomass burning decreased up to 1750 (arguably due to cooling), increased up to 1870 (due to increased population density), and then decreased again, arguably due to land use changes (Marlon et al. 2008).

With the onset of industrialisation, economies in western Europe started to produce aerosols. Time series of aerosol burdens since 1850 are shown in Fig. 3.59. For a long time, the atmospheric effects of aerosols may have been rather local as compared to those from biomass burning aerosols (von Danckelman 1884). However, industry and thus aerosol emissions grew rapidly, particularly after the Second World War (Sect. 4.4.2). The increase of sulphate aerosols was especially dramatic. In Europe and North America, aerosol concentrations peaked in the 1980s.

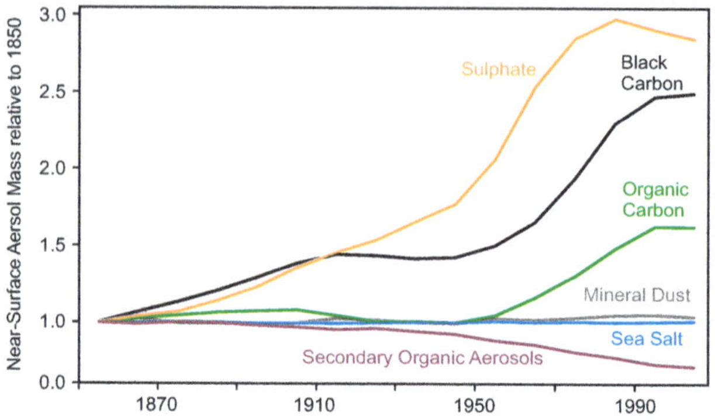

Fig. 3.59 Time series of global aerosol burdens relative to the 1850s from model simulations used in the IPCC Fifth Assessment Report (Data are from CAM3.5 simulations driven by CCSM3 (CMIP4) sea-surface temperatures and the 1850–2000 IPCC emissions, provided by J.-F. Lamarque)

Air quality measurements measures and the collapse of socialist economies led to a decrease of aerosols in some regions, while aerosols continue to increase in rapidly developing regions (specifically, East Asia).

3.3.4.4 Tropospheric Trace Gases

Besides aerosols, other tropospheric constituents also affect the radiation of the atmosphere. One of the most prominent among them is ozone. On the one hand, ozone is a greenhouse gas and contributes to the greenhouse effect. On the other hand, ozone photolysis (and subsequent reaction with water vapour) is the main source of OH radicals in the troposphere. Reaction with OH, in turn, is the major sink for most tropospheric trace constituents, including the greenhouse gas methane. Section 3.4.4 will discuss this aspect further.

The main natural source of tropospheric ozone is influx from the stratosphere. In the troposphere, ozone is photochemically produced from nitrogen oxides in presence of hydrocarbons or carbon monoxide, which are predominantly anthropogenic (although soils, vegetation, lightning, and biomass burning also contribute). Its main sink is dry deposition.

Ozone is highest in boreal summer over industrialised regions as well as in subtropical regions such as the Arabian Peninsula. Tropospheric ozone concentrations doubled from the 1930s or 1950s to the 1980s (Staehelin et al. 1994), and the increase continues on a global level, though now mostly in East Asia rather than Europe and North America (Hartmann et al. 2013).

The increase of tropospheric ozone since the 1950s affects local radiative forcing because ozone is a greenhouse gas (Shindell and Faluvegi 2002). The total radiative forcing due to tropospheric ozone corresponds to approximately 0.6 W m^{-2}.

3.3.5 *Land Surface*

The part of the atmosphere that is very close to the earth's surface is home to 7 billion people and a large part of the living world. This part of the atmosphere is strongly influenced by its lower boundary, the land surface. Properties of the land surface affect the radiation balance (via albedo), the mass balance (via evapotranspiration), and the momentum balance (via surface roughness, which relates to the mechanical generation of turbulence). Important properties of the land surface are vegetation fraction and type (or presence of snow), soil moisture, and albedo.

The land surface affects climate locally, but also on larger scales. Conversely, the atmosphere affects the land surface and thus the land surface and local climate evolve as a coupled system.

The land surface may change over time (e.g., changes in vegetation or the built environment) and its influence on the atmosphere also changes. Land surface changes were long ago recognised as a factor causing (perceived or real) climatic changes (see Glacken 1967; Grove 1995). Changes in rainfall after deforestation of tropical island colonies was a source of concern in the Renaissance and was studied by scientists in the 17th century (Halley 1694). In his book, Brückner (1890) gives a very extensive, critical review of research on the effects of land use changes on climate. In particular, deforestation (sometime reforestation) was blamed in the 18th and 19th century to have caused climatic changes. In North America, numerous publications claimed a climatic change due to deforestation. Some authors argued that the climate had become warmer, some felt it had become colder and winds more easterly (e.g., Jefferson 1787; Volney 1803; Williamson 1771). The climatic changes may or may not have occurred, but were perceived as such and were widely discussed.

In Europe, land cover–climate relations were discussed in the context of the presence or absence of climatic changes since the Roman period (Ideler 1832; Mann 1790) and in the context of extreme flooding in the 19th century. The latter were often blamed on deforestation and in many countries discussions reached political levels. Another emerging topic related to land use, climate change, and environment in colonial science in the late 19th century was desertification. The discovery of ancient pluvial epochs in the Sahara region fuelled plans for desert transformation (Lehmann 2013).

Today, land surface changes and their effect on climate are again an important research area (Seneviratne et al. 2010). Hence, one of the main topics of climate science during Brückner's time is again in the spotlight of research.

3.3.5.1 Mechanisms

A schematic view of land surface effects is shown in Fig. 3.60. Land surface properties including albedo, leaf area index, surface roughness, and soil moisture

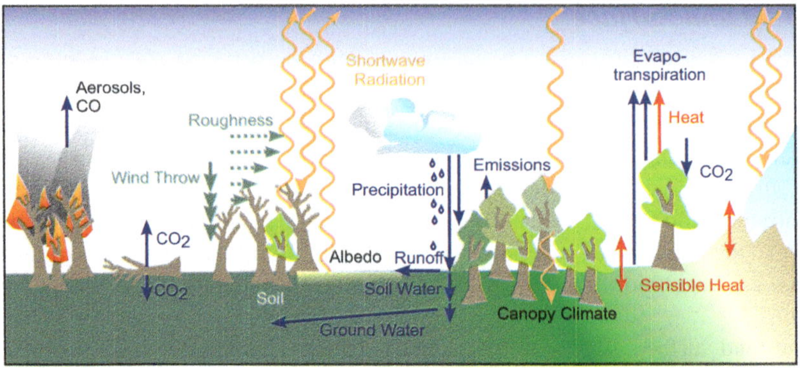

Fig. 3.60 Schematic of land surface effects. *Orange/red*, *blue*, and *green arrows* denote fluxes of energy, mass, and momentum, respectively

interact with the atmosphere in various ways. First, some of the properties affect the surface energy balance. The net radiation depends on the albedo, which is one of the most important land surface properties. The albedo varies from almost zero (tarred street) to almost one (fresh snow). The land surface also affects the partitioning of the heat fluxes into sensible and latent heat, the so-called Bowen Ratio. If the soil is dry, a larger fraction of the energy fluxes is put into sensible heat. Thus, the atmosphere is heated more strongly over dry surfaces. The roughness of the terrain or the vegetation affects wind speed and turbulence, which changes not only the momentum budget, but can also affect the turbulent fluxes of gases and heat. Evapotranspiration is the physical link between the energy balance and the (soil) water balance (see review by Seneviratne and Stöckli 2008).

For vegetated surfaces, further effects need to be considered. Plants affect radiation (via shading) and turbulent water vapour fluxes (via transpiration), and provide a storage term for water, energy, and carbon. Carbon assimilation is correlated with evapotranspiration, which thus is one of the most fundamental properties for describing land–vegetation–atmosphere feedbacks. Vegetation emits hydrocarbons, and when it burns it can become a source of aerosols and gaseous species. Conversely, vegetation can also react to changes in climate.

On time scales of decades, the land surface is an important component of the carbon cycle. Soils also need to be taken into consideration on centennial time scales. Spatially, land–atmosphere interaction is most relevant in transitional zones between wet (where evapotranspiration is not limiting) and dry climates (where evapotranspiration is not varying much). This includes the Mediterranean area, the Sahel, the Midwest United States, and India (Koster et al. 2004; Seneviratne and Stöckli 2008).

Land–atmosphere interaction is particularly relevant for droughts. Once a precipitation deficit has occurred and evapotranspiration is reduced, the land surface may also act as a feedback to the atmosphere and worsen or prolong the drought. Section 3.4.4 will go into more detail regarding climate feedbacks including the land surface.

Fig. 3.61 Land cover can be spatially very heterogeneous and with it, local climate. The image shows a delta, agriculture, and settlements near Rostov-on-Don, Russia (Near-natural colours, LandSat-5 image 2010 (LT51740272010161MOR00), provided by NASA)

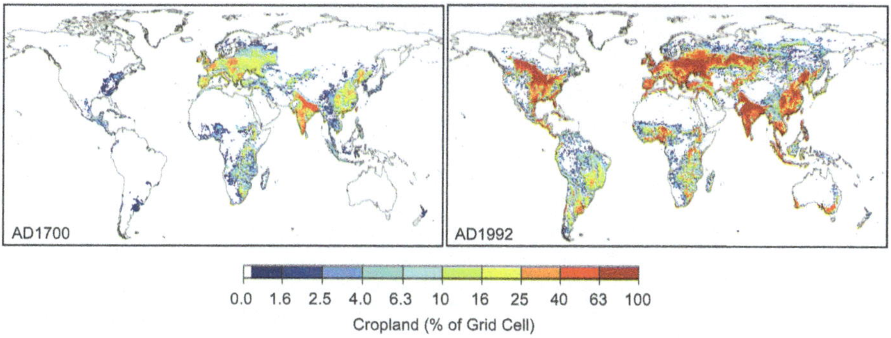

Fig. 3.62 Global cropland area in 1700 and 1992 from (adapted from Pongratz et al. 2008)

Land use is extremely spatially variable (Fig. 3.61) and land cover differences may induce changes in local circulation systems. Therefore, in addition to the energy balance, local-to-regional exchange processes need to be taken into account (see Fig. 1.6).

3.3.5.2 Climate Impacts and Development of Land Cover Since 1600

Over the past 400 years, the landscape has changed dramatically in many regions. Forests have diminished and increased, cities have been built, and deserts have grown. Historical land cover data were reconstructed by Pongratz et al. (2008). Figure 3.62 shows two time slices, 1700 and 1992, for the example of cropland.

This shows the intensification (in Europe, India, and Africa) and expansion (North and South America) of agriculture at the expense of pasture and forest area. Forests decreased rather strongly in the tropics and northern subtropics, but increased in the southern extratropics.

The surface roughness length decreased in all regions until the second half of the 20th century but arguably increased over the past three decades. Vautard et al. (2010) and McVicar et al. (2012) report widespread "stilling" of surface winds over large parts of the world's land masses since the 1970s, which they attribute to increased roughness due to increased biomass.

Land surface changes can be much larger and climatically much more relevant on regional-to-local scales. For instance, clearing of forests may have contributed to the "Medieval Climate Anomaly" in Europe (Goosse et al. 2006). The land surface is therefore a factor that we need to consider when studying climatic changes of the past 300 years.

3.4 Coupling Between Systems and Feedback Mechanisms

3.4.1 Interaction Within the Machinery

Many elements of the machinery interact to produce climatic variability, as Brückner noted in 1890. Sometimes the response emerges from interaction of seemingly separate parts of the machinery. For instance, the response of the climate system to volcanic eruptions or solar activity changes (see Sect. 3.2) emerges through the interaction of quite different mechanisms—atmospheric chemistry, radiative processes, and atmospheric dynamics. The notion of climate as the product of all interactions between different components of a system goes back at least to the German scientist Alexander von Humboldt who wrote in 1845:

> "Climate [. . .] depends on the perpetual interaction of the ocean surface, with its currents and temperature contrasts, and the dry, radiating land surface with its physical complexity, bare or with vegetation cover." (Translated from von Humboldt (1845), p. 304)

Interactions in the climate system can act to modify an initial perturbation. If they amplify the initial perturbation (until other restoring forces grow and balance the system in a new state), we speak of a positive feedback. If the system acts to damp the initial perturbation and return to the base state, the feedback is negative. There are many examples in the climate system for both types of feedback. For instance, the radiative forcing due to the increase of greenhouse gases is amplified by the climate system.

The possibility of positive climate feedbacks was very topical during Brückner's time because the existence of an ice age could hardly be explained without such mechanisms. One of the first scientists to discuss the role of feedbacks in the climate system was James Croll, who suggested that ice ages could have been caused by a rather small initial cooling that was amplified by the ice–albedo feedback (Croll 1875). If snow or ice accumulates on the Northern Hemisphere land masses, the planetary albedo is increased and less solar radiation is converted to heat. Thus, Croll discovered a key mechanism of the climate system.

Today, we know of many more feedbacks (many have already been mentioned by this chapter). In these last few pages of this chapter, I would like to highlight the particular role played by one constituent (while neglecting many other important feedback mechanisms) that participates in almost all coupling processes and feedback mechanisms: water.

Box 3.4 Negative feedbacks: The Daisy World, CLAW, and Gaia
Consider an earth-like planet in which the land surface is covered with flowers: daisy world. Assume that there are only two kinds of daisies: black and white. Both have the same growth response curves with an optimum growth temperature bounded on both sides by temperature regimes with

(continued)

Box 3.4 (continued)
suboptimal growth. Because black daisies absorb more solar radiation than white daisies, they heat up more strongly. If temperature is not too far above the optimal temperature, white daisies may still grow while black daisies heat up excessively and recede. As a consequence, the albedo decreases and so does the air temperature in the climate system of the earth-like planet. If temperature decreases, black daisies grow at the expense of white daisies, which suffer more from the cooling; air temperature increases. Through their albedo, the daisies themselves regulate temperature.

This simple model of a negative land surface–climate feedback was proposed in 1983 (Watson and Lovelock 1983) and was refined in the 1980s and 1990s. Though it may not be relevant for climatic applications per se, toy models of climate feedbacks are important heuristic tools.

A more complex negative feedback cycle was suggested in the 1980s, the CLAW hypothesis, named after the authors of the paper (Charlson, Lovelock, Andreae, and Warren) (Charlson et al. 1987). In essence, the hypothesis highlights a specific aspect of the global sulphur cycle. It starts with increased emissions of DMS from oceanic phytoplankton due to higher temperatures. DMS is oxidised in the atmosphere and forms aerosols (see Fig. 3.14). These change cloud properties and lead to a decrease of incident solar radiation and, thus, temperature. Though its importance has been overstated, CLAW has triggered attention as it may be seen as a particular aspect of James Lovelock's[20] overarching "Gaia" hypothesis (Lovelock and Margulis 1974), which considers the earth as a self-regulating body.

Another interesting aspect of CLAW is that it points to the crucial role of the sulphur cycle. Apart from its role in forming tropospheric aerosols, which then act as cloud condensation nuclei (Sect. 3.3.4), sulphur is particularly important in the stratosphere. It arrives in the stratosphere through volcanic eruptions (Sect. 3.3.1) or transport upwards from the troposphere. Volcanic sulphate aerosols interact with ozone, but first of all, they cool the surface. Nobel laureate Paul Crutzen[21] suggested in 2006 that injection of sulphur into the stratosphere may have to be considered to counteract global warming (Crutzen 2006), a much discussed "climate engineering" solution (see Box 4.5, p. 299).

[20]James Lovelock, born 1919, is a British chemist and earth system scientist. Apart from his work on the Gaia theory, Lovelock also developed instruments for environmental chemistry analytics.

[21]Paul Crutzen, born 1933, is a Dutch atmospheric chemist. Crutzen continuously contributed insightful concepts to the field of atmospheric chemistry. For his work on the chemistry of atmospheric ozone, he was awarded the Nobel Prize in chemistry in 1995.

3.4.2 Feedbacks Involving Water Vapour and Clouds

Water is everything—it is the key constituent of the climate system. Water is an agent, mediator, and response in the climate system. Water exists in the climate system in all three aggregate states. Water vapour is a powerful greenhouse gas. Water is a solvent and a chemical reactant and participates in gas phase, liquid, and heterogeneous chemical reactions. Water forms clouds, which are very important in the climate system. Water links the mass balance and energy balance. Water is the most important link between the atmosphere, ocean, cryosphere, land surface, and biosphere. Therefore, it is not surprising that many of the known feedback effects of the climate system involve water. At the same time, the distribution of water in the climate system is very complex (Fig. 3.63). In the following, I outline selected feedback mechanisms that revolve around water in the climate system.

The most well-known feedback involving water is arguably the greenhouse gas–water vapour feedback. Water is a very powerful greenhouse gas and its abundance depends upon temperature (assuming constant relative humidity; see Sect. 3.1.1). Because greenhouse gases increase surface and tropospheric temperature, this feedback is positive and hence water vapour amplifies the forcing by greenhouse gases (Boucher et al. 2013). It also amplifies other warming mechanisms and feedbacks. The water vapour feedback is more efficient in the tropics than in the extratropics and in the mid- and upper troposphere than near the ground.

Related to the water vapour feedback is the lapse rate feedback. If temperature increases faster at higher than lower altitudes, more radiation is emitted and thus the feedback is negative (Boucher et al. 2013). However, at the same time, the warmer air is able to contain more water vapour (in a region where water vapour is a particularly effective greenhouse gas) and, thus, less radiation is emitted to space. Therefore, water vapour and lapse rate feedbacks are often studied together.

Potentially powerful feedback mechanisms are those that are related to clouds, similar to those related to aerosols listed in Table 3.3. However, feedbacks involving

Fig. 3.63 The distribution of water in the climate system is amazingly complex and fascinating (Photo: S. Brönnimann)

water vapour and clouds are still poorly understood and are an important source of uncertainty in climate models (Boucher et al. 2013). Furthermore, feedbacks with between clouds and atmospheric circulation need to be considered (Bony et al. 2015).

3.4.3 Feedbacks Involving Freezing Water: Arctic Amplification and the Cryosphere

In polar or high-altitude regions, ice and freezing or thawing processes are important players involved in many feedback mechanisms. In fact, climatic changes are much larger in the Arctic than elsewhere, and temperature signals due to global forcings appear most strongly in the Arctic, a phenomenon known as Arctic amplification (Manabe and Stouffer 1980).

The Planck feedback (cooler regions radiate less energy) and the lapse rate feedback (warming occurs near the ground from where it is less efficiently radiated) both contribute to the Arctic amplification (Pithan and Mauritsen 2014).

Croll's ice–albedo feedback is arguably the most important contributor. However, changes in sea ice extent and snow cover affect more than just the radiation fluxes. They also affect sensible heat fluxes between ocean (or land surface) and the atmosphere, moisture fluxes into the atmosphere, and ocean salinity. Heat and moisture fluxes into the atmosphere may in turn affect cloud cover and thus longwave radiation fluxes. Furthermore, the changes may feedback on atmospheric circulation and thus affect the transport of heat and moisture (Serreze and Barry 2011), eventually affecting snow cover (Wegmann et al. 2015). Because the atmosphere over the Arctic is normally statically stable, feedbacks operating at the surface only affect the lowest layers of the atmosphere. Hence, the effect at the surface is stronger.

These Arctic feedback processes amplify recent greenhouse gas–induced warming. However, in the past, some of the mechanisms may also have contributed to Arctic warming episodes. Examples for both will be discussed in Chap. 4.

3.4.4 Water as a Reactant: The Role of the Hydroxyl Radical

Atmospheric chemistry and climate are linked in manifold ways. The climate effects of volcanic and solar variability and of tropospheric aerosols partially proceed through chemical mechanisms. Conversely, chemical reaction rates depend on ambient atmospheric conditions including temperature or radiation. Water plays a crucial role also in atmospheric chemistry. In the following brief discussion, I focus on gas phase chemistry in the troposphere.

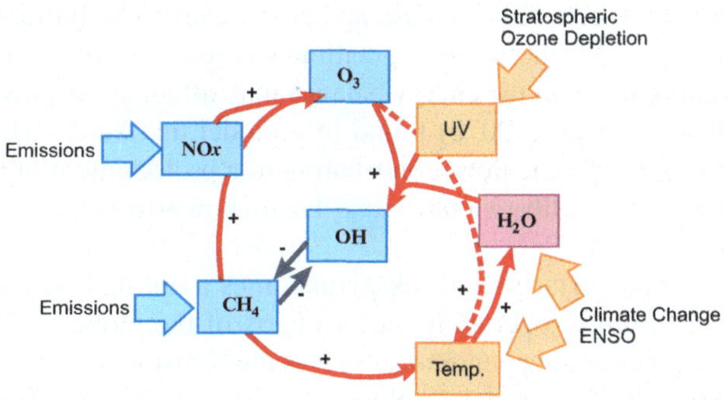

Fig. 3.64 Schematic of atmospheric chemical processes and their role in climate with a focus on the interaction between water vapour, temperature, ozone, and methane. *Red arrows* denote positive, *blue arrows* negative relations

The reaction of water vapour with atomic oxygen, which results from the photolysis of ozone by UV radiation, is arguably one of the most important reactions in tropospheric chemistry. It produces the hydroxyl radical (OH), which is also called the detergent of the atmosphere. Most atmospheric trace gases are eventually removed through oxidation by OH. Hence, OH determines their lifetime. This is also true for the greenhouse gas methane. A schematic view of the related chemical mechanisms is given in Fig. 3.64.

The formation of OH requires water, ozone, and UV radiation; thus, it depends on climate. Let us look at methane's lifetime in Fig. 3.64. Methane's dominant sink is its reaction with the OH radical and vice versa. Thus, if methane increases, OH decreases and methane increases its own lifetime. However, in a changing atmosphere and a changing climate, the sources of OH also change. The presence of nitrogen oxides allows an alternative oxidation pathway for methane in which ozone and thus OH are produced rather than consumed. In a warmer and hence wetter world, with everything else remaining the same, OH is expected to increase due to the increased water vapour. Hence, in a changing climate and a changing atmosphere, with increasing emissions of nitrogen oxides, the lifetime of methane also changes. Similar mechanisms may also operate on interannual time scales, e.g. during El Niños, when the global atmosphere is more humid than during La Niñas. Also, years with strong biomass burning may alter the chemical regimes. Section 4.4.2 will go into more detail with respect to changes in the lifetime of methane during the 20th century.

3.4.5 Feedbacks Involving Evaporating Water: Land Surface Feedbacks

One of the most important positive feedback mechanisms are land surface feedbacks on drought development. This mechanism also involves water, or more specifically, evapotranspiration. The starting point here is a change in rainfall. Once a precipitation deficit has led to the drying out of the soil, this will affect the heat fluxes.

Sensible heat will increase and heat waves (which often accompany droughts) will increase. Furthermore, vegetation may react to water stress and reduce transpiration. This change in the energy balance may affect atmospheric circulation. For instance, Fischer et al. (2007b) found in a model that land surface feedbacks may amplify the anticylconic flow, contributing to a prolongment of the drought conditions. Soil moisture feedbacks have also been shown to be relevant for heat waves (Hirschi et al. 2011).

Land surface feedbacks, sometimes amplified by land use practices, have been suggested to operate in various parts of the globe and have played a role in some major climatic phenomena such as the "Dust Bowl" droughts in the Midwest United States (see Sect. 4.3.3). Some regions are particularly susceptible to land surface feedbacks (Koster et al. 2004). In a changing climate, however, regions in which such feedbacks are not yet operating in today's climate may become susceptible in the future, including regions in central Europe (Seneviratne et al. 2006).

Land surface feedbacks that involve vegetation responses may also occur on longer time scales. They may proceed through the carbon cycle, or may lead to desertification. This brief overview of some feedbacks in the climate system highlights the importance of the water cycle not only as a component of the climate system that affects ecosystems and humans, but also as an ingredient in many feedback mechanisms. Past and future changes in the water cycle, especially at local to regional scales, may modify these feedbacks effects and may act to amplify or damp other perturbations.

With this, I finish my brief overview of climate processes. It is now time to focus on the main topic of this book: climatic changes since 1700.

Chapter 4
Climatic Changes Since 1700

With contributions from: Florian Arfeuille, Jonas Bhend, Jörg Franke, Jenny Grütter, Sina Lenggenhager, Renate Auchmann, Peter Stucki, Alexander Stickler and Martin Wegmann

Over the last 300 years, countless climatic variations at different places and times have been witnessed, some affecting millions of people and changing the course of history, some going largely unnoticed. In this chapter, I discuss selected climatic changes in European and global climate history from about 1700, the end of the Little Ice Age, to the present. The 20 events start with the cold Maunder Minimum and end with the global warming hiatus, covering different aspects of the climate system. We will encounter continental-scale temperature dips, hydroclimatic anomalies, perturbations of the stratosphere, and changes in atmospheric composition. Thereby, we will see the mechanisms discussed in Chap. 3 at work: oceanic modes and atmospheric variability, volcanic eruptions, and humans changing their environment.

Datasets and literature on these events are available and this book is largely based on these studies. In fact, the introductory section compares many of the different reconstructions over the last few centuries. However, a full review of literature and datasets is outside the scope of this book. Rather, for reasons of consistency, the main aim of this chapter is to trace the 20 events using a small set of data products—reconstructions, reanalyses, and model simulations (see Sect. 2.9)—using simple analyses. Some of the datasets are new and allow a more mechanistic perspective of past climate changes than hitherto possible.

The chapter is structured chronologically and is divided into four subperiods (1700–1890, 1890–1945, 1945–1985, and 1985 to present). Each of the corresponding section starts with an overview, followed by a description of selected climate events and ending with a brief synthesis. Before starting with the first subperiod, however, I will give an overview of climatic changes over the entire 300-yr period and even look further back, to cover the entire last millennium.

S. Brönnimann, *Climatic Changes Since 1700*, Advances in Global Change Research 55, DOI 10.1007/978-3-319-19042-6_4

4.1 Climatic Changes of the Past Centuries: An Overview

The year 1700 is an admittedly arbitrary starting point for this chapter. It roughly corresponds to the beginning of early instrumental observations; this was the reason why Brückner chose it for his title. It is not, however, the beginning of a specific climate phase. The year 1700 lies in the middle of a phase called the "Little Ice Age". Therefore, this section starts by taking a step back to look at the last 300 years in the context of the last millennium. Then, it will zoom in on climatic changes of the last several hundred years (our main datasets reach back to 1600) and address changes in global and regional temperature and precipitation, mostly in the form of time series. The machinery will be explored by analysing large-scale atmospheric circulation indices. The last parts of the section will cover the changes in extremes and the role of external drivers of global and hemispheric temperature—the forcings.

Section 4.1 presents climate reconstructions based on tree rings, ice cores, corals, and documentary sources (see Sect. 2.2). In addition, I use our GCM simulations (CCC400) and our assimilated data set (EKF400, see Sect. 2.9). The GCM simulations are helpful to address long-term and large-scale climatic changes—that is, the forced signal—while EKF400 more accurately captures interannual-to-decadal variability, at least in well-constrained regions. In some sense, climate proxies and early instrumental measurements provide "ground truth", while EKF400 provides a comprehensive climate state, though uncertain, which is physically consistent with the forcings and the sparse assimilated information and its errors.

4.1.1 The Last Millennium

According to recent reconstructions of Northern Hemisphere mean temperatures (Fig. 4.1; see Stocker et al. 2013; Masson-Delmotte et al. 2013), the 17th and early 18th centuries were the coldest phase of the last millennium. Temperatures were ca. 0.5 °C below those in the periods around 1400 and around 1000 and they were 1.5 °C cooler than today. A similarly cold phase (albeit much shorter) occurred in the early 19th century. Less is known about the temperatures of the Southern Hemisphere. However, recent reconstructions (Neukom et al. 2014a) suggest lower temperature variability in the Southern Hemisphere and asynchronous climate variations in the two hemispheres. Climate models fail to produce these findings. Common among the hemispheres is that both achieved their lowest temperatures around 1700 and have been experiencing pronounced warming over the last few decades.

Although Brückner and contemporaries addressed historical warm and cold periods in the northern North Atlantic region (Pettersson 1914; Brooks 1922; see Jansen et al. 2007),[1] our current view of climate of the last millennium dates back

[1]In fact, the demise of the Greenland Vikings was related to climatic changes already in 1824 (Ehrenheim 1824).

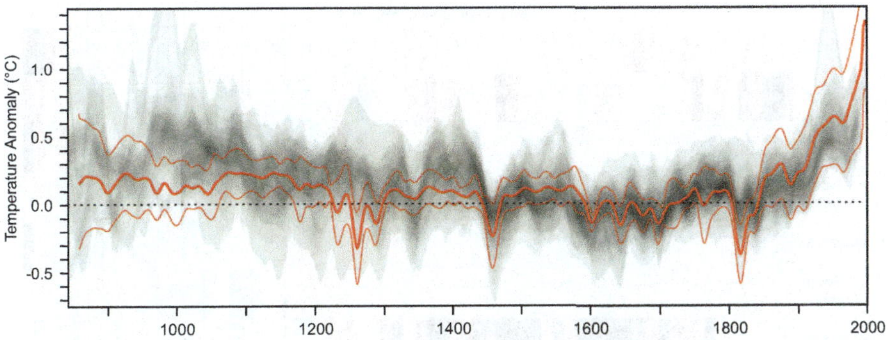

Fig. 4.1 Simulated (*red*) and reconstructed (*shading*) Northern Hemisphere temperature changes from 850 to 2000. The *thick red line* depicts the multi-model mean while the *thin red lines* show the multi-model 90 % range. The overlap of reconstructed temperatures is shown by *grey shading*; all data are expressed as anomalies from their 1500–1850 mean and are smoothed with a 30-year filter. Note that some reconstructions represent a smaller spatial domain or a specific season, while annual temperatures for the Northern Hemisphere mean are shown for the simulations (From Stocker et al. 2013)

to the 1960s, specifically to the work of two great historians of climate, Hubert Lamb[2] and Emmanuel Le Roy Ladurie.[3] Lamb analysed the climate history of the last millennium, mainly based on European sources, and found a warm period from around 1100 (or earlier) to 1300 or so, which he called "Medieval Warm Epoch" (Lamb 1965, 1972, 1977). The period was followed by a rather cool period that ended around 1850 and is now called the Little Ice Age.[4] After the end of the Little Ice Age, a period of global warming began. At least since the 1950s, this has been largely anthropogenically caused. Many other climate historians (e.g. Brázdil et al. 2005; Glaser 2008; Le Roy Ladurie 1971; Pfister 1982) have since refined Lamb's analysis for Europe (see the summary on historical climatology in Glaser (2008) and Mauelshagen (2010)).

Lamb's concept of a Medieval Warm Epoch (now commonly known as the Medieval Climate Anomaly (MCA)) followed by a Little Ice Age was very successful. In fact it was so successful that it was often applied indiscriminately to other regions. According to Diaz et al. (2011), the MCA conditions were warmer than those of the Little Ice Age over much of the northern mid- and high latitudes,

[2]Hubert Lamb (1913–1997) was a British climatologist who is known for his work on climate history and for his dynamical perspective of past climate. Lamb was the founder of the Climatic Research Unit (CRU) of the University of East Anglia, Norwich, UK.

[3]Emmanuel Le Roy Ladurie, born in 1929, is a French historian, known among other things for his work on climate history. He advocated for the separation of human and climate histories. Le Roy Ladurie is one of the leading figures of the French "Annales" school of historiography, which considers history from the point of view of mentalities.

[4]The term "Little Ice Age" was introduced earlier, but referred to a much longer period (the last ca. 5000 years), which is nowadays referred to as the neoglacial (Mann 2002).

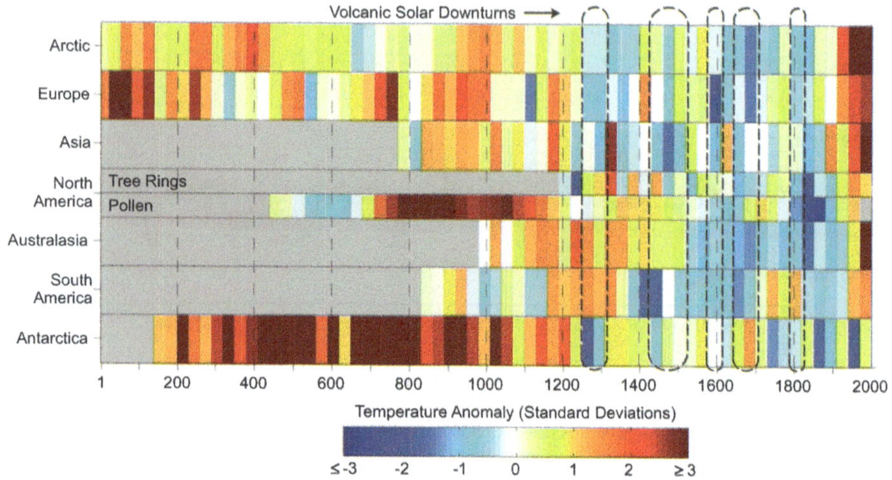

Fig. 4.2 Continental-scale temperature reconstructions. Thirty-year mean temperatures for the seven PAGES 2k network regions, standardised over the period of overlap among records (1190–1970). North America includes a shorter tree ring- and a longer pollen-based reconstruction. *Dashed outlines* enclose intervals of pronounced volcanic and solar negative forcing since 850 (From Ahmed et al. 2013, reprinted by permission from Macmillan Publishers Ltd: Nature Geoscience, copyright 2013)

but specific results depend on the season and region. Recent continental climate reconstructions (Ahmed et al. 2013; displayed in Fig. 4.2) suggest warm periods between 830 and 1100 on the northern continents (see also Bradley et al. 2003). However, no evidence for warming is found in South America and the Australasian warming was later, partly compensating for the centennial changes in the northern continents (see also Neukom et al. 2014a). During the MCA, the equatorial eastern Pacific may have been rather cool (more La Niña-like, see Graham et al. 2007). Moreover, there were widespread hydroclimatic anomalies (Graham et al. 2011), such as prolonged droughts in the western United States and northern Mexico (Seager and Burgman 2011), dry conditions in tropical Africa and partly dry conditions in Asia.

After the MCA, many regions exhibited a period of transition and cooling towards the Little Ice Age. The definition of the start of the Little Ice Age varies from author to author. Glaser (2008) uses the term for the period 1550–1850, Masson-Delmotte et al. (2013) for the period 1450–1850, and Le Treut et al. (2007) for the period 1350–1850. Although the term Little Ice Age was based on information from the North Atlantic domain, this may have been a phase of global cooling. Many regions exhibit their strongest cooling periods during the 14th–19th centuries, although not synchronously. The recent continental-scale climate reconstructions of the PAGES 2k initiative (Ahmed et al. 2013; see Fig. 4.2) confirm that there is little synchronous decadal-to-multidecadal variability across continents. All continents exhibit low temperatures in the period 1580–1880, punctuated with warm periods in the 18th century (but again not synchronous). In Antarctica, the cooling was least pronounced. Section 4.2 will go into more detail with respect to the last part of the Little Ice Age climate, the subperiod 1700–1890.

After Brückner's book was published, temperatures started to rise globally. A first 50-yr phase of increase ended in the 1940s and was followed by a 30-yr-long temperature plateau. Since the 1970s, global temperatures have continued their increase. Sections 4.3–4.5 cover these three phases and discuss their prominent climatic events.

The phase since 1850 is sometimes called the anthropocene—a time characterised by humans-driven environmental change. In fact, the changeover from a climate system controlled by natural variability to one increasingly dominated by anthropogenic factors is the main characteristic of the development of climate since 1700.

4.1.2 An Overview of Climatic Changes Since 1700: Brükner's View

It is much easier to study climatic changes of the past few hundred years than those of the medieval period. This is evident by the reduced spread of reconstruction preceding 1600 in Fig. 4.1. From the 17th century onward, far more sources of information—even instrumental data—are available (see Sect. 2.2.1 and Fig. 2.3), covering a larger part of the globe than earlier sources. The following text zooms in on this phase and analyses global- and large-scale climatic changes in the form of time series. For this phase we can rely on existing proxy-based reconstructions (which we discuss in Sect. 4.1.3), as well as our CCC400 simulations and the EKF400 assimilation (which we analyse in Sect. 4.1.4). Before we look at these datasets, however, it is apt to start with a brief summary of Brückner's book, "Climatic Changes Since 1700".

Brückner summarised his main findings in the form of a figure, which is reproduced in Fig. 4.3. It shows six time series, each in the form of 5-year averages. The series comprise the sunspot number, an instrument-based global temperature series (back to 1735), a global precipitation series (back to 1775), a series of European grape harvest dates, a chronology of European cold winters, and a series of ice conditions of Russian rivers. Remarkably, Brückner drew a global temperature time series (extended from Köppen 1881) and a global rainfall time series. For this purpose he used 280 temperature station series and 321 rainfall series, which he aggregated regionally. In a second step he formed an average of regional averages (although many regions were poorly covered or not covered at all). The global temperature series shows cold conditions around 1735, 1770, 1815, and 1840 and warm phases in between; features that are still found in current reconstructions. The series of cold winters in Europe, which is in rough agreement with the chronology of Glaser (2008), and the grape harvest dates point to cool conditions between around 1685 and 1715. This latter period is today considered the coldest phase of the Little Ice Age in Europe and coincided with the "Late Maunder Minimum" of solar activity (discussed in Sect. 4.2.2). In his summary, Brückner found cold and

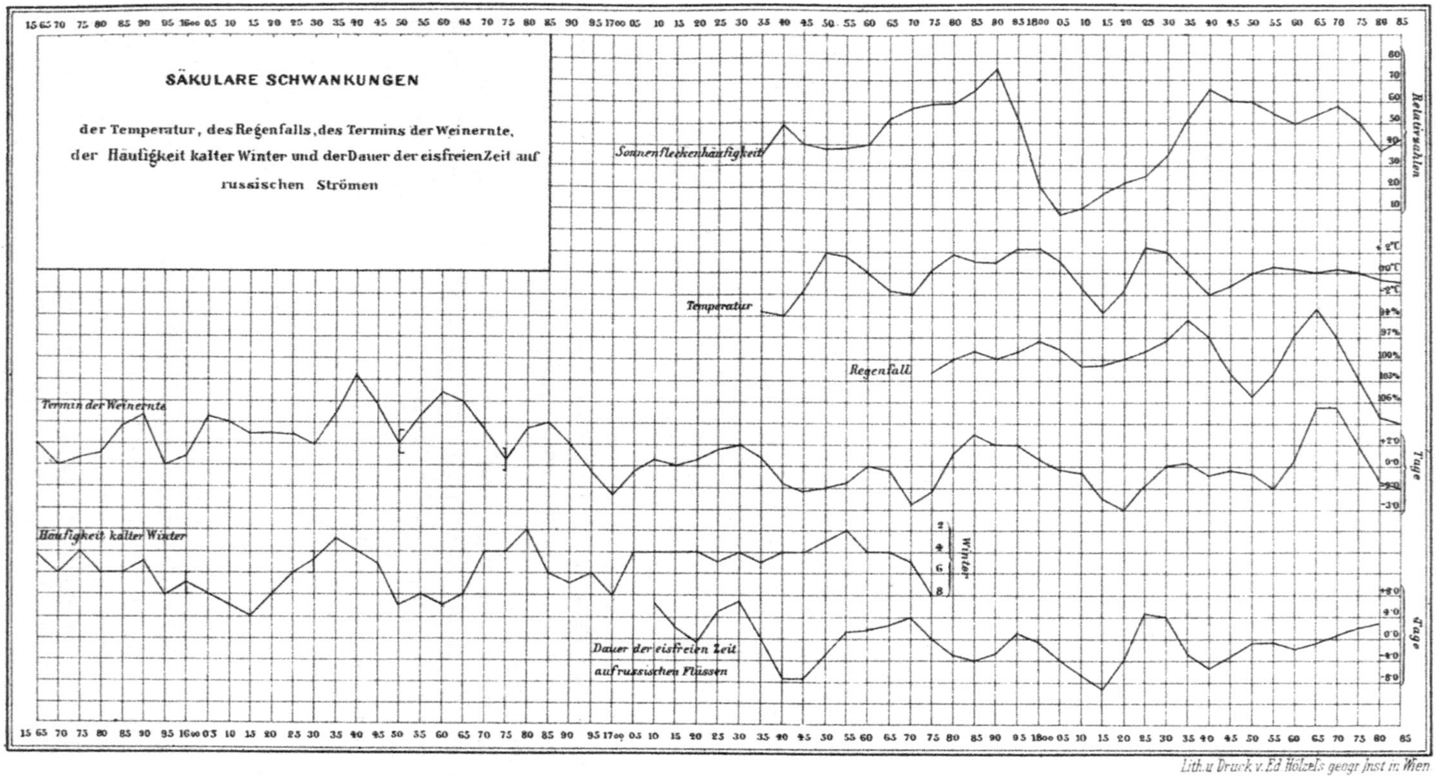

Fig. 4.3 Secular changes in climate as displayed by Brückner (1890). The *curves* show (from top to bottom) sunspot frequency (relative number), global temperatures, global rainfall, grape harvest dates in Europe, frequency of cold European winters (per decade; note the inverse scale), and the duration of the ice free period of Russian rivers. All *curves* are anomalies from their 1851–1880 mean value. Years indicate the end year of a 5-yr period. All *curves* except the cold winters are weighted averages

wet periods around 1700, 1740, 1780, 1815, 1850, and 1880; the warm and dry periods were centred around 1720, 1760, 1795, 1830, and 1860.

Analysing the timing of the maxima and minima, Brückner proposed a 35-yr cycle of global climate, or rather: a quasi-periodic variation of climatic continentality.[5] According to Brückner, this cycle, which was later largely rejected by palaeoclimatologists, could be of considerable relevance. Lines of equal precipitation move by several hundred kilometres over the course of a cycle. Since the data only covered land areas, Brückner argued that ocean regions would show the opposite sign. Indeed, this was confirmed by island and coastal stations. Dry periods on land (on the annual average) were also characterised by higher temperatures (globally, although with some exceptions) and a larger amplitude of the annual temperature cycle. Although Brückner's starting point was lake levels—an analysis that he then extended to precipitation—he saw the precipitation signal as merely an indication of a change in the mode of operation of the machinery. He argued that the precipitation changes were caused by changes in atmospheric circulation. Analysing long sea-level pressure time series (a rarity), he found that during dry periods the Azores high in boreal winter extended to central Europe and the Icelandic low was strong. Brückner argued that high pressure over the land masses inhibits the penetration of moist oceanic air to continental regions. Conversely, during "wet" periods, more moist air from the ocean is transported to the land and rains out. However, his explanation is partly at odds with the current view of the North Atlantic Oscillation (see Sect. 3.2.3), which finds a redistribution of precipitation rather than a continental increase or decrease.

In Brückner's view, the ultimate driver of circulation changes was temperature changes. In fact, he found a near-global signal of temperature variations concurrent with precipitation changes. The amplitude of the global temperature cycle found by Brückner was on the order of 0.5–1 °C.

How did Brückner explain these changes? In fact, he did not. Although he showed a sunspot record in this figure, he discarded an effect of sunspots. However, in his view the sun was the only possible pacemaker of the cycle. Hence Brückner speculated about a 35-year oscillation in solar activity that is not apparent in visual sunspot records. Interestingly, Brückner did not address volcanoes as a possible cause.[6]

In the second last chapter of his book, Brückner looked at the consequences of his climate cycle. He considered the effects on natural systems (e.g., glaciers), traffic, and economy. He even addressed effects on epidemics. This view towards practical applications of climatological knowledge (today we would say climate impacts) is typical for Brückner's work (Stehr and von Storch 2000).

[5]Climate or weather cycles were a typical, often-criticised perspective of 19th century science. Unlike many contemporaries, Brückner did not see his cycle as an exact, deterministic component; the cycle length of 35 years was an average length and could vary.

[6]Although the effects of the Krakatau eruption on the atmosphere and climate were actively discussed at the time Brückner wrote his book, volcanoes were only seen as possible drivers of climatic changes since the 1910s with the work of Humphreys (1913) and others, see Box 3.2, p. 124.

Brückner's cycle and the suspected mechanism has long been rejected. What remains outstanding is his perception of multidecadal climatic changes, his incentive to reconstruct global and regional temperatures for past centuries, and his strive towards emphasising practical application of climate science. Today, his efforts remind us that, in addition to a long-term climate trend, we should take into account the variability of the climate system on multidecadal scales. On these time scales, variations of continental-scale temperatures by 0.5°C or more may occur.

4.1.3 Climatic Changes Since 1700 in Climate Reconstructions

4.1.3.1 Global and Northern Hemisphere Temperature Changes

Do Brückner's reconstructions stand the test of time? They were certainly a good first try. Three recent reconstructions of global land temperatures, namely a multi-proxy reconstruction based largely on tree rings (Mann et al. 2008), a reconstruction based on glaciers (Leclercq and Oerlemans 2012), and a more recent multi-proxy reconstruction of land and ocean temperature (Crowley et al. 2014), are shown in Fig. 4.4 (top). Also shown is the ensemble mean land surface temperature from CCC400 (further discussed in Sect. 4.1.4) and observations from 1860 onwards (Jones et al. 2012). A first glance reveals a different character of the five curves; evidently they do not measure the same thing. The observations are arguably closest to the true variations in climate. The glacier-based reconstruction captures only multidecadal-to-centennial variability and does not exhibit the decadal dips or spikes. The reconstruction by Mann et al. (2008) captures decadal changes, but suppresses much of the interannual variability (see also Sects. 2.8.2 and 2.8.4 for the spectra of proxies and reconstructions). CCC400, which uses the latter reconstructions as boundary conditions, is not expected to show the full interannual variability of climate. However, the main reason for its smoothness is the fact that this figure shows the ensemble mean, thus averaging out atmospheric randomness. The reconstructions from Crowley et al. (2014) show somewhat more variability on interannual and decadal scales than the other reconstructions, but less than instrumental observations (which are for land only).

All curves (note that they are referenced to their 1900–1990 average) show warming over the period. This is slow at first and then accelerates. Global land temperatures from Mann et al. (2008) show little change during the 17th century, a dip around 1700, a slight increase during the 18th century, and then two further dips in the 19th century. The agreement with Brückner's curve (Fig. 4.3) is remarkable. The glacier-based temperature reconstruction shows very little warming before 1850. In both reconstructions, the main warming period started in the late 19th century and then occurred in two phases. In this respect both reconstructions agree with observations. The reconstruction by Crowley et al. (2014) shows pronounced decadal variability in the late 18th and early 19th centuries, with the lowest global temperatures reached in 1816.

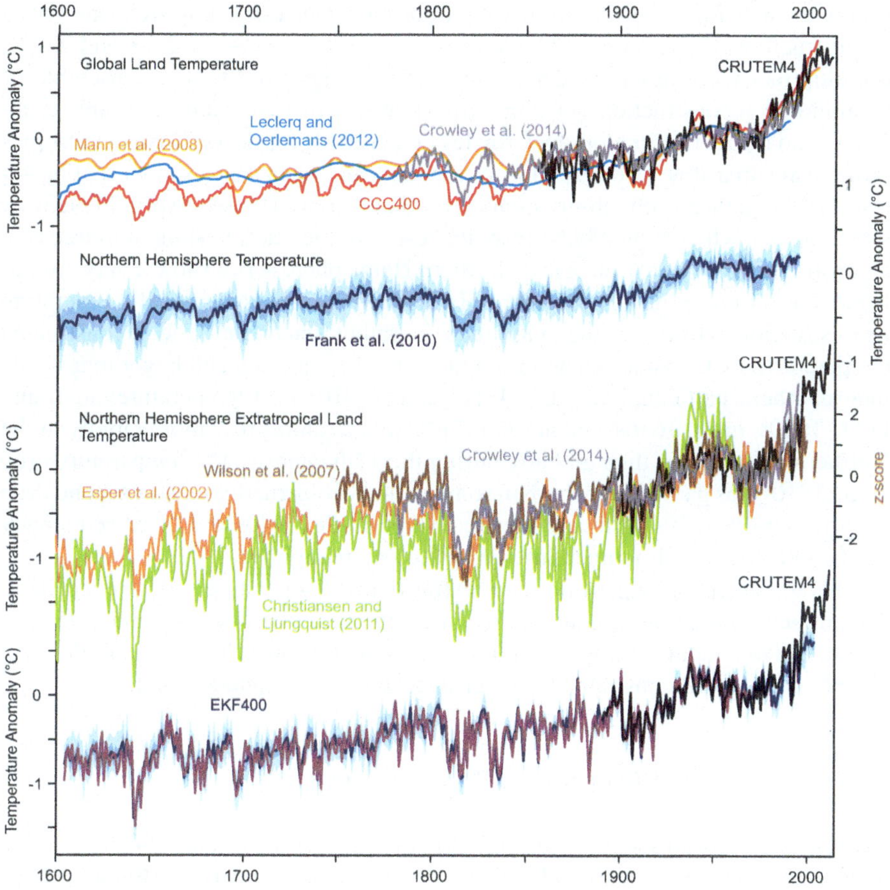

Fig. 4.4 (top) Global land surface temperature anomalies from glacier-based reconstructions by Leclercq and Oerlemans (2012), multiproxy reconstructions by Mann et al. (2008) and Crowley et al. (2014) (the latter is for land and ocean), CCC400, and observations (CRUTEM4). (second from top) Temperature anomalies for the Northern Hemisphere by Frank et al. (2010) obtained by rescaling existing reconstructions. (second from bottom) Temperature anomalies of the northern extratropics from tree ring-based reconstructions of Esper et al. (2002) (arbitrary units) and Wilson et al. (2007) (right y-axis; z-scores) and multiproxy reconstructions of Christiansen and Ljungqvist (2012) and Crowley et al. (2014) (left y-axis, °C). Also shown are April–September averages of temperature over land, 20°–65°N (CRUTEM4). (bottom) EKF400 ensemble mean April–September temperatures for northern extratropical land areas (*blue*; °C). *Light* and *dark shadings* denote the 50 % and 90 % ensemble ranges, respectively. The *purple* and *black lines* in the bottom denote the best ensemble member and observations, respectively. All anomalies are with respect to the 1901–1990 average

Do these reconstructions really represent global-mean temperatures? In fact, reconstructions of the Southern Hemisphere are difficult and only few attempts have been made (Neukom et al. 2014a). There is certainly more information available for the northern extratropics than for the rest of the globe. Therefore, the following will focus on the northern extratropics. Corresponding tree ring-based reconstructions (Esper et al. 2002; Wilson et al. 2007; see Sects. 2.8.1 and 2.8.3), a multiproxy reconstruction (Christiansen and Ljungqvist 2012), an ensemble of recalibrated reconstructions of Northern Hemisphere temperatures (Frank et al. 2010), and the reconstructions by Crowley et al. (2014) for the northern extratropics (note again that this reconstruction is for land and ocean) are shown in Fig. 4.4 (middle), together with observations over land areas (I used April–September temperatures, which arguably correspond best with the season captured in tree ring reconstructions). The extratropical Northern Hemisphere series show a very similar behaviour as the (arguably Northern Hemisphere dominated) global temperature reconstruction. However, the amplitude is rather uncertain. The reconstructed temperatures by Christiansen and Ljungqvist (2012) have a much larger magnitude than all others, particularly those by Frank et al. (2010). Low temperatures are found in the 1600s, in the 1640s, and around 1700. Temperatures rose more rapidly in the Northern Hemisphere than globally during the 18th century. The temperature drop in the 1810s was particularly sharp in some of the reconstructions, and a second drop occurred in the 1830s. Temperatures recovered in the 1840s and increased slowly until 1890. Afterwards temperatures increased more rapidly.

The agreement among all northern extratropical temperature series is excellent. The mutual correlations among the reconstructions are between 0.7 and 0.95. The timing of temperature variations in the northern extratropics since around the 17th century is thus well constrained from proxies, but the magnitude is not.

4.1.3.2 Continental Temperature Reconstructions

Brückner also constructed continental climate time series. In the following, I also address the continental scale and look at available temperature reconstructions from the recent PAGES 2k project together with selected other reconstructions (Fig. 4.5; the series from EKF400 are included. However, they are only discussed in the next section because they should not be interpreted as reconstructions but rather as constrained model results).

The *Arctic* exhibits a large magnitude of interannual temperature variability. The older reconstructions by Overpeck et al. (1997) and the recent PAGES 2k reconstruction agree very well. The coldest period was the early 19th century. The warming over the entire period amounts to 2 °C and is concentrated in the first half of the 20th century (this period is further discussed in Sect. 4.3.2). The second strong warming period since the 1970s, which is known from instrumental data (Sect. 4.5.8), is less well reflected in the proxies.

Climatic changes of the past few hundred years are arguably best studied in *Europe*. One selected multiproxy reconstruction of annual temperature (Xoplaki et al. 2005) is shown in Fig. 4.5 together with the PAGES 2k continental reconstructions. Both reconstructions indicate stable (slightly rising) temperatures throughout the 18th century. Then, after the drop in the 1810s and the subsequent recovery, stable temperatures are reconstructed from the 1820s to the 1890s. Brückner's

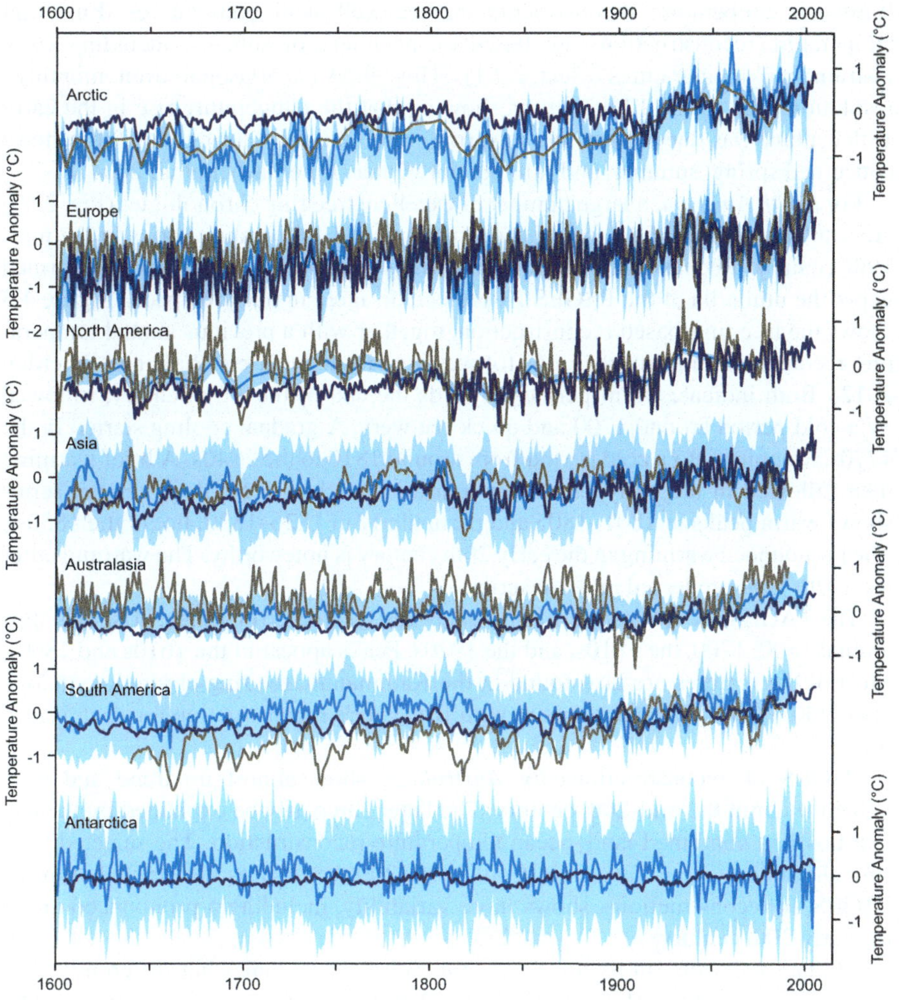

Fig. 4.5 Continental temperature reconstructions. Shown are the PAGES 2k annual reconstructions (*light blue*) with maximum and minimum (*light blue*). *Brown lines* denote other continental-scale recontructions: Xoplaki et al. (2005) for Europe, Wahl and Smerdon (2012) for western North America, Villalba et al. (2003) for South America, and Overpeck et al. (1997) for the Arctic. For Asia and Australasia, two tree-ring chronologies from Kathmandu (Cook et al. 2003) and Tasmania (Cook et al. 2006) are shown, respectively. *Dark blue lines* denote corresponding series from the ensemble mean of EKF400 (defined as averages poleward of 65°N and 65°S for the Arctic and Antarctica, respectively), and averages of land temperature over the Europe (35°N–65°N, 10°W–50°E), North America (20°N–65°N, 180°W–70°W), Asia (10°N–65°N, 50°E–180°E), Australasia (50°S–10°N, 90°E–180°E) and South America (60°S–20°N, 82°W–30°W). All series are anomalies with respect to the 1901–1960 period

European temperature reconstruction agrees well with both curves. European temperature reconstructions are based on a variety of sources, including early instrumental temperatures (Sect. 2.2.1). This allows a seasonal–even monthly–resolution back to the 17th century, showing that the temperature rise in the early 18th century was due to warming in spring, while the drop during the 19th century concerned spring, summer, and fall, but not winter (Luterbacher et al. 2004).

For *North America*, a large number of excellent tree ring chronologies (Fig. 2.30) are available prior to the onset of early instrumental temperature series in the mid-19th century. Reconstructions from tree rings and pollen, both were performed under the umbrella of the PAGES 2k project, disagree in many respects. Figure 4.5 shows the tree ring-based reconstruction, together with a previous annual temperature reconstruction from tree rings for western North America (Wahl and Smerdon 2012). Both indicate warm periods towards the end of the 17th century, followed by a cold period around 1700 and quick recovery. A gradual cooling started in the 1770s, dropping into a cold period from around 1810 to the 1840s. A brief warming then followed in the early 1850s (in contrast, Brückner's North American series shows warm peaks in 1801–1805 and particularly in 1826–1830; not in the 1850s). The pronounced warming in the early 20th century is noteworthy. The warming after the 1970s is not reflected in these series.

The PAGES 2k reconstruction for *Asia* has a large error range. It shows dips around 1600, 1700, the 1810s, and the 1920s. Peaks appear in the 1610s and 1800s. Most of the features are also found in the Kathmandu tree ring chronology (Cook et al. 2003), and a good agreement with Brückner's reconstruction for Asia is also found.

PAGES 2k reconstructions for *Australasia* show almost no trend and little variability until the mid-20th century. The latter can partly be explained by the fact that this is a combined land–ocean temperature reconstruction. The selected tree-ring chronology from Tasmania (Cook et al. 2006), which is also included in the PAGES 2k reconstructions, shows more variability, including a pronounced dip in the early 20th century.

Climate reconstructions for *South America* behave somewhat differently than those in the rest of the globe. In the PAGES 2k reconstructions, as well as those by Villalba et al. (2003), temperatures slightly increased in the 17th and 18th centuries, peaking near 1800. The subsequent dip is much more pronounced in the Villalba et al. (2003) reconstructions than in PAGES 2k.

Information from ice cores in *Antarctica* indicates that the peninsula was warm in the second half of the 18th century (Abram et al. 2013). West Antarctica was warm in the 1830s, when the tropical Pacific was in a more El Niño-like state (Steig et al. 2013). The latter warm period also appears in the continental PAGES 2k reconstruction. Overall, the PAGES 2k reconstruction shows large interannual-to-decadal variability and comparably little multidecadal-to-centennial variability. The proxies used for this reconstruction do not reflect the recent temperature increase in West Antarctica since the 1950s, which has been found in observation-based analyses (Bromwich et al. 2013; Steig et al. 2009).

Overall, the continental reconstructions show that little of the variability at interannual-to-decadal timescales is common to all regions, but there is some common variability among the northern continents. Some pronounced cold dips (such as those in the 1810s or 1830s), and episodes of warming are concurrent in most of the Northern Hemisphere series, others are not. The southern continents show a smoother temperature development (which is perhaps expected due to the larger ocean area) and variations are not synchronous with those in the Northern Hemisphere.

4.1.3.3 Precipitation and Drought Reconstructions

Precipitation generally varies on smaller spatial scales than temperature. Therefore a continental or even global perspective (as taken by Brückner) may have limitations. Towards the end of this section and in Sect. 4.1.4, I address precipitation reconstructions in central Europe and in the Alps. Here, an overview of some close-to-continental-scale changes is given. Large-scale reconstructions of hydroclimate are still very rare because, compared to temperature, more data are usually required to obtain skilful results at regional-to-continental scales.

In many or even most parts of the world, precipitation changes are more relevant to societies than temperature changes. One of these regions is Africa. Nicholson (2001), based largely on lake-level data, concluded that in most of *Africa*, including the Sahel region, it was dry in the early 19th century (Nicholson 2001). A pluvial period then set in after around 1850, concurrent with a positive AMO phase. Pronounced decadal variations in Sahel rainfall then occurred during the 20th century (see Sect. 4.4.3). A recent precipitation reconstruction for southern Africa back to 1796 (Neukom et al. 2014b) indicates high precipitation in the 1870s and 1880s in both the winter and summer rainfall zones and a drying (although with marked decadal variability in the winter rainfall zone) during the 20th century.

Several hydroclimatic reconstructions from Asia indicate pronounced historical droughts, some of which will be addressed in more detail in Sects. 4.1.4 and 4.2.6. The most comprehensive source of information is the Monsoon Asian Drought Atlas (Cook et al. 2010). Time series of droughts in this atlas, subdivided into the regions India, Southeast Asia, and China are shown in Fig. 4.6. Large drought events documented in this atlas and in the associated literature include the 1638–1641 drought in northeastern China, the 1756–1768 "Strange Parallels" drought in Myanmar, Vietnam, and Thailand (D'Arrigo et al. 2013; Lieberman 2003), the 1790–1796 eastern India drought, which may have been part of a larger-scale event (Rodysill et al. 2013; Sect. 4.2.3), and the 1876–1878 global drought (Sect. 4.2.6).

Hydroclimatic reconstructions from North America (Cook et al. 2004), reveal that from the 17th to the 20th century, the western United States was wetter than on the millennium average, suggesting that the preceding period of megadroughts had come to an end. Time series for western and eastern North America are shown in Fig. 4.6. A pronounced pluvial period occurred in the 1830s (Brückner's continental

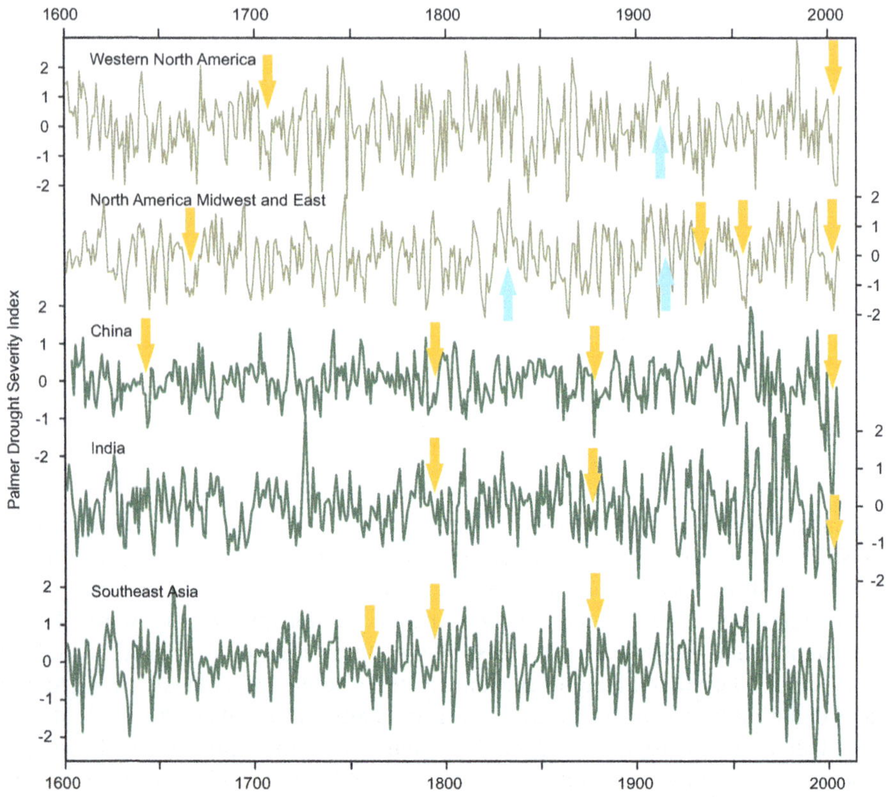

Fig. 4.6 Drought time series from the North American Drought Atlas (Cook et al. 2004) and the Monsoon Asian Drought Atlas (Cook et al. 2010) for the regions western North America (17.5°–62.5° N, 137.5°–110° W), the Midwest and eastern North America (17.5°–62.5° N, 110°–57.5° W), China (30°–56.3° N, 61.3°–143.8° E), India (8.3° S–30° N, 61.3°–90° E), and Southeast Asia (8.3° S–30° N, 90°–143.8° E). The *arrows* indicate those droughts (*orange*) and pluvials (*blue*) mentioned in the text

precipitation series for North America peaked in 1846–1850). Short dry periods occurred in the 1660s, the 1700s, the 20th century, and in the present. The "Dust Bowl" (Sect. 4.3.3), the 1950s droughts, the 1998–2004 drought (Sect. 4.5.6) and the 2010–2013 drought were the most prominent recent events. For southern South America, Neukom et al. (2010) found a long-term increase in summer precipitation and decrease in winter precipitation into the 20th century (not shown).

For Australia, Gergis et al. (2012) reconstructed a 206-yr hydroclimatic reconstruction that put recent droughts, such as the 1995–2009 "Big Dry" or "Millennium Drought" (see Sect. 4.5.6) into perspective. Thus, although reconstructions reveal some pronounced droughts, they are not nearly sufficient to provide an overview of large-scale hydroclimatic changes over the past several centuries.

4.1.3.4 Sea-Surface Temperature Reconstructions

Sea-surface temperatures are more difficult to reconstruct than temperatures over land. Few annually resolved, marine proxies such as corals are available. Therefore, coastal continental proxies such as tree rings are typically used. The only annual reconstruction of global sea-surface temperature fields, to my knowledge, is that of Mann et al. (2009a). These data were used as a boundary condition for our CCC400 simulations prior to 1870 (adding intraseasonal, ENSO-related variability as the reconstructions only provide annual means; see Sect. 2.9). The ENSO-augmented sea-surface temperature data (hereafter termed MC) are used in the following analyses. For individual oceanic indices, other reconstructions also exist and can be compared to the MC data.

Four different sets of indices derived from MC, observations, and other reconstructions are shown in Fig. 4.7. In general, the agreement among different reconstructions is not very good prior to around 1940. Observations are also uncertain prior to the Second World War (see Sect. 2.6.1). However, there are also common features among the reconstructions.

Using corals, tree rings, and a tropical ice core record, D'Arrigo et al. (2008) reconstructed *tropical sea-surface temperatures* (30°S–30°N) by building upon the coral-based reconstruction of Wilson et al. (2006). Both series (shown in Fig. 4.7) resemble the global temperature reconstructions (Fig. 4.4). D'Arrigo et al. (2008) found pronounced dips in the years following strong tropical volcanic eruptions. This includes the very low temperatures in the early 19th century. The corresponding MC curve agrees well with the reconstructions, although the individual dips are smoothed out. Entirely coral-based reconstructions have recently been published for different tropical oceans (Tierney et al. 2015) and indicate cooling rather than warming in the 18th century.

With respect to the tropical *Pacific Ocean*, ENSO indices (see Table 3.2) have been reconstructed based on various proxies (documentary data, tree rings, corals, and a combination of proxies; see Sect. 2.1). In Fig. 4.7, five reconstructions are shown Cook et al. (2000), Braganza et al. (2009), McGregor et al. (2010), Li et al. (2011), and Emile-Geay et al. (2013), along with the corresponding NINO3.4 index from MC (note that the last part in the Emile-Geay et al. (2013) series is from HadSST2 observations). Although most of these reconstructions share common proxy records, they differ widely. This is particularly the case in the 17th century. The recent reconstructions from McGregor et al. (2010) and Emile-Geay et al. (2013) agree that the 1750–1770 as well as the second half of the 19th century were rather La Niña like. Meanwhile, the 1720–1750, 1830–1860, the 1900s, the 1980s, and the 1990s were more El Niño-like. The MC NINO3.4 index fits well with other reconstructions, but clearly shows suppressed interannual variability, thus showing more of a PDO than ENSO signal. Prominent El Niño events that will be discussed occurred in 1877/1878, 1939–1942, and 1997/1998 (Sects. 4.2.6, 4.3.4 and 4.5.5).

The PDO (see Sect. 3.2.8 and Table 3.2) has been reconstructed by Biondi et al. (2001) and MacDonald and Case (2005). Figure 4.7 shows these reconstructions,

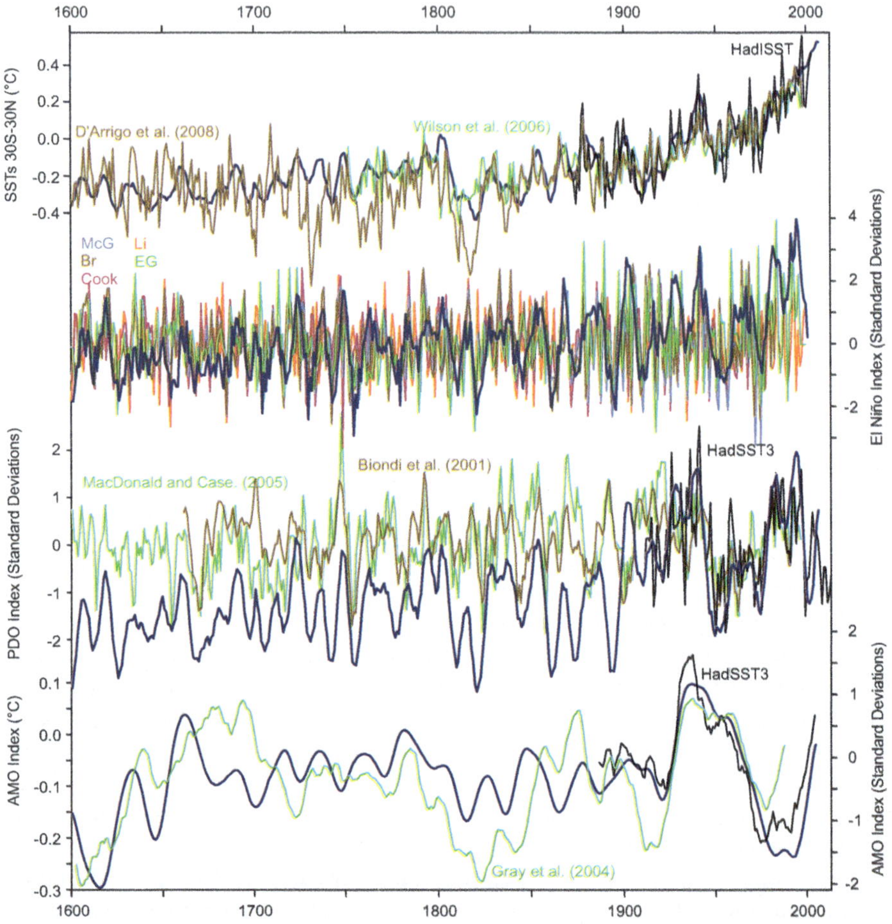

Fig. 4.7 Indices of tropical sea-surface temperature (30° S–30° N, from Wilson et al. (2006), D'Arrigo et al. (2008), and HadISST), ENSO (standardised in the 1650–1970 period), PDO (Biondi et al. 2001; MacDonald and Case 2005), and AMO (Gray et al. (2004); index on right y-axis) from various reconstructions and from MC data (*thick blue*). ENSO indices can be thought of as a measure of NINO3.4 (Cook et al. (2000), Braganza et al. (2009), version R5 with sign inverted (Br in the figure), McGregor et al. (2010) (McG in the figure), Li et al. (2011), Emile-Geay et al. (2013) with HadSST2 as target (EG in the figure))

together with the PDO derived from MC data and from HadSST3 observations (Kennedy et al. 2011). The pre-20th century part of these reconstructions—unlike the later part—does not show coherent multidecadal variability. In the 20th century (period with sufficient sea-surface temperature data), all series agree. The index was positive from the 1920s to the early 1940s, then negative from the 1950s to the 1970s, positive in the 1980s and 1990s and becoming increasingly negative since the 2000s.

For the *Atlantic Ocean*, the AMO index (see Sect. 3.2.8 and Table 3.2) was reconstructed by Gray et al. (2004) and can be calculated from MC data (Fig. 4.7). It was low in the 17th century and had dips in the early 19th century, early 20th century, and in the 1970s. The index was high in the second half of the 17th century, in the 1870s, around the mid-20th century, and again in the 2000s. The correlation between the two reconstructions is low, and a good agreement with observations only emerged after 1910.

There is still considerable uncertainty with respect to the past climate of marine regions. Large areas of the oceans are not covered by annual-resolution climate reconstructions. For basins with existing data, reconstructions often differ considerably.

4.1.3.5 Regional Climate: Central Europe and the European Alps

To end the discussion of 400 years of climate variability, I would like to zoom into Brückner's main region of work: central Europe and the European Alps. Summer temperature reconstructions for central Europe based on tree rings (Büntgen et al. 2011) and documentary data (Dobrovolný et al. 2010; Glaser and Riemann 2009 for Germany) are compared in Fig. 4.8. The agreement is good for the documentary series (correlation of 0.86 prior to 1880), but poorer with the tree ring-based reconstruction, which shows pronounced cooling around 1820. The tree-ring based series shows long-term warming, which is less obvious from documentary sources. There is a clear similarity to Brückner's series of grape harvest dates (Fig. 4.3). Prominent features are a drop in summer temperatures around 1700, another strong drop in the 1810s, and three warming periods: 1820–1880, 1910–1950, and from the 1970s to present.

Multi-proxy reconstructions of summer temperature for the Alps by Casty et al. (2007) and Trachsel et al. (2012) are also shown in Fig. 4.8, together with a tree ring-based reconstruction by Büntgen et al. (2006), an index based on documentary series representing the Swiss Plateau Pfister (1999) and early instrumental data from HISTALP (Auer et al. 2007; Sect. 2.9). Together, they provide an overview of summer temperature variations and their agreement is good (median correlation coefficients from 1780 to 1880 of 0.75). The tree ring-based reconstruction agrees well with the multi-proxy reconstructions, but less so with the other reconstructions. Summer temperatures in the Alps show an increase over time—especially during the last 50 years—and exhibit multidecadal variability. The recent hot summers, particularly in 2003, clearly exceed the temperature range of the last 400 years. Cold summers were basically absent since the mid-1980s. Likewise, the cold summers of the 1810s (see Sect. 4.2.4) were below the normal temperature range.

The temperature curves for central Europe and the Alps show very strong similarities and are highly correlated (some of the correlations exceed 0.9). However, the dip in the 1810s was stronger in the Alps. Although regional differences may be large on seasonal or shorter scales, on a decadal scale both regions agree and show strong similarities to the reconstructions for Europe and even the northern extratropics.

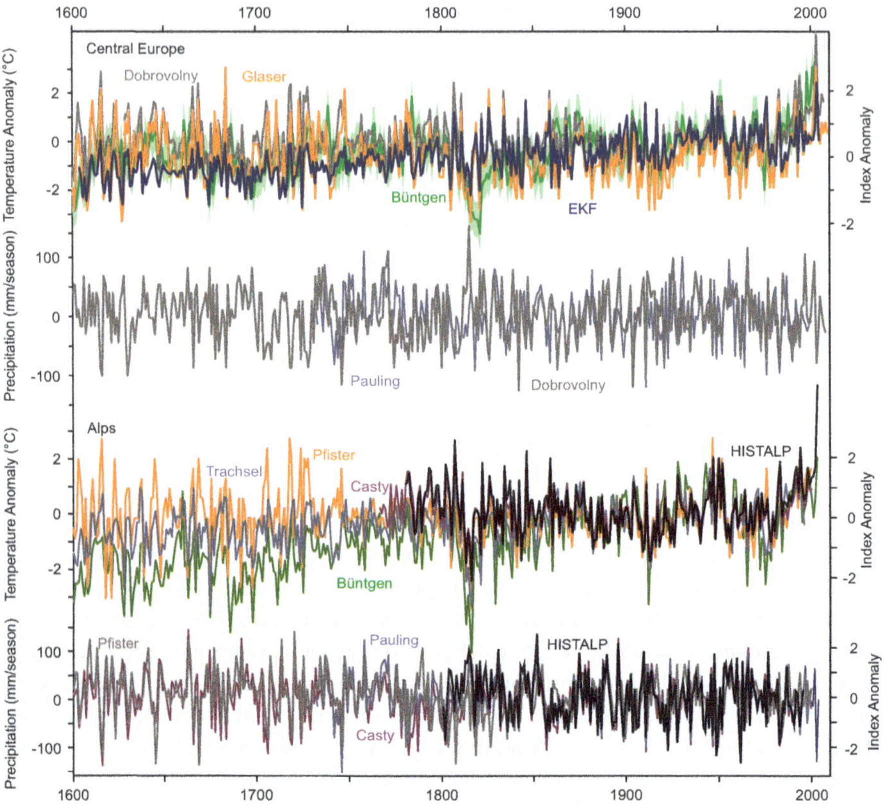

Fig. 4.8 Summer (Jun.–Aug.) temperature and precipitation reconstructions for (top) central Europe (average over 48°N–52°N, 6°E–12°E, or records within) and (bottom) the Alps (average over 44°N–49°N, 4°E–15°E, or records within) from observations (HISTALP, Auer et al. 2007), documentary data (Glaser and Riemann (2009) for Germany (smoothed), Pfister (1999) for Switzerland, both as indices, Dobrovolnỳ et al. (2010, 2015) for Central Europe, in °C and mm, respectively), multiproxy reconstructions (Casty et al. 2005, 2007; Trachsel et al. 2012), tree ring-based reconstructions (Büntgen et al. (2006) for the Alps and Büntgen et al. (2011) for Central Europe; *shading* indicates one std. dev.), and the ensemble mean of EKF400 (*blue*). All series are anomalies from their 1900–1990 average

For summer precipitation, Fig. 4.8 shows reconstructions for central Europe from multiple proxies (Pauling et al. 2006) and from documentary and early instrumental data (Dobrovolnỳ et al. 2015). Because the multi-proxy reconstruction uses documentary and instrumental information, they are not independent. The precipitation reconstructions exhibit interannual-to-decadal variability, but little multidecadal variability and no trends. This results in a "white" or "blue" spectrum.

In contrast to temperatures, summer precipitation in the Alps is not well correlated with central European precipitation. It is difficult to identify common features in all reconstructions for both regions. However, we might be able to address low summer precipitation values in the 1740s, the 1800s, and the period from the late 1850s to the early 1870s. We can also address high values before and after the latter period. This sequence partly agrees with Brückner's wet and dry periods and was associated with droughts, floods, and glacier advances. These are dealt with in Sects. 4.2.3, 4.2.5, and 4.2.7.

4.1.4 Climatic Changes in CCC400 and EKF400

From the available reconstructions, a picture of large-scale climatic changes emerges containing particularly detailed information about temperature. In the following, I add the modelled climates using the CCC400 and EKF400 data sets. The data sets are not better than the reconstructions but, in addition to providing climate on global and regional scales, they provide insight into the machinery, that is, changes in atmospheric circulation.

In all following figures based on CCC400 or EKF400, maps always show the ensemble mean or the best ensemble member (Sect. 2.9). Plots of time series include the ensemble mean, the 50 % and 90 % range of the members, the best member of EKF400 (in boreal summer), and occasionally the ensemble mean of CCC400. Annual time series of seasonal or annual means are displayed without smoothing or filtering.

4.1.4.1 Global, Hemisphere and Continental Scales

On the global scale, CCC400 temperatures over land (Fig. 4.4, top) agrees well with reconstructions and observations. A gradual warming during the 18th century peaks shortly before 1800 and is followed by a dip (more pronounced than in Mann et al. 2008) and subsequent recovery.

With respect to summer temperatures over northern extratropical land areas, existing reconstructions agree in terms of correlation but not in terms of amplitude. How does EKF400—which is well constrained by instrumental data, documentary data, and tree rings over this region and season—fit in the picture? Correlations of the different reconstructions with the EKF400 ensemble mean are around 0.7–0.8. With Frank et al. (2010), the correlation exceeds 0.9. Part of this agreement is because EKF400 and the reconstructions share some proxies. Another part is due to external forcings and sea-surface temperatures. The amplitude of EKF400 is somewhat higher than that of Frank et al. (2010), but lower than Christiansen and Ljungqvist (2012). The best ensemble member shows a similar amplitude of variability as observations and is well correlated with observations in the overlapping period.

Continental temperature series from EKF400 are shown in Fig. 4.5. Again, they are not independent from the reconstructions because both use tree ring series. The agreement is good for regions having observations available for EKF400 (e.g.,

Europe from the 17th century onward or other regions from the 19th century onward). The series also agree with the continental averages presented in Brückner (1890), who used many of the same early instrumental series for his analyses. The farther back the series go, the more they represent the pure model response. In the ensemble mean, this shows lower variability than the reconstructions on interannual-to-decadal time scales. This is particularly the case in polar regions, where sea-ice variability (an important boundary condition) is almost absent in CCC400 prior to around the Second World War. Therefore, EKF400 is more suitable for studying anomalies on interannual-to-multiannual scales in well-observed areas.

Conversely, EKF400 and CCC400 allow us to go beyond temperatures and to analyse the large-scale circulation. In the next two sections, we will have a closer look at circulation changes in EKF400 and CCC400.

4.1.4.2 Changes in the Zonal-Mean Circulation

A first overview of large-scale circulation variability can be gained from analysing the zonal-mean flow (Sect. 3.1.2). This paragraph addresses zonal mean circulation indices (see Table 3.2) since 1600 calculated from EKF400. For the most recent 140 years, additional data sets are available and corresponding indices are presented in Sect. 4.3.1. Note that the plotted series (ensemble mean) emphasise the forced component (external forcing and sea-surface temperatures) and damp unforced variability unless it is well constrained by observations.

Indices for the zonal-mean circulation (Fig. 4.9) mostly focus on the Northern Hemisphere in boreal winter. They encompass the strength of the northern Hadley circulation, the northern subtropical jet, the boreal winter polar vortex, and the position of the ITCZ in winter and summer (see Table 3.2 for definitions). Thus, the indices measure the pace and efficiency of the machinery. Expressions of climatic changes in variability modes (see Sect. 3.2) that are not zonally symmetric are discussed thereafter.

The *boreal winter Hadley cell* (here defined as the maximum zonal mean upward velocity at 500 hPa between 30°S and 30°N) shows a strengthening over the entire period that is interrupted by a strong dip after 1800. The Hadley cell is the motor shaft of the machinery. It directly shows how the machinery changed its pace during the past 400 years. Some of the dips broadly follow tropical volcanic eruptions, which can be explained by reduced net radiation and reduced evaporation in the tropics (see Sect. 3.3.2). However, minima in the strength of the Hadley circulation also coincide with solar minima. Generally, the Hadley circulation was weak when global temperatures were low, but not exclusively (see Schneider et al. 2010).

The position of the *boreal summer ITCZ* is related to cross-equatorial energy flux and equatorial ocean heat uptake (Schneider et al. 2014b; Ori et al. 2015) and severely affects large-scale hydroclimate. It shows little change prior to the late 19th century. There is a northward shift in the late 1940s followed by a southward shift into the 1970s. According to a study of lake sediments on Pacific Islands (Sachs et al. 2009), the ITCZ over the Pacific Ocean had a more southerly position during

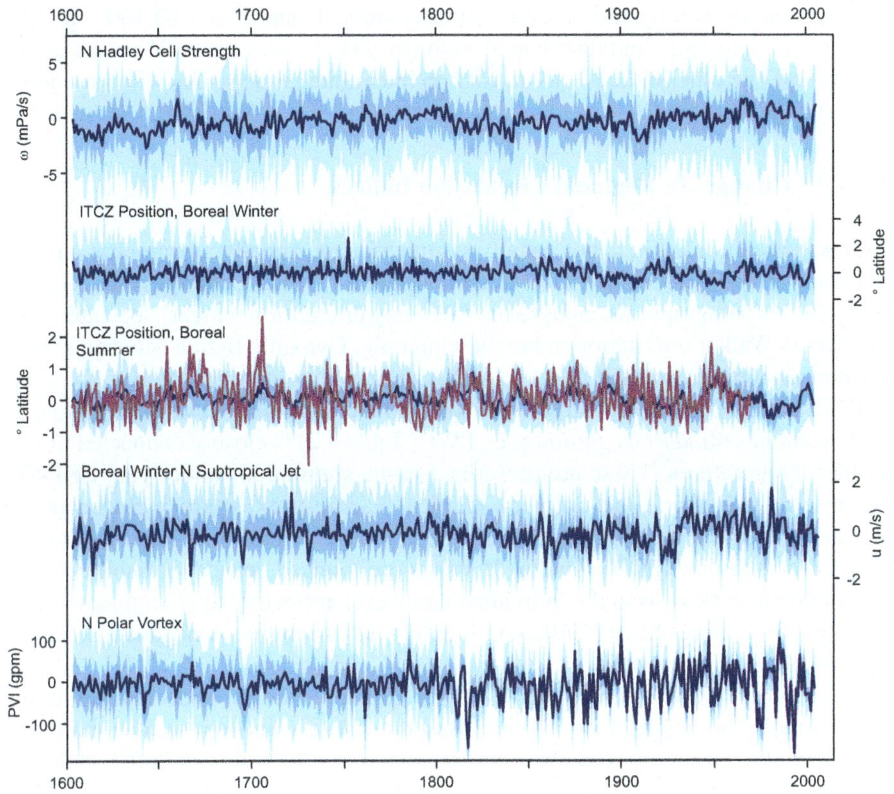

Fig. 4.9 Zonal-mean circulation indices during the last 400 years in EKF400 (ensemble mean is given in *blue* with 50 % and 90 % ranges shaded; *purple line* indicates the best member). All series are anomalies from their 1900–1990 averages. Shown are December–March averages (the position of the ITCZ also for Jun.–Sep.) of the strength of the northern Hadley cell, the position of the ITCZ, the strength of the northern subtropical jet, and the weakness of the lower stratospheric northern polar vortex, respectively

the Little Ice Age (1400–1850) than today. Conversely, Ridley et al. (2015) find a southward shift of the ITCZ over Belize since the 1850.

The *boreal winter subtropical jet* (i.e., SJ; see Table 3.2) hardly shows any long-term change. The weak jet in the 1920s is the most noteworthy feature. The year-to-year variability of the subtropical jet is strongly determined by the tropical-to-extratropical temperature difference and does not react strongly to climate forcings. However, it is related to internal modes of variability such as ENSO.

Similarly, the lower stratospheric *boreal winter Polar Vortex Index* (i.e., PVI; Table 3.2) does not show a clear long-term trend. Strong negative spikes (indicating a strong vortex) consistently follow volcanic eruptions. Pronounced positive spikes (indicating weak vortex events) occur more erratically. While these spikes appear in individual members, the ensemble mean does not show a strong imprint. Periods with a strong vortex occurred in the early 19th century and again in the 1990s. A weak vortex occurred around 1900.

In summary, among the zonal-mean circulation features in EKF400, only the Hadley cell strength and the boreal summer ITCZ show decadal-to-centennial variability. The other indices vary mostly on interannual time scales with little long-term change.

4.1.4.3 Changes in Regional Circulation Indices

Atmospheric circulation is not zonally symmetric, and zonally asymmetric circulation and variability (see Sect. 3.1.2) are important for climatic changes on a regional scale. In this paragraph, we analyse the following circulation indices (see Table 3.2): the Pacific Walker circulation in boreal winter, the Dynamic Indian Monsoon Index for boreal summer, and the NAO and PNA indices in winter (Fig. 4.10). For some of these indices, other reconstructions exist and are incorporated into the figures.

The *Pacific Walker circulation* (i.e., PWC; Table 3.2) is closely connected to sea-surface temperatures. These are prescribed in the simulations underlying EKF400. Because at the same time only few tropical Pacific proxies are included in the assimilation, the PWC largely exhibits the response of the model and is strongly correlated with the NINO3.4 index. It shows pronounced decadal variability in the ensemble mean (less so in the individual members), reflecting the fact that MC sea-surface temperatures have insufficient interannual variability. A negative excursion around 1900 is noteworthy.

Over the last 400 years, the PWC in EFK400 shows a decreasing trend, amounting to one-tenth of its strength. Based on observations, a weakening of the Pacific Walker circulation has also been suspected (see Vecchi et al. 2006). However it is offset by a strong, recent strengthening (Hartmann et al. (2013), see Sect. 4.3).

The *Dynamic Indian Monsoon Index* (i.e., DIMI; Table 3.2) is correlated with the PWC ($r = 0.69$ for the ensemble mean, $r = 0.21$ for the members). Some reconstructions (Anderson et al. 2002; Gong and Luterbacher 2008) suggest a strengthening of the Asian monsoon during the past four centuries, particularly through the 18th century and again from the 1850s to 1900. However, EKF400 only reproduces the latter increase. The most pronounced feature is the dip starting around 1900, concurrently with a drop in tropical (and global) temperatures, and continuing to the 1930s. This is found in reconstructions (Zhou et al. 2010), in 20CR (Fig. 4.42), and to some extent in Gong and Luterbacher (2008). In general, however, the DIMI in EKF400 does not agree well with the DIMI reconstructed by Zhou et al. (2010).

Reconstructions of indices of the *North Atlantic Oscillation* (i.e., NAO; see Sect. 3.2.3, Table 3.2) by Luterbacher et al. (2002), and by Cook et al. (2002) and an instrumental index by Jones et al. (1997) are shown together with the index from the EKF400 ensemble mean (the latter is plotted as pressure differences between the Azores and Iceland). Correlations among the reconstructed indices are high for the past 150 years. Significant correlations among the indices are also found in earlier periods. However, the agreement is poor for certain cases such as the first half of the 18th century. Note that a two-point circulation index constrained by only a few

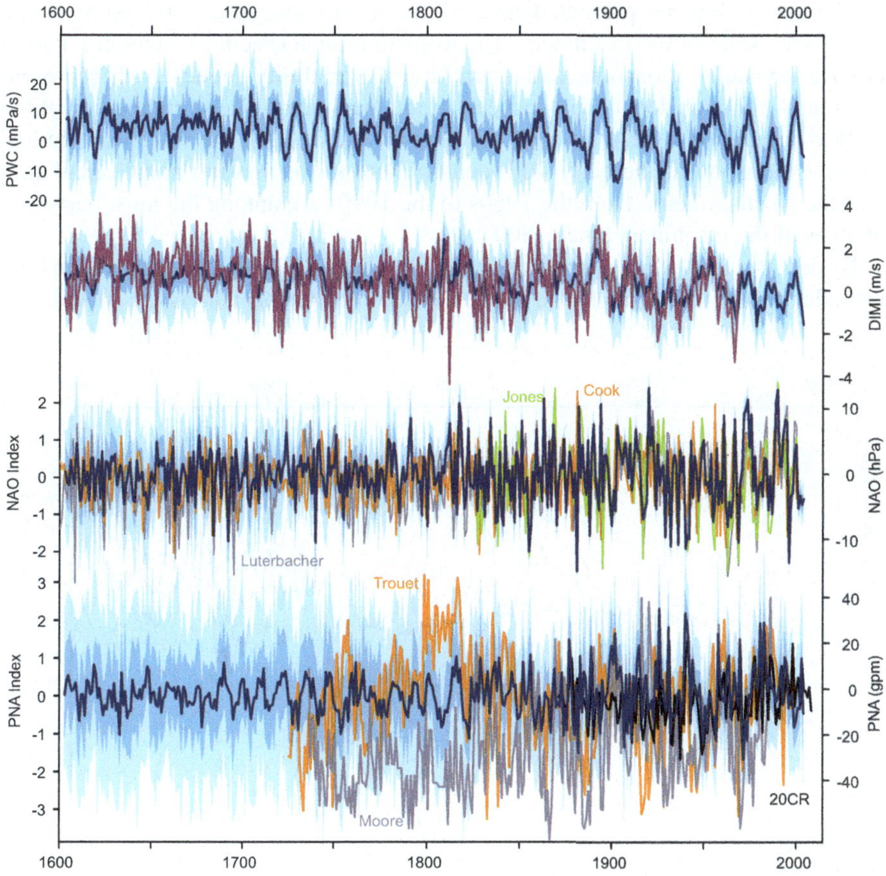

Fig. 4.10 Regional circulation indices in EKF400 (ensemble mean is given in *blue* with 50 % and 90 % ranges shaded; *purple line* indicates the best member) measuring the strength of the Pacific Walker circulation (Sep.–Jan.), the Dynamic Indian Monsoon Index (Jun.–Aug.), and the NAO and the PNA (Dec.–Mar.). For the latter two indices, existing reconstructions are also shown, and EKF400 is given in absolute terms (hPa and gpm). All series are anomalies from their 1900–1990 averages

proxies may easily be affected by slight changes in the centres of action (Lehner et al. 2012; Ortega et al. 2015). One of the most compelling features of decadal variability is the strong increase in the NAO from the 1960s to the 1990s (Sect. 4.5.4, see also Chap. 3) following an almost century-long phase of decrease.

For the *Pacific-North American pattern* (i.e., PNA; see Table 3.2) we can consult (in addition to EKF400 and 20CR) reconstructions of the Pacific climate based on accumulation rates in the Mt. Logan ice core (Moore et al. 2002) and a tree-ring based reconstruction (Trouet and Taylor 2010). The agreement among these series is obviously poor prior to the 1950s. Thus, one of the most important modes of variability is poorly constrained in the past.

The set of indices presented here allows us to diagnose—at least to some extent—the state of the machinery. The tropical branch (Pacific Walker circulation and Hadley cell), monsoon, and extratropical variability modes each represent different—although dependent—aspects of the machinery. The strengthening of the Hadley circulation over time (and the drop in the 1810s), the lull in the Indian summer monsoon in the early 20th century, and the increasing zonal circulation over the North Atlantic from the 1960s to the 1990s are among the most pervasive changes of the machinery since 1600.

By construction, a large part of the variability in the EKF400 ensemble mean (and a smaller part of the variability in the best ensemble member) in Figs. 4.10 and 4.11

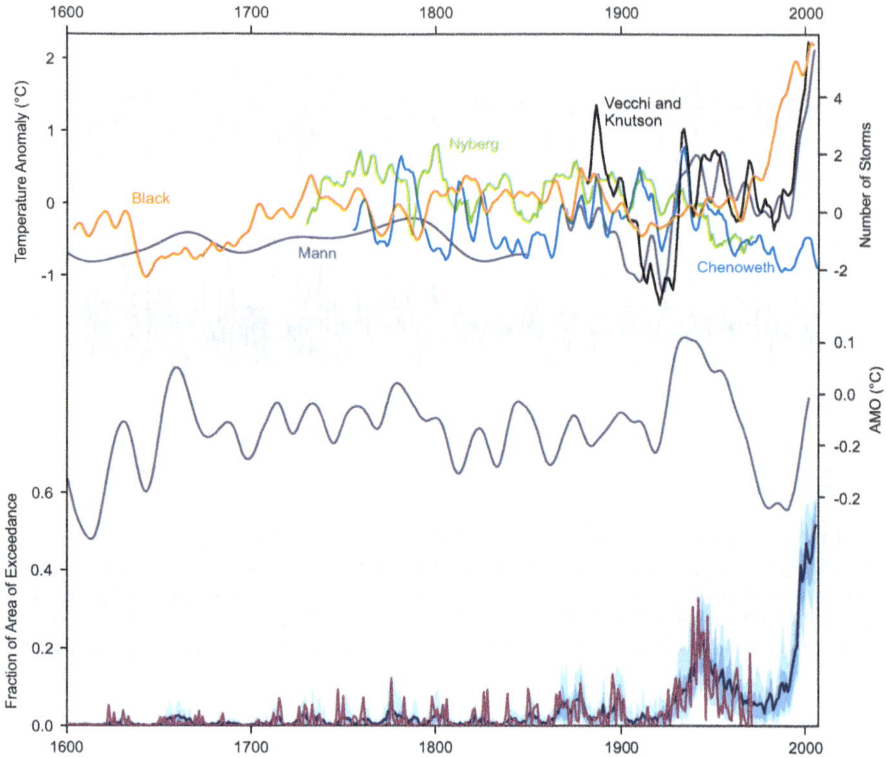

Fig. 4.11 (top) Reconstructions of tropical cyclone activity in the Atlantic. Land-falling hurricanes from Mann et al. (2009a), number of hurricanes from Nyberg et al. (2007), tropical storms in the vicinity of the Lesser Antilles from Chenoweth (2003) (multiplied by 2), reconstruction of sea-surface temperatures in the Carioca basin from Black et al. (2007), and adjusted HURDAT storm count from Vecchi and Knutson (2008); the latter three series were filtered with a 9-point triangular filter. (middle) AMO index from MC data. All *curves* are anomalies from the 1901–1970 mean value. (bottom) Fraction of area for which the annual maximum temperature exceeds the 90th percentile of the 1901–2000 period in CCC400, calculated for the United States (30° N–50° N, 124° W–70° W) and Europe (42° N–52° N, 0° W–30° W), then averaged. *Blue line* and *shadings* denote the ensemble mean and 50 % and 90 % ranges. The *pink line* is the best ensemble member (by Sina Lenggenhager)

is forced externally or by sea-surface temperatures. Thus, the series represent a model response constrained by observations rather than a reconstruction.

Sections 4.2–4.5 will revisit prominent excursions, steep drops, and increases found in these series. In the remainder of this section, we turn to another important aspect, namely changes in extreme events, and then very briefly analyse the role of external forcing factors for global temperature changes.

4.1.4.4 Changes in Extremes

Extremes are increasingly in the spotlight of climate change discussions. Can we learn from past extremes? Certainly! Extreme weather events and their impacts have been well documented by past societies. These data allow precise (and sometimes quantitative) assessments. Certain archives (e.g., sediments) preserve records of storms and/or floods, which can be used to explore extremes (e.g., de Jong et al. 2006; Martin-Puertas et al. 2012; Pfahl et al. 2009; Stewart et al. 2011). Obviously, long instrumental series can also be used to acquire information (Dangendorf et al. 2014). However, the spatial representativity of conclusions based on proxies and instrumental series might be limited. In the following, I focus on two examples: hurricanes in the Atlantic, and heatwaves in Europe and North America. Other extremes (e.g., floods and storms) will be addressed in Sects. 4.2 and 4.5.

Certain aspects of Atlantic hurricane statistics have been reconstructed (Fig. 4.11). These statistics describe different phenomena, which makes it difficult to obtain a clear picture of past Atlantic hurricane activity. Based on lacustrine sediments in eastern North America that capture land-falling hurricanes, Mann et al. (2009a) found relatively low hurricane activity during the Little Ice Age, an intermediate peak around 1790, and a sharp increase after the 1850s. Chenoweth (2003) and Chenoweth and Divine (2008, 2012) found no long-term trend in tropical storm activity based on documentary data from the Lesser Antilles. The reconstruction by Nyberg et al. (2007), derived from corals and marine sediments, shows a lower number of hurricanes in the mid-20th century than previous centuries. Black et al. (2007) reconstructed sea-surface temperatures in the Carioca basin from a foraminiferal Mg/Ca record, which also serves as a proxy for hurricane activity (Fig. 4.11). This record shows peaks in hurricane activity in the 1890s, the 1930s, and the present. Troughs are found in the mid-19th century, the 1920s, and the 1960s. Observation-based data from Mann et al. (2009a) and Vecchi and Knutson (2008, 2011) are shown in Fig. 4.11. These series only partly share the trends seen in Nyberg et al. (2007) and Chenoweth (2003), even though the latter is based on a subset of the same data. Estimates of hurricane frequency based on observations are biased low in the early days, when not all hurricanes might have been noticed, and need to be corrected. The effect of this correction is large prior to 1910 (Neu 2008).

Atlantic hurricanes are triggered by so-called African Easterly Waves—instabilities in the African Easterly Jet (Sect. 3.1.4). Their preconditioning and development depends, among other things, on sea-surface temperature (high temperature favours their development) and wind shear (strong vertical wind shear inhibits growth) over the tropical Atlantic. The number of Atlantic tropical storms that grow to hurricanes is larger during warm phases of the AMO (e.g., Zhang and Delworth 2006). Correspondingly, there is a positive correlation between the AMO index and hurricane number over the instrumental record.

In contrast to hurricanes, heatwaves are resolved in CCC400. A statistic of heatwave area based on 6-hourly CCC400 data from all ensemble members (the area of exceedance of the 90th percentile of 20th century annual maximum temperatures, averaged over the United States and central Europe) is shown in Fig. 4.11. According to CCC400, few heatwaves occurred before the late 19th century. In the 1940s, the heatwave area increased. However, the most striking result is the dramatic increase in the area affected by heatwaves over the last decades, which coincided with increasing global temperatures. Analyses of observations revealed an increase in heatwave severity in most regions over the last decades (Hartmann et al. 2013); for Europe, this occurred over the last century and possibly beyond (Della-Marta et al. 2007). This is further discussed in Sect. 4.5.7.

In addition, the area affected by heatwaves shows decadal-to-multidecadal variability. The resulting series shares close similarities with the AMO (see also Fig. 4.72). Thus both hurricane activity and heatwaves in North America and Europe—although unrelated from a dynamical point of view—vary with slow modes of the climate system such as the AMO.

Because heatwaves here are defined as the exceedance of a fixed threshold, their frequency is expected to increase with higher mean temperatures. A shift in the temperature distribution (assuming no change in the shape of the distribution) could provide an explanation. However, a higher rate of exceedance could also occur if there is a change in the variability or shape of the distribution without a change in the mean, or a combination of these factors (see Katz and Brown 1992). Changes in the tails of the temperature distribution can occur through feedback processes. Heatwaves can be amplified by land–atmosphere interaction at midlatitudes (see Sect. 3.4). Conversely, temperature maxima in polar, marine, or high-altitude regions can be damped by the melting of ice or ocean heat uptake. A warmer climate with more water vapour (thus more latent heat) can also affect other extremes. Storm systems can become more energetic and precipitation events more intense.

The question, whether increasing temperature variability due to a warming climate causes more frequent heat extremes was raised by the 2003 heatwave in Europe (e.g., Schär et al. 2004). To analyse how extremes relate to the mean temperature, we can study CCC400. Let T_{max} and P_{max} be the maximum 6-hourly temperature and precipitation sum, respectively, per ensemble member and year, averaged over all ensemble members. Variable T_{mean} is the annual mean temperature. We now express $T_{max} - T_{mean}$ and P_{max} as a function of T_{mean} (Fig. 4.12). For temperature, a stronger change of maxima than change in mean (red colours) appears over midlatitude land regions. However, a weaker change (blue colours) occurs in polar regions, as expected from the above considerations. It is important to note that land cover trends are misrepresented in our simulations (Sect. 2.9).

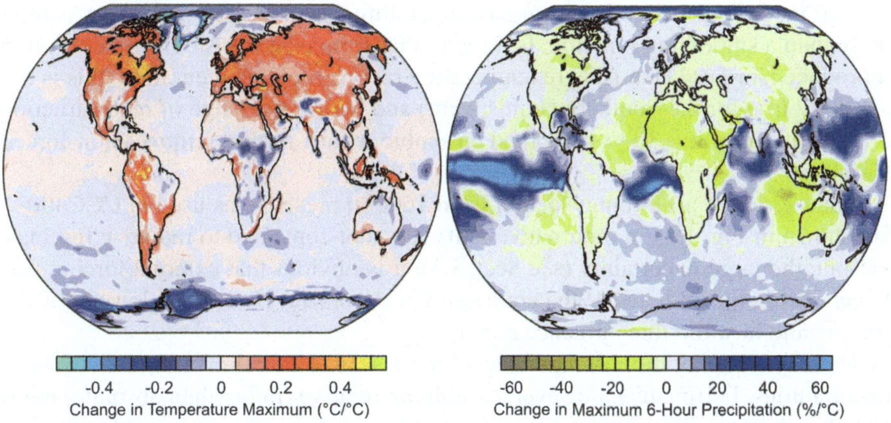

Fig. 4.12 Change as a function of annual mean temperature in (*left*) the ensemble average of the difference between annual maximum and mean temperature and (*right*) the ensemble average of the ratio of annual maximum precipitation and long-term mean annual maximum precipitation. The figure shows coefficients of a least squares regression (Figure by Sina Lenggenhager)

Maximum precipitation rates (Fig. 4.12, right) increase with rising annual mean temperatures. This is especially the case in the ITCZ region. Meanwhile, decreasing precipitation rates are found in the outer tropics. Although intensities increase over the North Atlantic storm-track region, a slight decrease is seen over the northern midlatitude continents. Trends in precipitation intensity over the past few decades suggest an increase in some regions and seasons (Fischer and Knutti 2014; Hartmann et al. 2013). In contrast, analyses of flood frequency from Alpine lake sediments from the past 2,500 years suggest more floods during cold periods (Glur et al. 2013; Stewart et al. 2011). Although other factors (e.g., the snow line altitude) also matter for floods, reconciling the two results is not straight forward.

Overall, CCC400 suggests that a warmer annual mean temperature is characterised by amplified temperature extremes in the midlatitude land regions (but damped temperature extremes in polar regions) and intensified rainfall over the equatorial oceans and storm-track regions (but damped rainfall extremes in subtropical arid zones).

4.1.5 Drivers of Global Climate Since 1600

What are the main drivers of global climate since 1600? In this section, I will focus on global-scale forcing mechanisms. These may differ from the regional drivers that are addressed in later sections. On a global scale, the response of the machinery to climate forcings may be fairly direct. For instance, climate models are basically sophisticated smoothers when it comes to global-mean temperatures. They convert incoming energy fluxes into a temperature signal and the machinery is little more than a smoothing function. Is it really that simple? Surprisingly, models often reproduce Northern Hemisphere temperature reconstructions accurately (this is not similarly true for the Southern Hemisphere) and within the range of reconstruction uncertainty (plotting 30-yr-smoothed ensemble means further removes non-forced variability).

Time series of all forcings factors from 1600 to present—as used in CCC400—are shown in Fig. 4.13 in their native units (i.e., not converted to radiative forcing). Despite the large uncertainty (see Sect. 3.3), it is obvious that external forcing was lower during the 18th and 19th centuries than at present. How much lower was the forcing and how did this influence climate?

Unfortunately, both the magnitude of the forcings and climate sensitivity have uncertainties. Fortunately, however, combining relative (rather than absolute) numbers for the forcings time series with simple statistics, we can roughly estimate the contribution of forcings to the temperature development. In Sect. 3.3.2, I used a simple regression model to separate volcanic and solar contributions to tree ring width. We can use the same model, extended beyond volcanic forcing, to assess the role of forcings (note that this type of fitting is overly simplistic and does not yield, for instance, reasonable estimates of climate sensitivity; see Wigley et al. (2005), see also Box 3.3, p. 149):

$$T_t = c_0 + c_1F_1 + c_2F_2 + c_3F_3 + c_4F_4 + T_{t-1}e^{-c_5\Delta t} \tag{4.1}$$

Here, T_t stands for anomalies in global—or Northern Hemisphere—averaged land temperatures at time t. Variables F_1–F_4 stand for the forcings exerted by greenhouse gases (we use the CO_2 concentration for simplicity), tropospheric aerosols (Muthers et al. 2014b), solar variability (total solar irradiance by Lean 2000), and volcanic activity (global- or hemispheric-averaged aerosol optical depth by Crowley (2000)). In a simple form, this model expresses the response of the "machinery" to external forcings. That is, it acts as a smoothing function. Here, the model is fitted to the observed global land surface temperatures from 1856 onward (see Fig. 4.4) and the fitted series is then extended backward (Fig. 4.13).

The resulting curve can now be compared with reconstructions or CCC400 land temperatures. The latter are also shown in Fig. 4.13. Not surprisingly, agreement between the fit and CCC400 is good because the modelled global temperatures (particularly in the ensemble mean) are strongly determined by the forcings. However, it also agrees with the independent reconstructions shown in Fig. 4.4. The most notable deviation is the warming in 1910–1940 (Sect. 4.3) and subsequent stagnation. This multidecadal signature in the observed and reconstructed temperatures is absent from the fit to the forcings. Throughout the remainder of this book, I will make use of this fitted curve and its decomposition to obtain a rough estimate of the forced components of the temperature signal in a given period. However, the reader should be aware that this is not a proper attribution approach (the model is very simple; it will not capture the correct autocorrelation in time and does not account for the extent to which responses are uncertain due to climate variability).

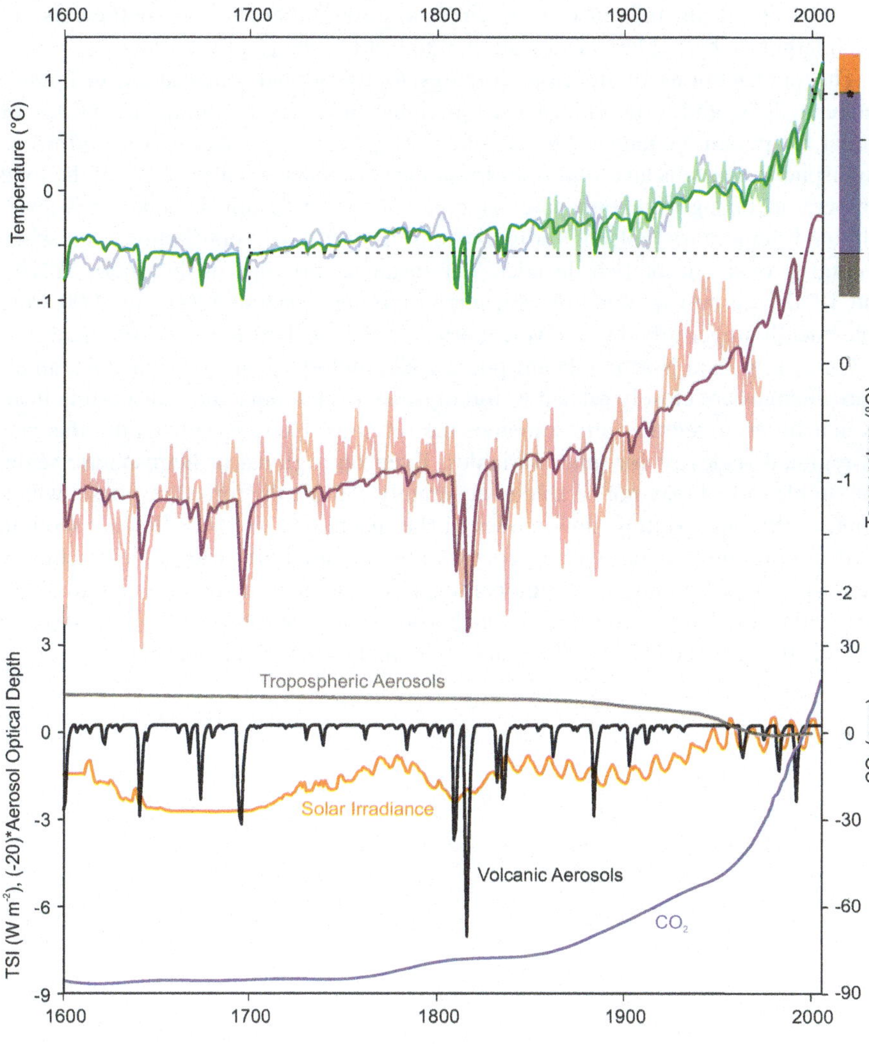

Fig. 4.13 Temperatures of (top) the global land areas (*green*) fitted to observations (*pale green*) with Eq. 4.1 and (middle) the northern extratropical land areas (*purple*) fitted to reconstructions (Christiansen and Ljungqvist 2012; *pale purple*) with Eq. 4.1 and extended backward and forward. The *pale blue line* is the global land temperature in CCC400. (bottom) Climate forcings (in native units; aerosols are scale-inverted; all series are expressed as anomalies from the 1985–2005 average). The scale (top, right) shows the contribution of the different forcings (color coded as (bottom)) to the fitted temperature difference (asterisk) between 2000 and 1700

This method can be applied to an extreme reconstruction such as the northern extratropical temperatures by Christiansen and Ljungqvist (2012). This reconstruction does not include the recent warming and ends with a cooling period from the 1940s to 1970s. The fit to this curve (Fig. 4.13) shows very similar results (including a highly significant influence of greenhouse gases) although the coefficients are much larger. In particular, the deviation in the early 20th century is large.

The contributions of different forcings to the global temperature difference between 2000 and 1700 (Fig. 4.13, top right) indicates a similar magnitude of global temperature changes and the effect of greenhouse gases. Contributions of tropospheric aerosols and total solar irradiance are much smaller (20 % of the total temperature change) and opposing. All coefficients are significant at the 95 % level. Although temporary, volcanic eruptions can have a large contribution even when averaging over one or two decades. For instance, the dips in the 1690s, 1810s, and 1830s can be explained by volcanic eruptions. Because 1700 and 2000 were volcanically unperturbed years, volcanoes do not contribute to their differences.

On a global and Northern Hemisphere scale, multidecadal-to-centennial temperature changes are closely related to forcings. By exchanging and transporting heat, the machinery integrates and smoothes the forcings. However, this is not the case on regional scales or for other variables. Even on a global or hemispheric scale, our simple model may not be able to accurately capture the forced response, which emerges through exciting slow modes of the machinery. In the following section, I will go into more detail with respect to internal variability within the machinery and responses to forcings on a regional scale by looking at subperiods. Within these subperiods, I will also explore individual events. The first subperiod is the last phase of the Little Ice Age (1700–1890), and I will analyse six climatic events.

4.2 The Period from 1700 to 1890: The Little Ice Age

The subperiod from 1700 to 1890 coincides with the last phase of the Little Ice Age (Note that although most authors define the last year of the Little Ice Age as 1850, the climate in many parts of the globe remained cool for another 30–40 years (see Ahmed et al. 2013; Figs. 4.4–4.6). A characteristic feature of the Little Ice Age was the glacier advances in the European Alps. A detailed view of these advances emerges from the rich historical documentation of glaciers and climate (e.g., Holzhauser 2010; Zumbühl et al. 2008) as well as from abundant dendro-climatological, geomorphological, and geological evidence (e.g., Holzhauser 2010; Nicolussi and Patzelt 2001). From the 18th century onwards, glaciers triggered the interest of painters. Some of these paintings (see Fig. 4.14) are sufficiently accurate to reconstruct glacier lengths at an almost annual resolution (Nussbaumer et al. 2012; Zumbühl et al. 2008, see also Sect. 4.2.5).

Glaciers also advanced outside of Europe during the 17th–19th centuries (Oerlemans 2005); for instance, in New Zealand's Southern Alps (Lorrey et al. 2013). This is reflected in glacier-based global temperature reconstructions (Leclercq and Oerlemans 2012). However, advances on other parts of the globe were not necessarily synchronous with those in the European Alps.

The Little Ice Age is often associated with frozen rivers and canals in western Europe. Paintings of winter landscapes by Dutch artists of the 16th and 17th centuries are often used to illustrate the different impacts of the Little Ice Age on Europe, from everyday life to politics. Some authors claim that the climate had a large impact on political history during this time (e.g., Behringer 2007; Fagan 2000). Meanwhile, climatic changes have always affected agrarian societies. However, societies have always had strategies to cope with climatic hardship and to overcome crises. Therefore, climatic changes rarely suffice as a monocausal explanation ("climate determinism") for social or political changes.

Fig. 4.14 View of the Glacier du Bois and the hamlet "Des Pras" (Painting by Jean-Antoine Linck, 1813. Swiss Alpine Club, Geneva, reproduction by H. J. Zumbühl)

The subperiod from 1700 to 1890, which is also the focus of Brückner's book, was not constantly cold. It consisted of several particularly cold phases punctuated by warm phases on all continents (Ahmed et al. 2013). In this section, I will give an overview of climate during the 1700–1890 subperiod, starting with a summary of climate during the 18th and 19th centuries from a large-scale perspective. Then, I will go into more detail for several prominent variations: the Late Maunder Minimum around 1700, the high temperatures and regional droughts around 1800, the "Year Without a Summer" of 1816, the Alpine glacier advance around 1850, the global drought of 1876–1878, and the frequent floods in central Europe in the 1860s to 1880s. Many other periods could have also been selected. Examples include the "Strange Parallels" drought in Southeast Asia (1756–1768), the dry conditions in Africa in the early 19th century (Nicholson 2001), the West Antarctic warming of the 1830s (Steig et al. 2013), the climate anomalies following the Krakatau eruption of 1883, and the European storminess peak in the 1880s (Cornes and Jones 2011; Matulla et al. 2008).

4.2.1 Climate of the 18th and 19th Centuries

4.2.1.1 Mean Climate of the 1700–1890 Period

Brückner's study period 1700–1890 represents the last phase of the Little Ice Age. Figure 4.15 shows maps of the mean temperature anomalies of this period (with reference to 1985–2005; following IPCC 2013) from the CCC400 simulations and reconstructions (Mann et al. 2009a). This figure suggests that globally, the 1700–1890 subperiod was approximately 0.5–2 °C cooler on average than present. Reconstructions and model simulations closely agree. CCC400 is driven by the

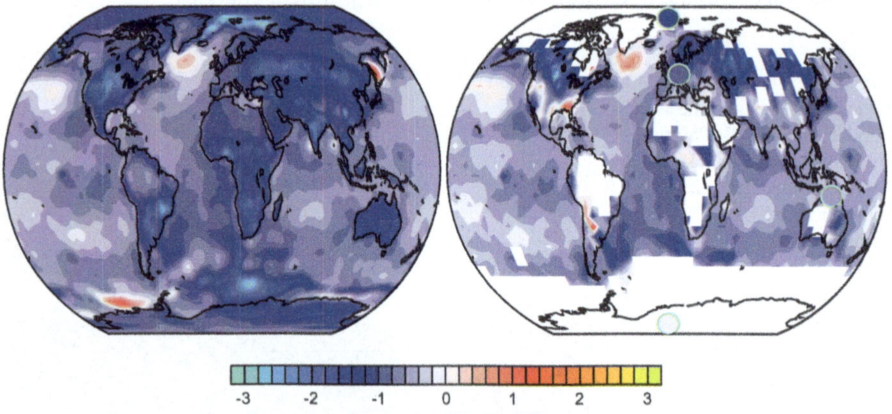

Fig. 4.15 Global map of the annual mean temperature difference in 1700–1890 relative to 1985–2005 in the (*left*) CCC400 simulations and (*right*) reconstructions (Mann et al. 2009b). The *dots* indicate temperature anomalies from the PAGES 2k reconstructions (Fig. 4.5; only shown if they reach at least to 2001)

Mann et al. (2009a) sea-surface temperatures. Hence, the agreement over the oceans is good by construction. However, both datasets also agree that the cooling was strongest over the continents. Furthermore, the cooling over North America and Eurasia in both data sets agrees with other reconstructions (e.g., Luterbacher et al. 2004). The mean temperature anomalies from the PAGES 2k reconstructions (dots) suggest a good agreement also for Australasia and the Arctic, but not over Antarctica where—according to the PAGES 2k reconstructions—temperatures were similar to those at present.

This figure implies regional changes in the meridional temperature gradient (e.g., over the North Atlantic or North Pacific), as well as the land–sea contrast. Both might have affected circulation systems such as monsoons or storm tracks (see Sect. 3.1). In fact, in terms of atmospheric circulation with 1985–2005 as a reference, the 1700–1890 subperiod featured a decreased strength of the northern Hadley cell in boreal winter, but an increased strength of the Pacific Walker cell. In simpler terms, there was a slightly different mode of the machinery.

4.2.1.2 Triggers of the Little Ice Age

What could have triggered the Little Ice Age? In Sect. 4.1.5, we addressed the drivers of global temperatures over the past 400 years using a simple regression model. In the following, I will briefly look even farther back—to the start of the Little Ice Age.

For a long time, a decrease of solar irradiance was considered the main culprit for a cold climate. In fact, this may be true (a schematic of the suggested mechanisms is given in Fig. 4.16). Solar activity was low from 1260 to 1330, from 1390 to 1560, and from 1645 to 1715 (Steinhilber et al. 2009). However, the magnitudes of these minima are difficult to quantify (see Sect. 3.3). Several studies show agreement

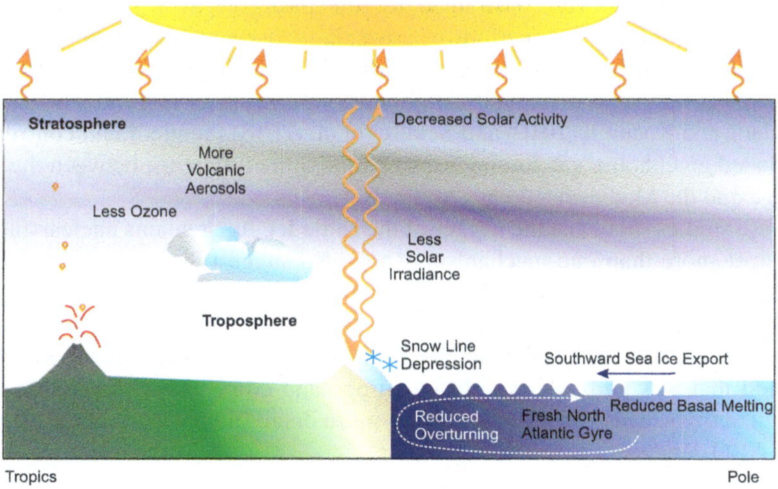

Fig. 4.16 Schematic view of the possible processes triggering and sustaining the Little Ice Age

between temperature proxies and solar activity (e.g., Eichler et al. 2009b). However, reconstructions are also compatible with smaller solar effects (see also Schurer et al. 2014). Models can reproduce the Little Ice Age climate for a large range of solar forcing conditions (Ammann et al. 2007), but not necessarily for an extremely large forcing (Feulner 2011).

Thus, other factors likely contributed to—or even dominated—the climate response during the Little Ice Age. The most important factor is volcanic eruption. The temperature response to an individual eruption lasts only 2–3 years. However, periods of high volcanic activity may possibly affect the Atlantic meridional overturning circulation (Stenchikov et al. 2009, see Sect. 3.3.1) and ocean heat content in general. In this way, volcanoes can influence the North Atlantic and European climate on longer time scales. Miller et al. (2012) showed that volcanic eruptions are sufficient to trigger a Little Ice Age in model simulations. They posit a sustained sea ice–ocean feedback mechanism in the North Atlantic that causes cold summers and a snow line depression. In their simulations, cool summers following volcanic eruptions led to increased southward sea ice export. In turn, this caused an anomalously cold and fresh North Atlantic subpolar gyre. Reduced basal sea ice melt led to enhanced sea ice extent. Furthermore, poleward oceanic heat transport was reduced in the model (Miller et al. 2012). In this way, the Samalas eruption of 1257 (or a sequence of strong eruptions) could have triggered the Little Ice Age (see also Sects. 3.1.4 and 3.3.1).

Solar minima, which last longer than the negative forcing from volcanic eruptions, can also induce changes in the ocean. Feedback processes can further amplify the effect of either forcing. Lehner et al. (2013) proposed that the Little Ice Age climate was maintained by a feedback between sea-ice and ocean circulation through salinity changes in the Nordic seas. Moffa-Sánchez et al. (2014) suggested that the Little Ice Age was linked to solar irradiance changes amplified via a feedback of the Atlantic meridional overturning circulation. A major obstacle in determining the role of solar or volcanic forcing is that the magnitudes of both forcings, particularly solar forcing, are poorly constrained for the past (see Sect. 3.2). Furthermore, during the last millennium, periods of low solar activity coincided with periods of high volcanic activity, which makes it difficult to distinguish the two. However, other factors may help to explain temperature changes prior to 1900. For instance, Jungclaus and Koenigk (2010) and Schurer et al. (2013) found a small contribution of reduced greenhouse gas forcing to the temperature difference between the Little Ice Age and the Medieval Climate Anomaly.

These studies show that the trigger of the Little Ice Age remains unclear and may consist of more than one mechanism. They also suggest that the response in the North Atlantic region may be different from the global response due to the complex response of the machinery.

A starting point for analysing the Little Ice Age is the Maunder Minimum, a period that was particularly cold in Europe and coincided with a minimum in sunspot activity. This period is studied in more detail in the next section on behalf of EKF400 data.

Fig. 4.17 Drawing of the procession on frozen Lake Constance in 1830 (Wikimedia Commons)

4.2.2 The Maunder Minimum: A Compelling Case for Solar Forcing?

The winter of 1694/1695 was harsh in central Europe (Glaser 2008; Pfister 1999). Lake Constance froze over, and in a solemn procession, a wooden statue of John the Apostle was brought from Hagnau on the northern shore to Münsterlingen on the southern shore. This tradition continues today. Every time the lake freezes and the ice is sufficiently thick to carry a procession, the statue crosses the lake and remains there until the next freezing event (Fig. 4.17 shows the procession in 1830 from Münsterlingen to Hagnau. Presently the statue is in Münsterlingen).

The winter of 1694/1695 was not the only harsh winter around 1700 (Glaser 2008; Pfister 1999). Johann Laurenz Bünti, a Swiss chronicler (Bünti 1973), wrote:

> "In January 1709 it was so cold for three weeks that no old man has ever seen such a long cold spell. (…) Many people perished because of the cold, got sick and died etc. The large rivers were frozen and ice covered; when the ice broke the floating ice caused undescribable damage (with destruction of bridges, also houses, flooding of many villages and fields, also drowning of many people and cattle)." (translated from Bünti 1973, p. 180)

In the mid- and late 17th century, European lakes and rivers froze frequently and temperatures were particularly low. This marked the peak of the Little Ice Age in central Europe.

These cold winters have drawn the attention of climate scientists for many decades. Based on historical observations of the sun, aurorae, and ^{14}C, Eddy (1976) confirmed that solar activity was very low and called this period the "Maunder Minimum" of solar activity (1645–1715). Is this proof of the important role of the sun on earth's climate? Eddy's paper revived old discussions on sun–climate relations. However, volcanic activity was also high (Fig. 4.13). In the following, I will focus on the last part of the Maunder Minimum—the "Late Maunder Minimum" (1685–1715). This was a particularly cold period (note that one of Brückner's cold phases was centred around 1700) with abundant documentary evidence and even sparse instrumental observations (see Sect. 2.8.1). However, most of the documentary and all of the instrumental information is from Europe, and only few proxies provide reliable winter temperatures.

4.2.2.1 The Late Maunder Minimum: Rock Bottom of the Little Ice Age Climate

The Late Maunder Minimum climate has been the target of many studies. Luterbacher et al. (2001) analysed reconstructed temperature fields in Europe. They found a strong cooling in central and eastern Europe and a weak cooling along the Atlantic coast. Winter sea-level pressure was increased in northeastern Europe but decreased in central Europe and the western Mediterranean, with more frequent blocking situations. In spring, the Atlantic storm track was displaced to the south. In summer, there were wet conditions across all Europe (Luterbacher et al. 2001). In the following I subdivide the Late Maunder Minimum into the periods 1686–1700 and 1701–1715 (see also Zorita et al. 2004).

Several model-based studies were able to reproduce some of these features. Shindell et al. (2001) obtained a similar temperature anomaly pattern with an atmospheric model coupled to a mixed layer ocean and a parameterised response to ozone changes. They attributed the climatic conditions to solar irradiance changes. These operate via changes in stratospheric ozone, a slowdown of the stratospheric circulation, and downward propagation (see Sects. 3.1.3 and 3.3.2, Langematz et al. 2005). Zorita et al. (2004) were also able to reproduce the temperature anomaly pattern in a coupled climate model with reduced solar irradiance. They found a temperature drop and reduced high-latitude salinity in the North Atlantic, which they attributed to reduced wind-stress forcing. So, how did the sun affect climate? Via the stratosphere or the oceans? Or did it not?

The difference in total solar irradiance between the Maunder Minimum and the present is not known. Estimates range −6 to 0 W m^{-2} (Shapiro et al. 2011; Steinhilber et al. 2012; see Sect. 3.3.2). The spectral pattern of the decrease, which might be relevant, is also unknown—further adding to the uncertainty. Several strong volcanic eruptions also contributed to a negative forcing (especially Serua in 1694). However, these data are also burdened by large uncertainties. Our simple forcing–temperature regression (Eq. 4.1) indicates a forced contribution (relative to 1700–1890) to global land surface temperatures in 1686–1700 and 1701–1715 of −0.24 and −0.13 °C, respectively. The main contributors were variations in solar irradiance (−0.12 and −0.10 °C) and greenhouse gases (−0.08 °C). The difference (0.09 °C) between the two sub-periods is due to volcanic forcing. However, the two subperiods are short and internal variability might have played an important role.

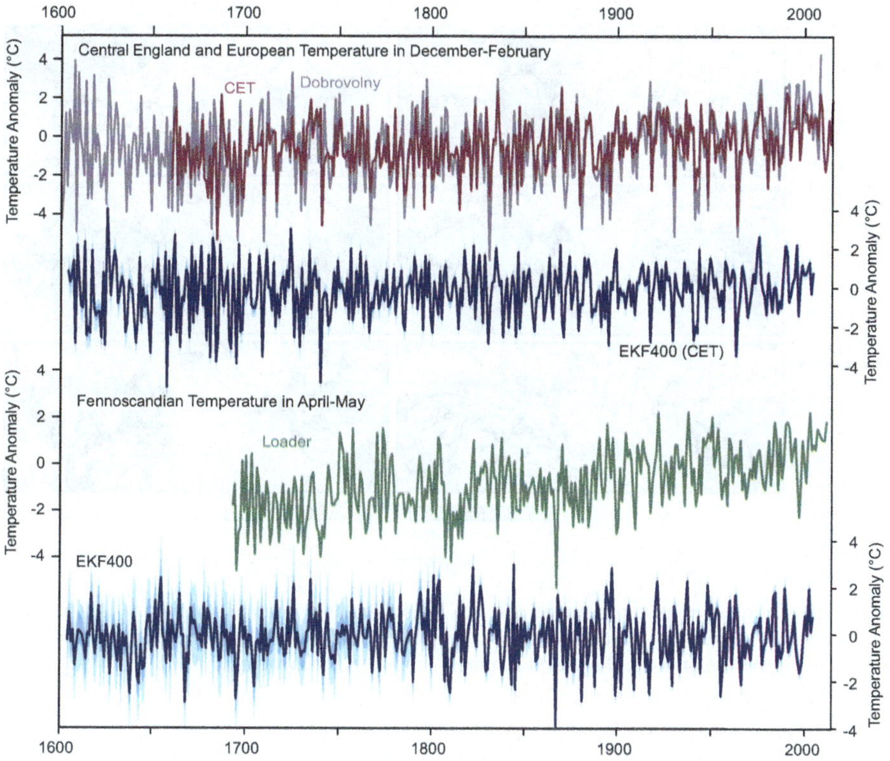

Fig. 4.18 Temperature anomalies in (top) central England in winter from instrumental data (Parker et al. 1992), (middle) in the Czech land from documentary data (Dobrovolný et al. 2010) and (bottom) Fennoscandia in April and May from reconstructions by Loader et al. (2011) based on the freezing of rivers, and EKF400 (ensemble mean is given in *blue* with 50 % and 90 % ranges shaded). All series are anomalies from their 1900–1990 average

The PAGES 2k reconstructions show that the Late Maunder Minimum was a large-scale event that appears in the polar reconstructions as well as in the series for Europe, Asia, and North America. However, it was not reflected in reconstructions for Australasia or South America. For Europe, the Late Maunder Minimum also appears in Brückner's figure (Fig. 4.3) as a period with cold winters and late grape harvest dates. However, cold winters were even more frequent in the early and mid-17th century. Reconstructions of summer temperatures in central Europe (Fig. 4.8) also show a slight dip during the Late Maunder Minimum. Winter temperatures were low in central Europe (e.g., in the Czech lands, Fig. 4.18) and in central England. Ice duration in Fennoscandia (a series that starts in 1693) shows frequent cool springs, but also warm ones. EKF400 and proxies are consistent (although the Late Maunder Minimum appears somewhat less pronounced), allowing us to explore this event in more detail.

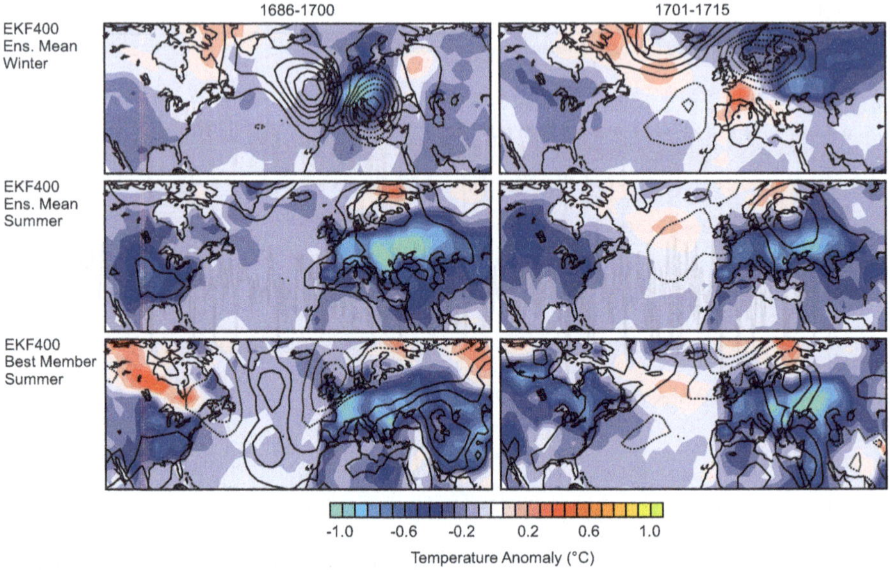

Fig. 4.19 Differences in temperature and sea-level pressure (contour interval is 2 hPa, 0 hPa contour is not shown, and negative *contours* are *dashed*) in EKF400 (best member and ensemble mean) in boreal winter (Dec.–Feb.) and summer (Jun.–Aug.) relative to 1700–1890

4.2.2.2 The Late Maunder Minimum in EKF400

North America–Europe is the region best covered by tree ring, documentary, and instrumental data (Sect. 2.9, note that winter data are only available over Europe). The EKF400 results for this sector imply that summers in both periods were cool over North America and Europe between 30°and 60°N (Fig. 4.19). Winter conditions were cold over North America in both periods. The winters 1686–1700 were cold in western and central Europe, with light positive temperature anomalies in northeastern Europe. Positive pressure anomalies to the west of France indicate frequent blocking; negative anomalies appear over the Adriatic Sea. The winters 1701–1715 were cold over northeastern Europe, but warm over western Europe (despite the cold winter of 1708/1709). This feature also appears in other reconstructions (Luterbacher et al. 2004). The NAO in the 1701–1715 winters was negative.

Do these results favour either solar or volcanic forcing? During 1686–1700, which featured more volcanic eruptions, the winter NAO index was neutral – not typical for volcanic forcing. Alternatively, during 1701–1715, which had fewer eruptions, the winter NAO index was negative, as expected from low solar activity. The cool summers could have been the result of either forcing. Hence, results are not clear and internal variability of the machinery must have been at work.

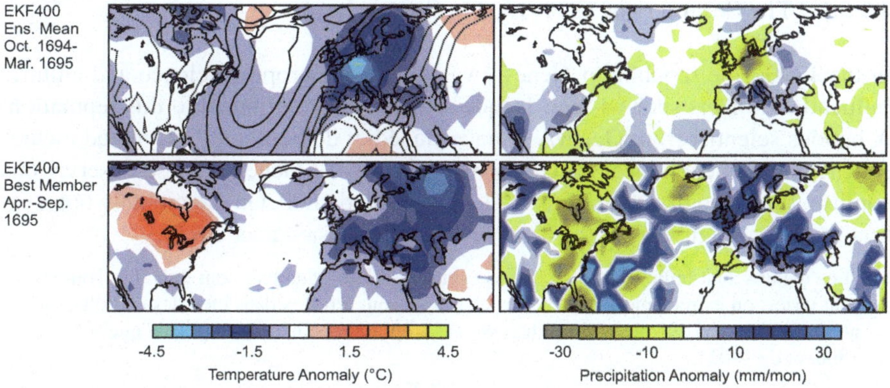

Fig. 4.20 Cold (Oct.–Mar.) and warm season (Apr.–Sep.) anomalies of temperature, sea-level pressure (contour interval is 4 hPa, 0 hPa contour is not shown, and negative *contours* are *dashed*), and precipitation in 1694/1695 in the EKF400 best ensemble members relative to 1700–1890

Let us return to the winter of 1694/1695 and the following summer, which were cold in Switzerland and southern Germany (Fig. 4.20). EKF400 is in good agreement with documentary information; there was a cold winter over Central Europe. The summer (shown with the best ensemble member) indicates cool and rainy conditions over large parts of Europe, including the south-central region. The summer followed the Serua eruption of 1694 and the described features are in good agreement with the expected effects of volcanic eruptions. At least for this specific year, volcanic forcing might have played a role.

4.2.3 *The 1790s: Temporary Warmth and Strong Tropical Circulation*

After the end of the Maunder Minimum, solar activity increased, and only few tropical volcanic eruptions occurred. Tropical sea-surface temperature increased, and the boreal winter Hadley circulation strengthened. Both peaked around 1800. Global temperatures also reached an intermediate peak in the 1790s.

Concurrently, the enlightenment transformed the way people perceived weather and climate. Observations and measurements became much more frequent (Sect. 2.2). Major changes to societies also were underway. The United States obtained independence and the French revolution brought the rise of democracy and nationalism. In the following we take a look at the 1790s, a period of temporary warmth.

4.2.3.1 The Summers of 1800–1804 in Switzerland

In the 1790s, the republic of Geneva was a hub of European science and culture. Influential thinkers such as Jean-Jacques Rousseau[7] contributed to this reputation. A famous scientist from Geneva, Horace Bénédict de Saussure,[8] promoted meteorological and phenological observations in the city of Geneva. These observations confirm warm and dry conditions in Geneva around 1800. On the monthly observation sheet for August 1803, the observer Marc-Auguste Pictet wrote

> "Les regains sont brûlés par la sécheresse constante. Les pâturages sont nuls. Les sources sont taries, ou extrêmement basses. Les charrures sont arrêtées dans bien des endroits. On ne se souvient pas d'avoir vû de sécheresse aussi opiniâtre: elle est plus grande que celle de 1800, et de 1802."

Twice daily meteorological data from Geneva were recently digitised by our group. The data allow constructing daily weather types back to the 18th century. As the station moved several times, we used the homogeneous record during 1799–1821 (see Auchmann et al. 2012) to analyse warm summers around 1800. The classification uses pressure, pressure tendency, and wind direction to distinguish 16 weather types (see Sect. 4.2.4 for more details). The 16 types were grouped into 5 main types (high and low pressure, northeasterly and southwesterly winds, and the rest) and 5-yr periods were analysed, as done by Brückner. Results for the summer half-years 1800–1804 (predominantly warm and dry summers) and 1805–1809 (cool and wet) are shown in Fig. 4.21. Clearly, the warm pentad featured fewer low pressure and more high pressure situations than the cold period. Also,

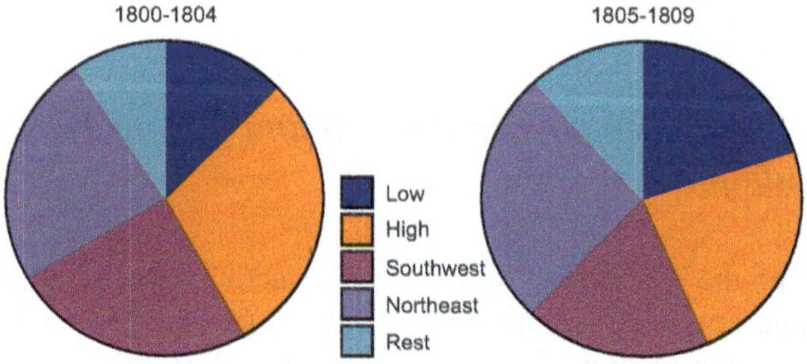

Fig. 4.21 Frequency of weather types in Geneva in April–September (*left*) 1800–1804 and (*right*) 1805–1809. The classification is based on Auchmann et al. (2012) and summarising high and low-pressure classes (including frontal passages) as well as (among the remaining classes) those with southwesterly or north-to-northeasterly winds

---

[7]Jean-Jacques Rousseau, 1712–1778, was a political philosopher whose work influenced the French revolution and the romantic movement.

[8]Horace Bénédict de Saussure, 1740–1799, was a physicist and traveller. He was among the first to study the vertical structure of the atmosphere by climbing Mont Blanc (4,808 m asl) with a barometer.

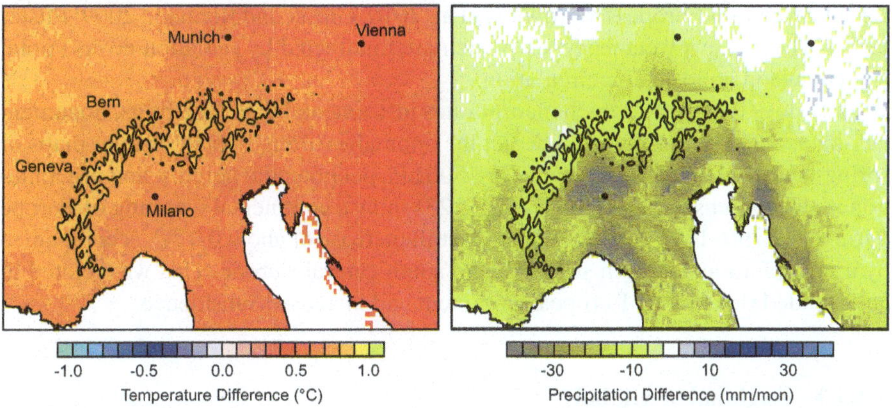

Fig. 4.22 Differences in April–September (*left*) temperature and (*right*) precipitation between the periods 1800–1804 (1801–1804 for precipitation) and 1805–1809 (Data source: HISTALP). Topography above 2 km is marked

southwesterly flow situations, which usually bring warm air to Geneva, were more frequent during the summers of 1800–1804 than in subsequent years. Thus, the differences in temperature (1.3 °C) and precipitation (15 %) between the periods can be partly explained by a shift in the weather types. Conversely, the weather types independently confirm the anomalous warmth.

4.2.3.2 Warming in Europe

The anomalous weather was not restricted to western Switzerland. Gridded temperature and precipitation data from HISTALP reach back to 1781 and 1801, respectively. Although the first drought year is missing in the case of precipitation, the data provide a spatial overview (Fig. 4.22). The difference fields clearly show warm and dry conditions in the entire domain relative to the subsequent 5-yr period (Auer et al. 2007). The strongest drying is found in the southern Alps and northern Mediterranean. However, high summer temperatures are not supported by local proxy records. This phenomenon is known as early the instrumental paradox (see also Frank et al. 2007; Hiebl 2006). In fact, some instrumental data from Austria were later revised due to a suspected radiation bias (Böhm et al. 2010).

Warm summers around 1800 are also known from farther away sites, such as Prague (Brázdil et al. 2013) and Scandinavia (Moberg et al. 2003). In Alpine summer temperature reconstructions (Fig. 4.8), slightly elevated temperatures appear around the late 1780s to the 1800s and are roughly comparable to the 1930s but far from present-day values. Can we trust the instrumental series or are they all affected by radiation biases? Warm summers around 1800 are independently confirmed by documentary data (Fig. 4.8). Warm springs are indicated by early thawing dates of Fennoscandian rivers (Fig. 4.18). Documentary data also indicate dry summers and springs in central Europe (Dobrovolný et al. 2010; Glaser and Riemann 2009; Pfister 1999). The Geneva weather types (which are unaffected by radiation errors) point to a change in atmospheric circulation.

How does EKF400—which incorporates instrumental data, tree rings, documentary data, and model forcings—depict the anomalies around 1800? Spatial maps of spring-summer temperature anomalies, 500 hPa geopotential heights, and precipitation in the best ensemble member (Fig. 4.23) show a continental warming of Europe during the 1790–1804 period, with an overall increase of about 0.4 °C, accompanied by a precipitation deficit in southern and south-central Europe. This was related to an extended ridge to the European continent and increased subsidence.

4.2.3.3 Regional or Global?

Anomalous climatic conditions were not restricted to Europe. Reconstructed anomalies of the Palmer Drought Severity Index for North America and Asia are shown with summer precipitation anomalies for Europe in Fig. 4.23. Although the results are rather noisy, several regions with a higher frequency of drought can be found. Literature provides further evidence of drought or, in the western U.S. and central Asia, a pluvial. Cook et al. (2010) found droughts in Asia and Rodysill et al. (2013) reported droughts in Mexico during this period. Were these events connected?

Zonally, EKF400 exhibits a fairly symmetric picture. This implies that regional anomalies can be seen in a global context. Precipitation increased in the Inner Tropics whereas a decrease occurred in subtropical regions. Specifically, drought conditions are seen in Asia, parts of Europe, and Mexico in the best ensemble member. However, the pattern remains noisy. Precipitation in EKF400 is almost identical to CCC400, confirming that the pattern is reproducible from the sea-surface temperatures and forcings.

The pattern for temperature is more uniform than for precipitation. The warming in EKF400 encompasses Europe. A global and hemispheric temperature maximum is also seen in many reconstructions (Figs. 4.1 and 4.4). In fact, one of Brückner's global warm phases was centred around 1795.

4.2.3.4 Causes of the 1790s Warmth

Can the temperature difference from 0.1 °C (global land surface temperature) to 0.4 °C (European temperature) between 1790–1804 and 1700–1890 be explained by external forcing? Based on our simple forcing-temperature regression (i.e., the fit to global land surface temperature observations; Fig. 4.13), the contribution of external forcings was only around 0.03 °C. This is mostly due to the lack of volcanic forcing. Though warmer than all previous decades in the 18th century, the 1790s temperature in the fitted curves was quickly surpassed in the 19th century. However, CCC400 ensemble-mean land temperatures deviate from the fitted curve and quantitatively capture the global warming. They also reproduce an enhanced summer warming over Europe of around 0.15 °C. This indicates that sea-surface temperatures may

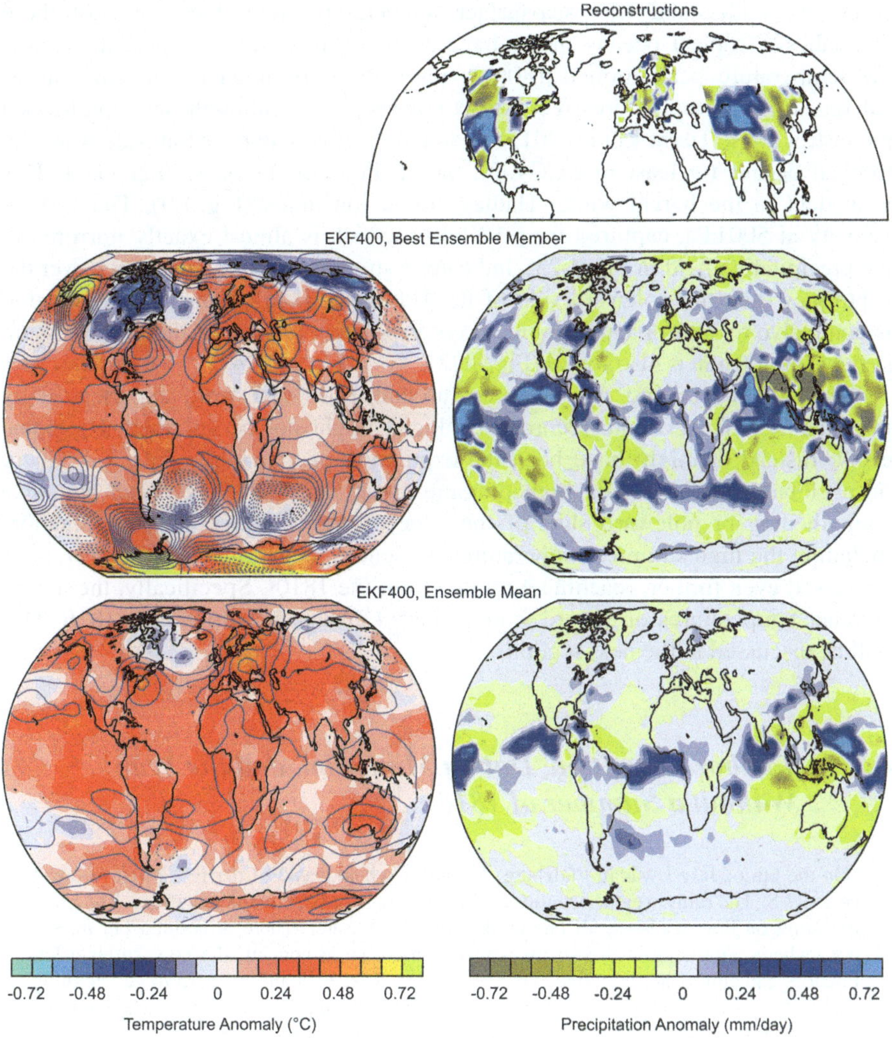

Fig. 4.23 Anomaly fields of March–August (*left*) surface air temperatures overlaid with 500 hPa geopotential heights (contour spacing is 3 gpm, zero contour not shown) and (*right*) precipitation in 1790–1804 relative to 1700–1890. These were derived from (*top*) reconstructions, (*middle*) the EKF400 best ensemble member, and (*bottom*) the EKF400 ensemble mean. Reconstructions for North America and Asia give the Palmer Drought Severity Index (same scale as precipitation)

have played a role. However, the remaining 0.2–0.3 °C of summer warming found in reconstructions must be explained by the internal variability of the atmosphere or a response to forcings that is not captured (e.g., local responses or feedbacks) in CCC400. Note that this analysis carries large uncertainties on all sides (forcings, model response, observations, assimilation, and attribution statistics).

Do we understand why sea-surface temperatures reproduce some of these anomalies? Tropical oceans might have been warm during this period. In fact, the temperature peak around 1800 is particularly pronounced in some tropical sea-surface temperatures reconstructions (Fig. 4.7), although new coral-based reconstructions (Tierney et al. 2015) do not show this feature. Alongside warming tropical oceans (at least in CCC400), the Hadley circulation strengthened. This is evident in the boreal winter Hadley circulation index (Fig. 4.9). The vertical velocity at 500 hPa, captured from EKF400 (which is almost exactly mirrored in the precipitation field in Fig. 4.23), indicates a strengthening of subsidence over the subtropics (i.e., the poleward side of the Hadley cell) in summer. This situation is conducive to subtropical drought. A corresponding change of the high pressure systems at 500 hPa is also found (Fig. 4.23).

Thus, the warm and dry summers around 1800 that were experienced by inhabitants of Geneva and recorded by Pictet might not be an isolated regional event. Instead, it might have been the expression of a change of the machinery on a global scale. Frequent high pressure situations might have been part of a strengthening or poleward shift of the edge of the Hadley cell. This situation, including the high summer temperatures in Geneva, ended in 1805. Temperatures decreased even further, reaching a minimum in the 1810s. Specifically, the lowest summer temperatures on record were achieved during the summer of 1816. This will be discussed in the next section.

4.2.4 Tambora Eruption, Dalton Minimum, and the Year Without a Summer of 1816

> "On the 6th of July I was in Pfalzburg. Yesterday wheat was 23 f per Frl, a loaf of bread costs 27 S. The rainy weather continues. The hay has not been made anywhere. The grass is rotting on the meadows, all mountains are full of water. There is nothing but misery everywhere. The beggars are so frequent that there is no council, the poor suffer a lot. Potatoes can almost not be found. The sester costs 30 to 34 S and are hard to get. Nothing can grow, it is always too cold, I do not know what will happen until harvest which will be very late. Fortunately, as to what concerns ourselves, there is no paucity."

This quote is from the unpublished diary of Johann Peter Hoffmann (1753–1842), a farmer and local magistrate in Alsace. It shows the hardship endured in central Europe during the summer of 1816. Consequently, 1816 is known as a "Year Without a Summer". Adverse weather caused widespread crop failure and severe famine (Pfister 1999; Luterbacher and Pfister 2015) and people reportedly resorted to eating grass from the meadows (Fig. 4.24). Malnutrition caused an increase in mortality. However, demographic data indicate pronounced local differences. This implies that governance was at least as important as climate (Krämer 2015). Nevertheless, the Year Without a Summer of 1816 highlights the societal importance of abrupt climatic events.

Fig. 4.24 Painting from Anna Barbara Giezendanner (1831–1905), showing people eating grass during famine in Switzerland

4.2.4.1 Anatomy of the Year Without a Summer

There is never a good time for a climate disaster. However, the Year Without a Summer of 1816 came at a particularly bad time—in the wake of the Napoleonic wars and the midst of politically troubled times—which might have worsened the societal impacts of the climate event (Oppenheimer 2015). It also affected the quality of observations. Many instrumental series have gaps in these years. Nevertheless, based on early instrumental measurements, documentary data, and proxies, it is possible to obtain an overview of the event. Temperatures were very low in central and western Europe, eastern North America (while in Europe several Years Without a Summer are known, 1816 is the only one for North America), and Asia. The summer of 1816 was particularly cold in the European Alps (Fig. 4.8). Tree ring width was reduced in eastern North America and across Eurasia, but normal or even enhanced in Scandinavia or North Africa (see also Briffa et al. 1998). Little evidence is found for an imprint in the Southern Hemisphere. Interestingly, Brückner does not mention the year 1816 in his book because he exclusively considered 5-yr averages. However, he listed 1815 as the centre of a cold and wet period.

In central Europe, cold conditions persisted throughout the entire summer (Harington 1992; Oppenheimer 2003). In western Europe, rainfall was much higher than normal, as expressed in the opening quote of this section. Snow fall repeatedly continued to reach Alpine valleys until early summer (Pfister 1999). Harvest yields were very low and late; grapes in northwestern Switzerland were harvested on 10 November (Meier et al. 2007), the latest date on record. In contrast, on the

Iberian Peninsula, precipitation was low in some places (Trigo et al. 2009). In Scandinavia, temperatures were near normal. In Russia, conditions were warmer and drier than normal. The west–east difference was noted in contemporary European newspapers. In the eastern U.S., storms, cold surges, and late snowfall (Chenoweth 2009) characterised the summer of 1816 (see also Klingaman and Klingaman 2013).

4.2.4.2 Causes

What caused the Year Without a Summer? At that time, the causes of cold and rainy summers were unknown. Newspapers speculated about sunspots and others blamed lightning conductors (see Bodenmann et al. 2011). Obscure astronomical theories were advocated, while chroniclers pointed to historical precedents. In the following summer, large masses of drifting ice were found in the North Atlantic and were held responsible for the cold weather of 1816 in Europe (Barrow 1819). Today, the Year Without a Summer of 1816 is mainly attributed to the April 1815 eruption of the volcano Tambora in Indonesia, which injected huge amounts of sulphur into the stratosphere (e.g., Stommel and Stommel 1979; Stothers 1984).

Other mechanisms or factors may have contributed to the Year Without a Summer of 1816. For instance, a strong eruption in 1808/1809 (Guevara-Murua et al. (2014), see also Sect. 2.9) may have set the stage for a sustained ocean cooling (Stenchikov et al. 2009). Furthermore, the Tambora eruption fell into a secular phase of low solar activity called the "Dalton Minimum". This may have contributed to cool summers around 1816 (Anet et al. 2014). Another sequence of two major volcanic eruptions over a short time period occurred in the 1830s and resulted in similar—though smaller—imprints in continental-to-hemispheric climate series (Figs. 4.4–4.7). In this sense, the Tambora eruption and the unknown 1808/1809 eruption were typical for the first half of the 19th century.

Do we understand the mechanism? Sea-level pressure reconstructions (Luterbacher et al. 2002) indicate that during the winter of 1815/1816 the NAO was neutral. The summer of 1816 was characterised by negative pressure anomalies off of Great Britain (Trigo et al. 2009). In the following, I add to this by analysing, on the one hand, EKF400. On the other hand, I will explore daily weather from early instrumental data.

4.2.4.3 The Year Without a Summer of 1816 in EKF400 and Early Instrumental Data

Let us return to Geneva where Marc-Auguste Pictet was continuing his observations and growing concerned about a negative temperature trend over the preceding 20 years in his observations (Pictet 1818). Geneva is a suitable place to start our analysis for several reasons. In addition to having weather observations, Geneva lies near the heart of the 1816 climatic anomalies. Comparing the summer of 1816 weather data from Geneva with a contemporary homogeneous reference period

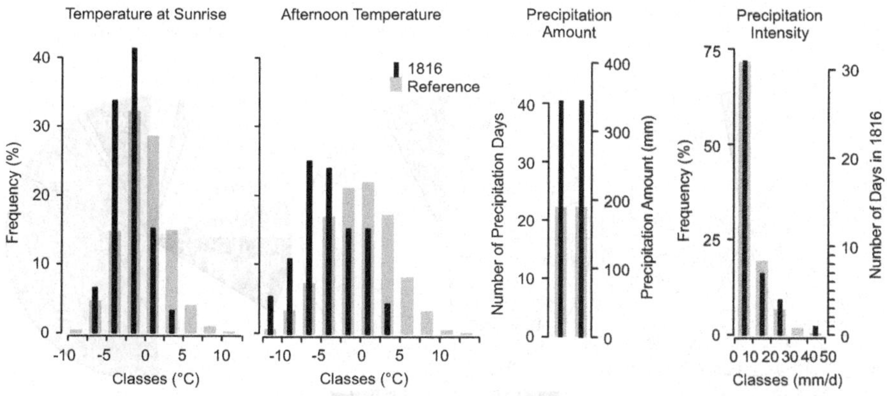

Fig. 4.25 Statistics of daily weather from Geneva from June to August 1816 relative to the reference period 1799–1821. (*left*) and (second from left) Histograms of temperature anomalies (with respect to the 1799–1821 mean seasonal cycle) at sunrise and 2 p.m. local time (second from *right*) Precipitation amount and frequency, and (*right*) intensity (precipitation on days with >0.1 mm of precipitation) (Auchmann et al. 2012)

(1799–1821), we found that the low daily mean temperatures were mainly due to the afternoon observation. In the sunrise observation, a much smaller signal was found (Auchmann et al. 2012). With respect to variability, the afternoon measurement (Fig. 4.25; second from left) shows a shift in the distribution, while the early morning measurement (Fig. 4.25; left) shows an absence of warm mornings, but no increase in extremely cold mornings. Also, there is an overall reduced variability. These features were related to an increase in cloud cover, which is found in the observations. Increased cloud cover reduced incoming radiation during the day but prevented clear, cold nights.

With respect to precipitation, the 80 % increase in the summer of 1816 can be explained by an increase in the frequency (Fig. 4.25; second from right). However, the intensity distribution (Fig. 4.25; right) did not change.

We also analysed the daily weather type classification in Geneva with 16 classes (Auchmann et al. 2012) for the reference period and for the summer of 1816. High pressure situations were almost completely absent in 1816, whereas some cyclonic situations tripled or even quadrupled. This explains a considerable fraction of the temperature and precipitation differences and becomes apparent when colour coding the classes according to their temperature anomalies in the reference period (Fig. 4.26; middle). In Fig. 4.26 (right), colour coding is according to the observed anomalies and shows that temperatures in each weather type were lower in 1816 than the reference period.

This suggests that a general cooling as well as a change in atmospheric circulation occurred. Both can be seen in EKF400, which shows a low pressure anomaly over Britain (Fig. 4.27) in the summer of 1816. The weather types suggest that Switzerland was in the track of depressions moving across western Europe from around April to July, which is confirmed by European subdaily pressure data

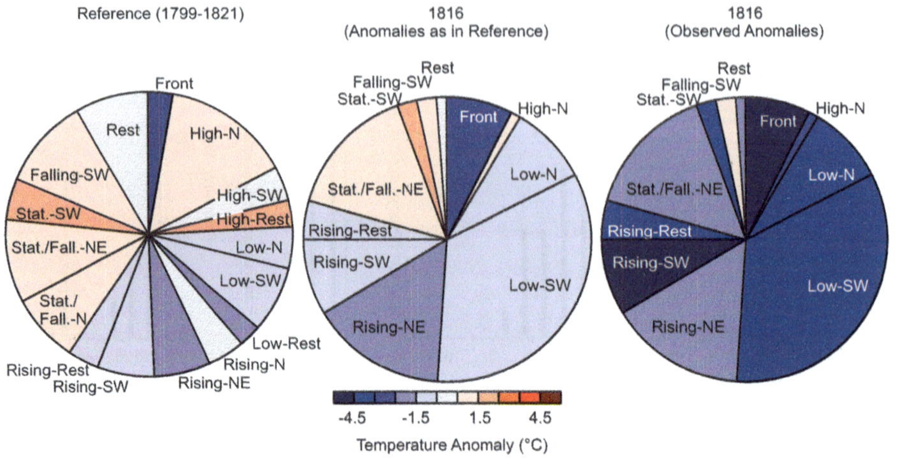

Fig. 4.26 Frequency of weather types in Geneva from June to August (*left*) 1799–1821 (excluding volcanic years) and (*middle*), (*right*) 1816. The colours in the (*left*) and (*middle*) denote the average 2 p.m. temperature of each weather class in 1798–1821. The colours in (*right*) uses the average temperature per weather class for 1816 (based on Auchmann et al. 2012)

(Brugnara et al. 2015). Only August brought a brief relief. With respect to cooling, CCC400 already shows a moderate cooling and a precipitation surplus (i.e., these features are forced). However, after assimilation, EKF400 shows much stronger anomalies.

The relation of this anomalous circulation and rainfall to the eruption is unknown. CCC400 simulations suggest that rainy summertime conditions in south-central Europe following tropical volcanic eruptions might be a consequence of weakened African and Asian monsoons (Sect. 3.2.1), which in turn is related to reduced short-wave radiation following an eruption (Wegmann et al. 2014). However, further evidence is needed to support this hypothesis for the case of the Tambora eruption.

4.2.4.4 The Mistery Continues

Recent work has shed some light on the causes of the Year Without a Summer of 1816. However, there are still several major uncertainties, for instance with respect to the tree-ring signature (Anchukaitis et al. 2012; Büntgen et al. 2015). Another uncertainty is the aerosol amount in the Northern Hemisphere. Because the eruption occurred in April a large fraction of the aerosol cloud may have moved to the Southern Hemisphere. This is also the case in our model-based reconstruction of volcanic forcing (Arfeuille et al. 2014) as displayed in Fig. 4.28 (right). However, temperature reconstructions for the Southern Hemisphere (Neukom et al. 2014a) show little cooling in 1816 or during many other major volcanic eruptions. Therefore, this topic needs further research.

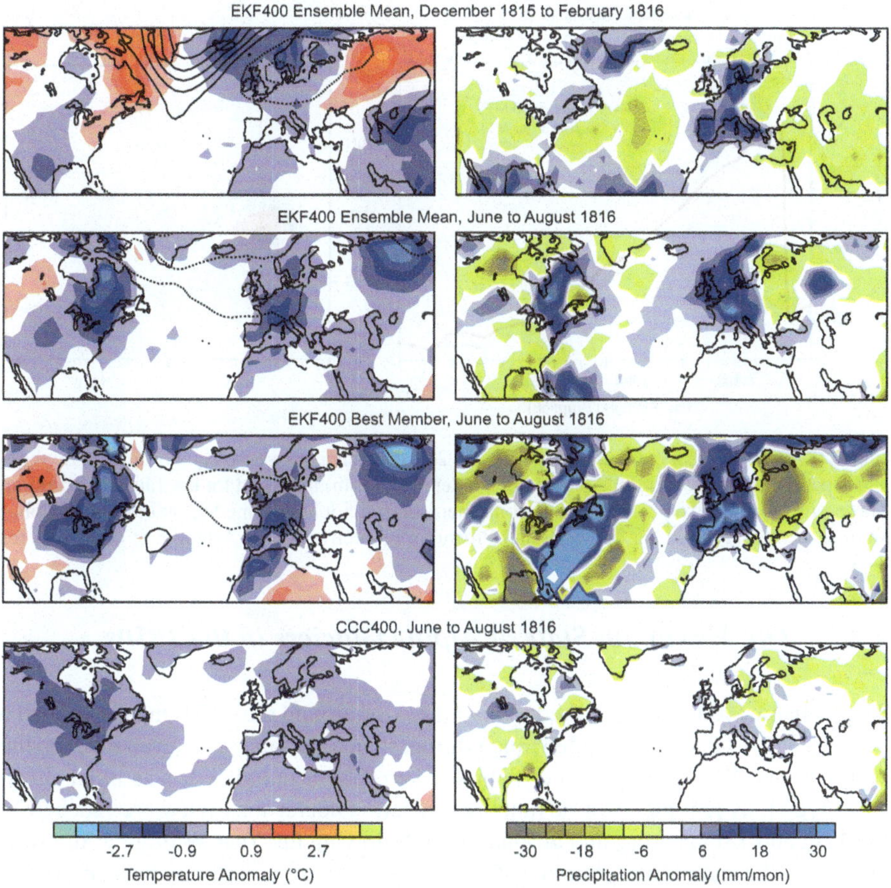

Fig. 4.27 Anomaly fields of temperature, sea-level pressure (contour spacing is 2 hPa, 0 hPa contour is not shown, and negative contours are dashed) and precipitation in EKF400 (best member and ensemble mean) in boreal winter (Dec.–Feb.) and summer (Jun.–Aug.) of 1816 relative to 1700–1890

Another uncertainty concerns the mass and size distribution of the volcanic aerosols. In Fig. 4.28 (left), modelled aerosol extinction for Tambora and Pinatubo are compared (note the logarithmic scale). Although aerosol extinction immediately following Tambora reaches higher values than following Pinatubo, the decrease occurs more rapidly due to coagulation (large particles scatter less efficiently, see Sect. 3.3.1). Around 2.5 years later, Tambora aerosol extinction falls below Pinatubo values because gravitational settling removes particles faster. Do we expect the effects of a large eruption to be shorter lasting than those for a small eruption? Perhaps Tambora could provide answers. More than 200 years after the eruption, there are still opportunities to learn from this event.

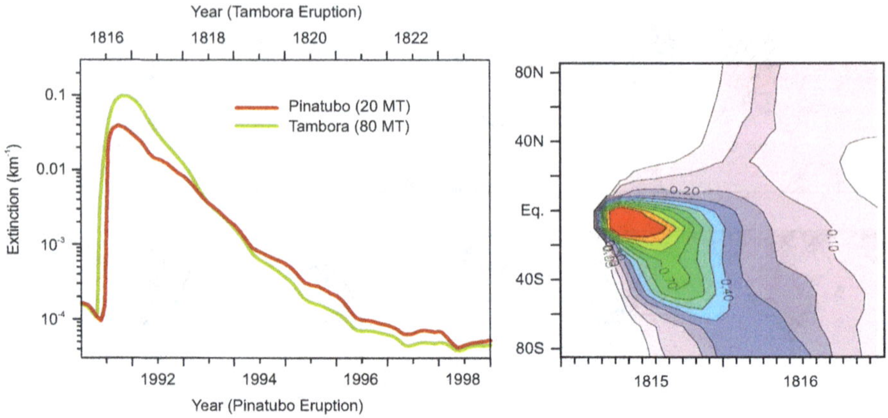

Fig. 4.28 (*left*) Aerosol extinction (1.02 μm) at 20-km altitude above the equator for the Tambora and Pinatubo eruptions. (*right*) Aerosol optical depth (500 nm) modelled for the Tambora eruption. These are obtained from a two-dimensional aerosols microphysics model, assuming a sulphur amount of 80 and 20 Mt for Tambora and Pinatubo, respectively (See Arfeuille et al. (2014) for details)

4.2.5 *The Maximum State of Alpine Glaciers in the 1850s*

The British travellers who started to visit the Alps around the 1850s must have had a majestic view. Glaciers extended much farther than today and some even reached the valley floors. Glacier ice was not debris covered, as it is today. It was blue, shiny, and ragged, as is typical for advancing glaciers. Glacier maxima were reached in the 17th and 18th century and around 1820, following the Year Without a Summer. However, the maximum around 1850 was the last substantial advance of Alpine glaciers and marked the final phase of the Little Ice Age.

After 1850, Alpine glaciers started to recede to their present positions, sometimes halting for brief phases of stagnation or temporary advances.

4.2.5.1 A 19th Century View of Glacier Changes

Although glaciers in the mid-19th century attracted the first tourists, they were traditionally seen as a threat. It was only in the late 18th century that the Alps and the glaciers became a matter of interest to science. One of the first to study glaciers was Jakob Samuel Wyttenbach[9] (see Zumbühl 2009). Glacier length changes were seen as an indication for climatic changes since the 1800s. Following the Year Without a Summer of 1816 (see Box 2.2, p. 22) evidence for historical changes in

[9]Jakob Samuel Wyttenbach, 1748–1830, was a Bernese priest and natural scientist. Wyttenbach travelled the Alps together with painter Caspar Wolf and published his observations. He was a co-founder of the Bernese and Swiss Natural Sciences Societies.

Fig. 4.29 (*left*) Aletsch Glacier from the "Topographische Karte der Schweiz" (Dufour Karte), 1:100000, Blatt XVIII, 1854 (updated in 1866 and 1876) (*right*) Aletsch Glacier as seen from space in 2001 (ASTER on Terra). NASA/GSFC/METI/ERSDAC/JAROS, and U.S./Japan ASTER Science Team

glacier extent was compiled (Venetz 1830). Eventually, evidence for pre-historical glacier maxima was found in Switzerland; one of the origins of the ice age theory (see also Krüger 2013).

With the lively discussion of ice age theories in the 19th century, glaciers and their changes were more intensely studied. To understand why glacier length changes, one must understand how a glacier's mass balance and flow responds to climatic changes. Sonklar (1858) helped to establish the relationship between climate variables and glacier growth. He suggested that precipitation from October to April and temperature from May to September were the decisive factors for glacier growth in the Alps, in addition to local wind conditions. Brückner continued Sonklar's work in the eastern Alps in the 1880s. Around the same time, systematic monitoring of glaciers started in Switzerland and internationally. First this involved length measurements, later it included mass balance measurements. In fact, Brückner published the reports of the International Glacier Commissions on glacier monitoring from 1908 to 1911. Since 1986, the World Glacier Monitoring Service at the University of Zurich has monitored numerous glaciers around the world based on in-situ measurements and remote sensing data.

It is appropriate to give Alpine glaciers a special emphasis in this book. After all, Brückner devoted a substantial part of his scientific career to the study of Alpine glaciers. In this section we analyse climatic conditions during the last maximum of

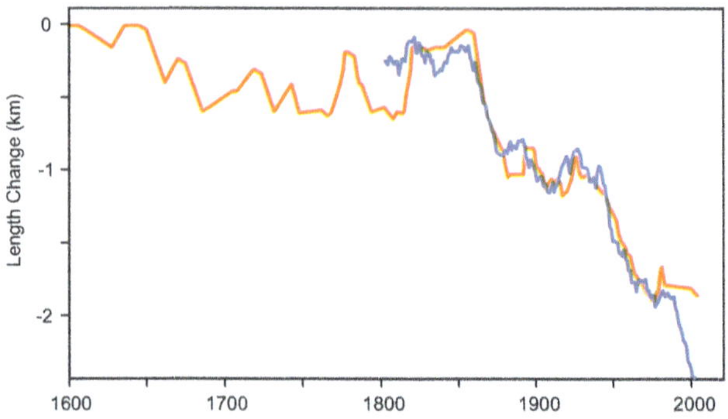

Fig. 4.30 Time series of the length of the Lower Grindelwald Glacier relative to the 1600 extent. *Orange*: reconstruction based on documentary evidence (Zumbühl 1980; Zumbühl et al. 1983, 2008), *blue*: accumulated glacier growth index following Sonklar (1858), calculated from HISTALP data (47° N, 7° E, the offset and scaling factor of the index were chosen to best match the long term trend in the observed glacier length)

Alpine glaciers. The glacier advance in the European Alps in the 1850s—the centre of a cold and wet phase in Brückner's cycle listing—is well documented because this coincided with the first tourist exploration of the Alps. Early photographic documents are available from several glaciers. Topographical and geodetic surveys also provide information for some of the glaciers (Fig. 4.29). Even more information is available for the receding phase, which occurred during the high phase of Belle Époque tourism.

4.2.5.2 Causes of the 1850s Glacier Advance

The phases and causes of the glacier advance in the Alps from the 1600s to 1850s were studied in Sonklar (1858); a paper that greatly impressed Brückner. Sonklar found a widespread glacier advance in the Alps in the 17th and early 18th century, local advances or stagnation in the second half of the 18th century, and a phase of renewed advance in the first half of the 19th century. Soon after this publication glaciers started to recede.

Some of the phases outlined by Sonklar are reflected in more recent reconstructions of the length of the Lower Grindelwald Glacier (Fig. 4.30; Zumbühl 1980; Zumbühl et al. 1983, 2008). The time series shows earlier advances around 1600 and the 1640s. The advance from 1820 to 1860, peaking in 1855, clearly stands out. This was also successfully modelled (through a so-called "neural network" fed with reconstructed climate data; Steiner et al. 2008), although it slightly underestimated magnitudes. Steiner et al. (2008) identified the same influencing factors as Sonklar (1858)—winter precipitation and summer temperature.

With respect to the 1850s advance, Sonklar (1858) found a sequence of years that were particularly conducive to glacier growth. From 1837 to 1845, summers in the Alps were cool. Meanwhile, precipitation was abundant in winter. This is confirmed by the Alpine summer temperature reconstructions shown in Fig. 4.8 as well as Alpine precipitation data from HISTALP and Pauling et al. (2006) (see also Gimmi et al. 2007). However, not all available precipitation series from the region reflect this increase.

4.2.5.3 Climate and Glacier Growth Assessed from HISTALP Data

In the following, we will analyse the climatic causes of the glacier advance using HISTALP data. Glaciers advances resulted from cool summers and wet (albeit relatively warm) winters. Following Sonklar (1858), we can calculate a simple "glacier growth index" using the weighted average of seasonal temperature and precipitation (I used HISTALP data, standardised within the 1800–1890 period, with the weights determined by Sonklar 1858). The index was subsequently scaled and an offset was added to match, when accumulated, the long-term trend in the historical series of the length of the Lower Grindelwald Glacier (Fig. 4.30). The growth and retreat phases of the glaciers are very well reproduced in this curve, as in Steiner et al. (2008), though the maximum in 1856 was slightly underestimated.[10] Glacier length follows the glacier growth index with a lag of a few years. Although the dynamics of the glacier also need to be taken into account, this comparison suggests that the climatic drivers dominated the glacier length changes. The series also shows that the climate in the early and mid-19th century was conducive to glacier growth.

Sonklar (1858) suggested that the years from 1837 to 1845 were crucial for glacier growth. Anomaly fields of temperature and precipitation for these years (Fig. 4.31, note that 1871–1900 was used as a reference) confirm that the machinery produced a sequence of cold summers and wet winters, particularly in the western Alps. A map of the glacier index anomalies during these years indicates that conditions were favourable for glacier growth, especially in the western Alps. Meanwhile, in the eastern Alps, climatic conditions were average due to lower precipitation.

4.2.5.4 Retreat

> "Of all the Alps' characteristics, none appeals more to the people of the plains than the everlasting snow covering the highest mountain peaks, none attracts more than the distant glowing glaciers." (from Brückner, 1893, in (Stehr and von Storch 2000), p.193)

[10]The overestimation of the recent retreat might be due to the fact that the glacier tongue has remained stationary since the 1980s as a block of dead ice in a narrow gorge, buried under rockslides.

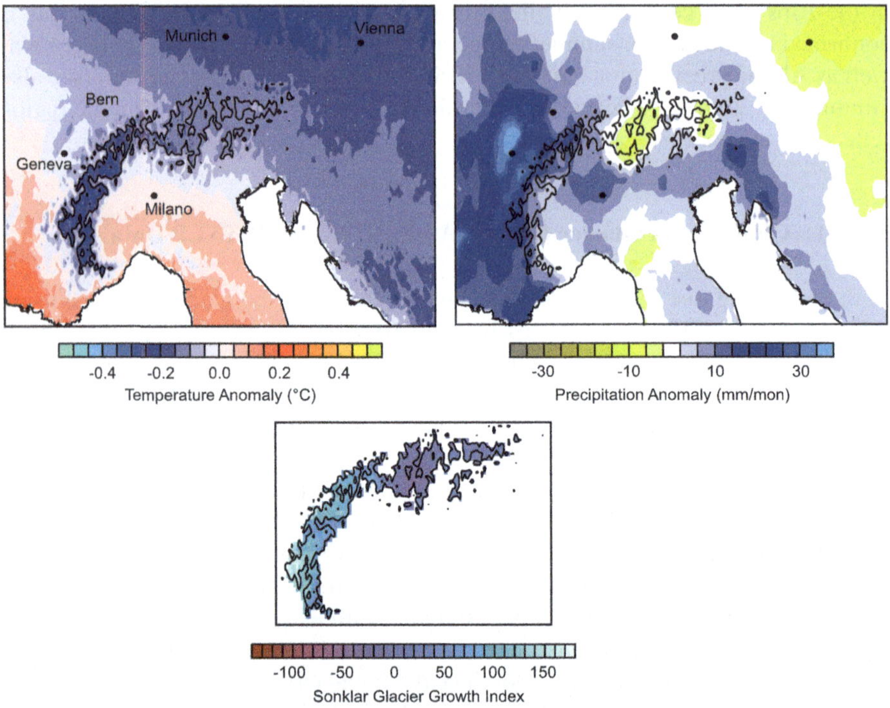

Fig. 4.31 Anomaly fields of temperature and precipitation from HISTALP for (*top, left*) May–September and (*top, right*) October–April during 1837–45 and expressed as anomalies from 1871–1900. (*bottom*) Average glacier growth index according to Sonklar (1858) calculated from HISTALP data (see text for explanation) from 1837–45. Topography above 2 km is marked

Brückner, a scientist from the plains (he grew up in the Baltic region), was fascinated by glaciers—their advances and retreats, and their dynamics over time. When Brückner wrote his book, the retreat had already started. He wrote that in the eastern Alps, glacier area had already diminished by 20 % between the 1850s maximum and the 1880s. However, during Brückner's time glaciers were still large compared to the present. This can be seen in the photograph of the Eiger Glacier around 1900 (Fig. 4.32). Since then, Alpine glaciers have receded with only short phases of advance. Particularly strong negative growth rates occurred in the 1940s and since the 1990s (see Fig. 4.30).

The Alpine glaciers will continue to retreat in the future. Many of them will be lost for a long time to come and future generations will never have the exciting views the first British tourists had in the 19th century.

Fig. 4.32 Eiger Glacier, with Eiger and Mönch in the background (*top*) around 1900 (Photochrom, courtesy of S. Nussbaumer) and (*bottom*) on 12 July 2009 (Photograph by S Nussbaumer)

4.2.6 Global Droughts During 1876–1878 and El Niño

The Year Without a Summer of 1816 and related widespread famine may have been perceived as a major catastrophe by the population in Switzerland. However, this catastrophe is dwarfed by the 1876–1878 global drought studied in this section. This was one of the most devastating climate events in history (see also Cook et al. 2010). Monsoon failure in India (Fig. 4.33) and China and drought in South America cost around 20 million lives (Aceituno et al. 2009; Davis 2001). Some authors call this the "late Victorian Great Drought" of 1877/1878 (although in some Asian regions it started in 1876).

The regions affected by drought are shown schematically in Fig. 4.34 and include much of India (where it is called "Great Famine"; Kripalani and Kulkarni 1997), northern and eastern China ("Great North China Famine";

Fig. 4.33 Governance worsened the consequences of drought. This image shows grain bags stacked for export in Madras (Chennai) during the drought (Digby 1878)

Hao et al. 2010; Kang et al. 2013; Zhang and Liang 2010), Southeast Asia, and northeastern Brazil (Aceituno et al. 2009). Were these droughts related to each other?

Some research suggests that they were related. The cause of the droughts—at least of some of them—might have been an El Niño event that started in late 1876 and matured in 1877/1878 (e.g., Aceituno et al. 2009; Kripalani and Kulkarni 1997). In India, however, the drought began a year earlier.

4.2.6.1 The 1877/1878 Drought in EKF400, CCC400, and 20CR

Let us look at this event in tree ring-based reconstructions (Cook et al. 2010), 20CR, EKF400, and CCC400. To some extent, this will allow us to address the contribution of sea-surface temperatures.[11] For the summers of 1877 and 1878, drought conditions are reconstructed across large parts of India and China, and 20CR shows a precipitation deficit over India and China. However, the two datasets disagree over Southeast Asia, where 20CR indicates increased precipitation. The best ensemble member of EKF400 reproduces a precipitation deficit over the Korean

[11]The sea-surface temperature field used to prescribe the GCM simulations contains atmospheric influences (internal variability). Hence, the resulting response of the atmosphere in the model cannot be addressed solely as an effect of the ocean on the atmosphere.

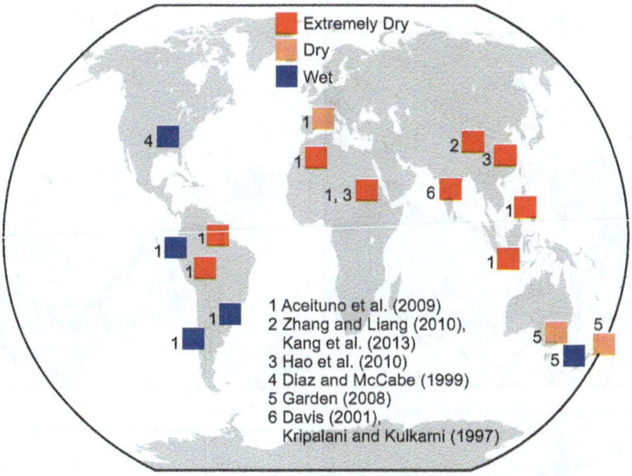

Fig. 4.34 Map of regions affected by drought in 1877/1878 (based on the work of Andrea Gammeter and Carmen Minder)

peninsula, China, Malaysia, and Indonesia. However, hardly any deficit is indicated over India, resulting in another disagreement over South Asia.

Analysis of 500 hPa geopotential height and vertical velocity (Fig. 4.35) as well as 850 hPa wind in 20CR shows anticyclonic conditions over China with associated subsidence (red solid contours) and a weakening of the East Asian monsoon (not shown). Over India, the monsoon circulation was similarly weakened.

If the drought was caused by a weak monsoon, what caused the monsoon to weaken? No major volcanic eruptions are reported. Hence, the drought must have come from internal variations in the machinery. A likely candidate is anomalous sea-surface temperatures. In fact, CCC400 simulations reproduce dry conditions in China, which explains part of the drought. However, the simulations also produce a zonally elongated wet band further south, covering all of India, which contradicts the well-known droughts there. Because this band also appears in 20CR, it might represent the response of models (CCC400 and 20CR) to the imposed sea-surface temperatures. Hence, either the models are inadequate or the drought in India was due to unforced atmospheric variability.

Reverting to statistics, the negative precipitation anomalies over China and India in 1877/1878 are in agreement with well-known El Niño teleconnections (e.g., Larkin and Harrison 2005). Precipitation over maritime areas of Southeast Asia is not well known from observations. Thus, the signal is difficult to interpret. The reported drying in parts of Africa and the Amazon basin, and increased precipitation along the Andes, in Patagonia, and in parts of North America are in agreement with El Niño teleconnections. In summary, a strong El Niño might be partly responsible for the droughts in 1877/1878. Its teleconnections were expressed very strongly and brought a lack of rainfall to densely populated regions, some of which had already suffered from antecedent dryness. Thus, the droughts were produced by the most

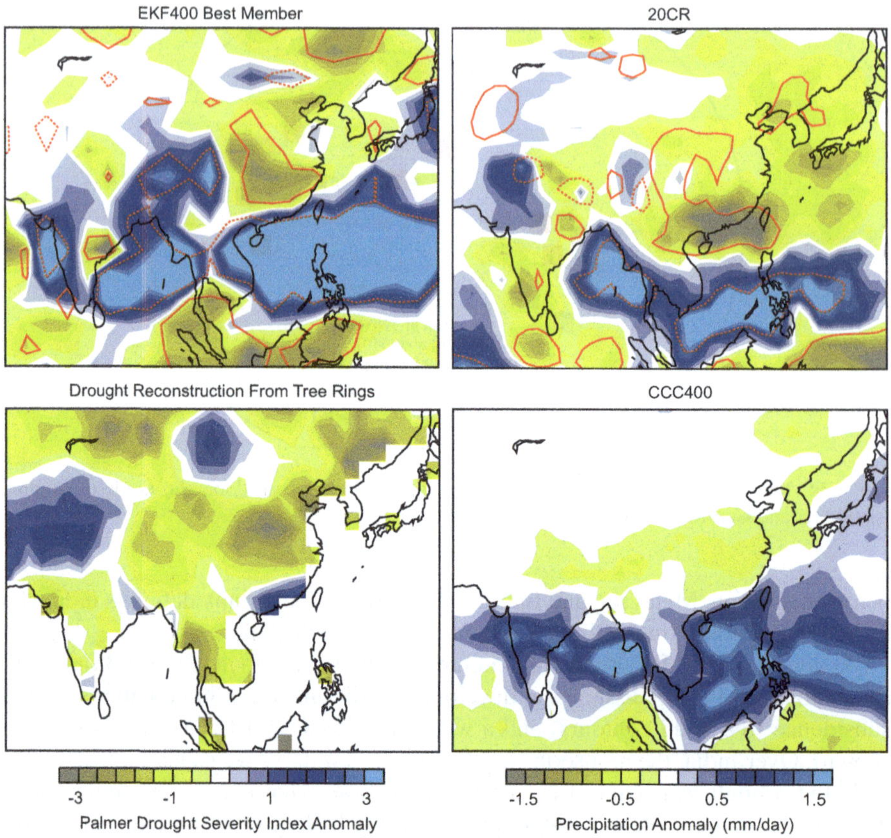

Fig. 4.35 The drought summers of 1877 and 1878 in (*bottom left*) tree ring-based reconstructions of the Palmer Drought severity index (Cook et al. 2010) and precipitation from July to September from (*top left*) the best member of EKF400, (*top right*) 20CR and (*bottom right*) CCC400. Anomalies are relative to 1871–1900. *Solid* and *dashed red contours* denote positive and negative vertical velocities at 500 hPa of 10^{-3} Pa s^{-1}, respectively

forceful mode of the machinery. However, additional atmospheric variability played a role, particularly for the drought in India.

Box 4.1 Francis Galton, Gilbert Walker, and Felix Exner: Statistics in climatology

The Great Famine in India of 1876–1878 also affected the development of meteorology and climatology. Specifically it impacted the use of statistical methods. The India Meteorological Department had just been founded when

(continued)

Box 4.1 (continued)
the famine struck. In its wake, director Henry Francis Blanford collected all available data from the Indian Ocean and Australasian region to study this event and its relation to climate. He noted anomalous pressure conditions over Asia, studied relations between the monsoon and other variables and performed experimental monsoon forecasts. Among other things, Blanford (1884) also related the variability of the Indian summer monsoon to Eurasian snow cover changes—a theory that is still being studied today (see Sect. 3.1.4). In the subsequent decades, Blanford and his successor aimed to strengthen the empirical work, but sought a more mathematical background.

In the early 1900s, the mathematician Gilbert Walker joined the India Meteorological Department to continue the work on monsoon forecast. He introduced the concept of correlation (a method developed by Francis Galton in 1888 (Galton 1888)) to climatology. He produced point correlation maps, but also analysed correlations across variables and with different leads and lags (e.g., Walker 1909). Walker's concept was successful. Using correlations, he discovered the main global- or large-scale circulation variability modes (see Sect. 3.2 and Table 3.2): the Southern Oscillation, the North Atlantic Oscillation (Fig. 4.36), and the North Pacific Oscillation (Walker 1924). Thus, Walker paved the way to modern climatology.

Other scientists adopted this new concept. Felix Exner (1913) performed similar analyses for the Northern Hemisphere. He discovered the Arctic Oscillation (Sect. 3.2.3 and Table 3.2) and quantified month-to-month predictability using lag correlation. The concept of correlation was far superior to analysing tables or visual inspection of time series—the method used by scientists such as Hann (1890) or Hildebrandsson (1897) working on the same topics.

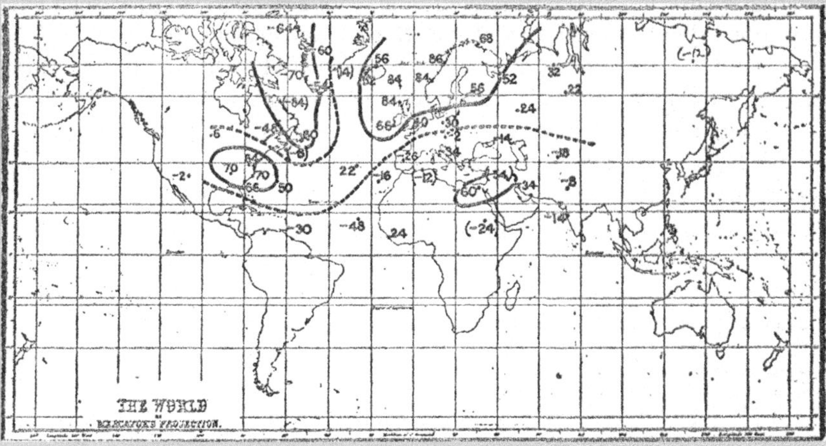

Fig. 4.36 Correlation map from Walker and Bliss (1932). Shown is the correlation (×100) of an NAO index with contemporary temperature for the December–February average. Copyright RMetS

Fig. 4.37 Contemporary depictions of the 1868 flood in the Alpine section of the Rhine River (Source: Staatsarchiv St. Gallen)

4.2.7 Frequent Flooding in the Central European Alps: 1830–1880s

Few flood events are still remembered after 150–200 years. One of such events, located along the Alpine section of the Rhine River, was the flood of 1817. This was caused by the melting of an enormous snow pack, which at higher altitudes consisted of the snow of two winters and the summer in between (Wetter et al. 2011): a consequence of the "Year without a Summer" of 1816 (Sect. 4.2.4). Another remembered flood event in Switzerland occurred in 1868 and affected the Valais and Ticino, destroying many bridges in the Alpine section of the Rhine valley (Fig. 4.37). This was deliberately used by lobbyists to influence opinions in the debate on the consequences of deforestation because floods in the lowlands were seen as a result of deforestation in the mountains. This paved the way to the Swiss forestry law, which was accepted by the Swiss parliament a few years later and entered into force in 1876 (Pfister and Brändli 1999). Sometimes an event at the right time makes all the difference in a decision-making process.

The 1868 event was typical for this period. During the 1850s–1880s, Switzerland experienced an increased frequency of severe floods (Fig. 4.38; e.g., Pfister and Brändli 1999; Schmocker-Fackel and Naef 2010a,b; Stucki et al. 2012). These floods not only caused economic losses, but soon became the focus of political discussions, as shown above for Switzerland. The paradigm of floods in the lowlands

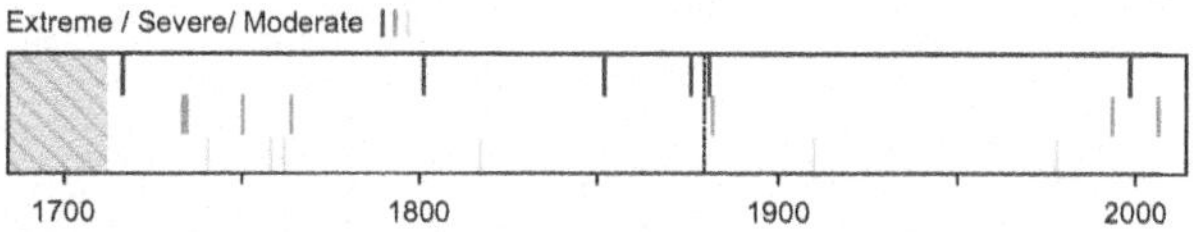

Fig. 4.38 Floods of the Rhine in Basel since 1714. Moderate, severe, and extreme floods were defined as runoff exceeding 4000, 4500, and 5000 m^3/s. Prior to 1878, the completion of a major hydroengineering project in the Aare subcatchment, thresholds were defined as 4500, 5000, and 5500 m^3/s (Data from Wetter et al. 2011)

being caused by mountain deforestation was also prominent in other countries and affected national policies. In fact, Brückner devoted several pages of his book to the political implications of this debate.

The increased frequency of floods was not restricted to Switzerland but arguably also affected surrounding countries. However, a consistent picture is difficult to obtain (Bichet et al. 2014). Most of the floods occurred from late summer to early fall and were related to heavy precipitation (Stucki et al. 2012). Atmospheric circulation patterns played an important role because few distinct weather patterns trigger floods in Switzerland and central Europe. A particularly flood-prone weather type is the Vb cyclone track whereby Mediterranean cyclones move east and north around the Alps to bring moisture to central Europe and the northern slope of the Alps. Based on an analysis of daily sea-level pressure fields, Jacobeit et al. (2004) suggested that this weather type was more frequent in these decades, consistent with more frequent floods.

4.2.7.1 The 1868 Flood in 20CR and Instrumental Data

The 20CR reanalysis allows us to study the flood of 1868 in depth (see Stucki et al. 2012, for more details on the following). As with any other flood in the largest watersheds of Switzerland, the 1868 event was triggered by heavy precipitation. However, heavy precipitation alone is not sufficient to produce a flood of that magnitude. Most floods are preconditioned by other factors such as saturated soils, a rapid temperature rise and associated snow melt, and high lake levels (Stucki et al. 2012). The 1868 case was no exception.

In the weeks leading up to the event, precipitation was excessive. Soils were saturated and the level of the Lago Maggiore (in southern Switzerland) was high (Stucki et al. 2012). Fresh snow had fallen in the mountains. The heavy precipitation triggering the 1868 flood occurred in two phases within about a week (from the end of September into early October 1868). The precipitation maximum for the entire event was measured at San Bernardino (1118.2 mm over 8 days). An analysis of 20CR (see Fig. 4.39; top, left) reveals that the first precipitation phase was triggered by a cutoff low over Ireland and southwesterly flow toward the Alps. In the second phase, a few days later (Fig. 4.39; top, right), an elongated upper-level trough moved southward with lifting on its eastern flank. Moist air was transported from the south,

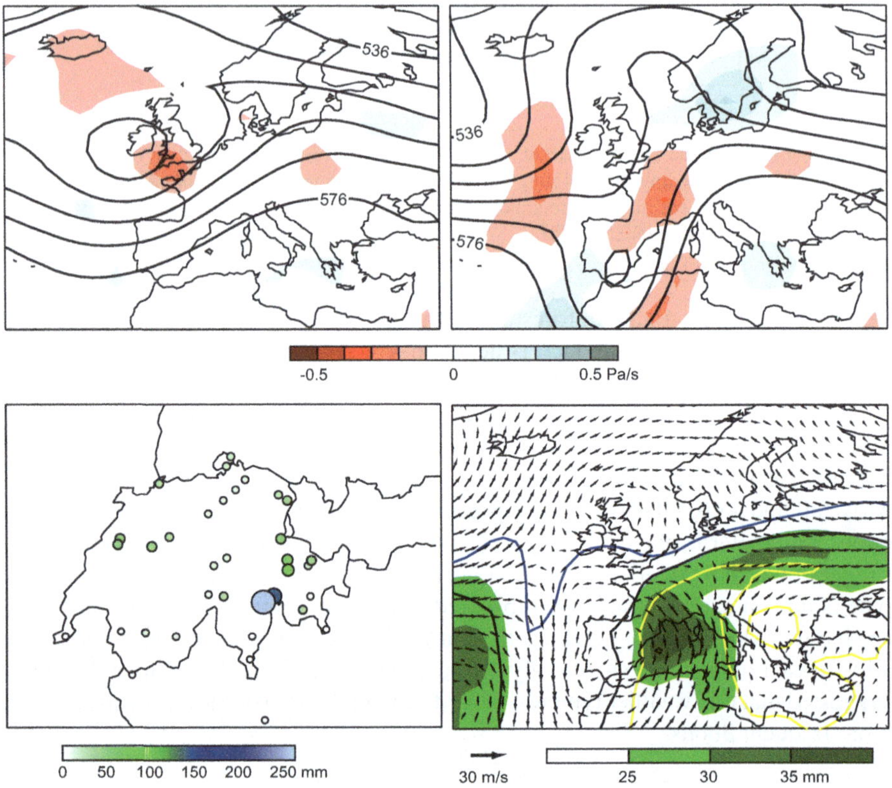

Fig. 4.39 Meteorological conditions during the heavy precipitation triggering the 1868 flood. (*top*) 500 hPa geopotential height and vertical velocity on (*left*) 28 September and (*right*) 2 October, (*bottom left*) precipitation on 28 September and (*bottom right*) equivalent potential temperatures (*blue*, *black*, and *yellow* isolines show 16, 24, and 32 °C), wind at 850 hPa (*arrows*) and precipitable water (*green shading*) on 2 October 1868. All data except (*bottom, left*) are from 20CR (Stucki et al. 2012)

across the Mediterranean, towards the Alps. At the same time, high temperatures led to the melting of snow. This resulted in flooding on both sides of the Alps. At the Lago Maggiore, it was the second largest flood since 1177. Fifty casualties were registered in Switzerland (Stucki et al. 2012).

What was the cause of the increased flood frequency? More flooding situations can be expected when the Azores high retreats due to a narrowing or weakening Atlantic Hadley cell. This allows more frequent disturbances from the Atlantic and the Mediterranean to reach central Europe and the Alps. However, CCC400 does not provide evidence for increased summer or autumn precipitation during these years (in fact, it even suggests a decrease). Then again, observations also show no clear peak (Fig. 4.8). Recently, Bichet et al. (2014) reproduced increased European summer precipitation in these decades from a climate model simulation prescribing sea-surface temperatures. They suggest a forcing from the tropical Pacific. However, further research remains to be done. First, it needs to be established whether the increased flooding was due to precipitation events or due to changes in the river beds.

After the 1880s, flood frequency decreased. For a long period, floods became extremely rare (Wetter et al. (2011); see also Fig. 4.38). The next big event (apart from an event in 1910) occurred in 1987, when a massive flood affected the central parts of Switzerland. Pfister (2009) calls the period between these events the "disaster gap" because floods, wind storms, and other disasters were rare (see Sects. 4.4 and 4.5.4). While the climatic factors leading to the disaster gap are unknown, it did have lasting effects on society. The disaster gap led to a loss of traditional disaster memory. Therefore, when the frequency of disasters increased in the 1980s, society had to relearn how to deal with them.

4.2.8 The Broad Lines: The Little Ice Age

Let us recapitulate the main lines of climatic changes in the 1700–1890 subperiod. This represents the last part of the Little Ice Age, which was a globally cool period with some exceptions. In particular, the 18th century was punctuated by warm episodes in many places. The Late Maunder Minimum 1685–1715 as well as the sharp temperature drop in the 1810s were the coldest phases of the subperiod in Europe and the Northern Hemisphere continents. The latter has been mostly attributed to the eruption of Tambora, which caused the Year Without a Summer of 1816. Volcanic forcing also contributed to the Late Maunder Minimum cooling. Reduced solar irradiance likely contributed to both cold phases. Another temperature drop, related to volcanic eruptions, occurred in the 1830s. These drops in summer temperatures initiated the growth of Alpine glaciers, the materialisation of the Little Ice Age.

In contrast, around 1800 was one of several comparatively warm phases at subtropical and midlatitudes that was well documented (though not well quantified) for Europe and visible on the global scale (Crowley et al. 2014). The warmth was accompanied by drought in subtropical regions due to changes in the tropical circulation. The Hadley circulation had been strengthening since the 17th century and reached an intermediate maximum in these years, before dropping off during the 1810s cold phase. Whether remote tropical influences also were responsible for the Alpine glacier growth and increased flooding in the mid-19th century remains to be studied.

Several pronounced decadal climatic features within this period were analysed in this section, indicating the importance of interannual and multiannual variability such as the El Niño of 1877/1878. In the 1880s, when Brückner wrote his book, the Little Ice Age came to an end. In central Europe this end was not only marked by the retreat of the glaciers (which Brückner documented), but also by a sequence of severe flood events, which even affected national policies in several European countries.

4.3 The Period from 1890 to 1945: Out of the Cold

Toward the end of the 19th century, temperatures finally increased in many areas of the globe. A period of global warming started around 1890 (somewhat later over the oceans), slowly at first, and then accelerated during 1920–1945. At the same time, societies, economies, and nations—and thus the livelihood of people—changed rapidly in Europe and North America. From a phase of industrial transformation, nationalism, imperialism, and scientific positivism, the world plunged into two World Wars, genocides, political divisions, and economic crises.

Several climatic events impacted society during the crises. The "Dust Bowl" droughts affected the U.S. in the midst of the Great Depression and extremely cold winters struck eastern Europe during the Second World War. An even more prominent climate event, a rapid warming of the Arctic, occurred in the 1930s—a time when the geostrategic importance of the Arctic was increasing. In this section, I summarise these three climate events based on the literature and trace the respective findings in the datasets and model simulations described in Sect. 2.9. In addition to reconstructions, 20CR (EKF400 is no longer used in this section), and CCC400, I also address the stratosphere based on the CASTRO chemistry-climate model simulations and the global historical ozone dataset HISTOZ.

With respect to climate change science, the period around 1900 saw an engaged discussion on ice ages and related theories (including those involving greenhouse gases; discussed in Box 4.2, p. 256), in which Brückner was a central figure (in fact, he co-authored the standard textbook on the topic; Penck and Brückner 1909). However, as to what concerns shorter time scales, many scientists had the opinion that the climate was varying about a fixed mean state and thus was essentially stable (Stehr and von Storch 2000).

This view changed in the 1930s and 1940s, when it became increasingly clear that the climate was changing. Global temperatures rose rapidly (Groissmayr 1949; Scherhag 1939a,b; Wagner 1940). However, the causes were unclear. When Callendar (1938) ascribed it to an increase in CO_2, his work was not well received. However, nor were competing theories (Fleming 2007; Weart 2008; see Box 4.2, p. 256).

Today's climate scientists looking back at this period find an increase in the amount and quality of data (though distinct from the "modern" weather and climate data that start with the International Geophysical Year 1957/1958). National weather services had been established and data had been globally compiled (see Sect. 2.2.2; Edwards 2010). These data have been analysed more or less continuously, leading to the current state of climate science. In addition to conventional weather observations, other geophysical observations started to become available, such as upper-level observations, radiation measurements, or spectroscopic data (Sect. 2.3). These provide information about the climate system that goes beyond temperature or precipitation and include quantitative data on volcanic eruptions or solar forcing. Previously, only indirect information was available for these variables.

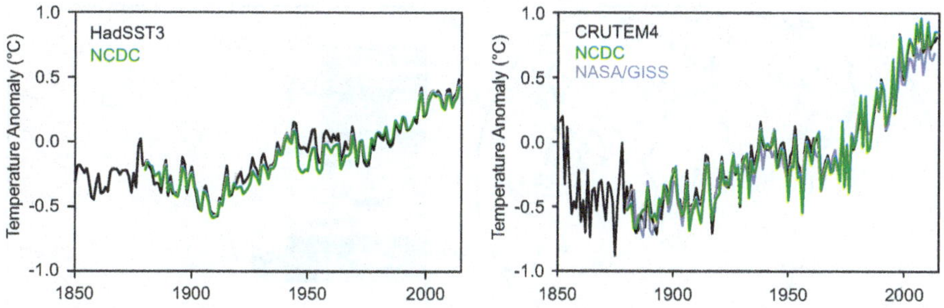

Fig. 4.40 Global (*left*) sea surface and (*right*) land surface temperatures from different data sources expressed as anomalies from 1961–1990

4.3.1 Global Drivers and Global Changes

Global-mean temperatures are available from several datasets that reach back to ~1850. Sea surface temperatures (Fig. 4.40, left) and land surface temperatures (Fig. 4.40, right) both show an overall warming pattern that occurred in two phases: one from 1900 to the 1940s, and one from the 1970s to the present. The phases are separated by a stagnation. Can we trust these global datasets? Coverage errors can still be large in global datasets and affect trends (Cowtan and Way (2014); also see Sect. 4.5.9). In addition, it has been argued that biases (e.g., due to urbanisation) remain in the data and distort the trend. The IPCC Fifth Assessment Report states that urbanisation effects are very small at the global scale (Hartmann et al. 2013). Triggered by debates of this sort, the "Berkeley Earth Surface Temperature" initiative recreated a global temperature dataset from raw observations using completely different approaches and arrived at practically the same result (Rohde et al. 2013). Further support for the correctness of the global temperature trend during the past 150 years comes from a recently constructed, non-instrumental global temperature index based on proxies (Anderson et al. 2013).

In the following, I will focus on the warming in the first half of the 20th century. Figure 4.41 (left) shows a spatial map of the annual-mean surface air temperature trends from the HadCRUT4 data. A clear warming appears from 1890 to 1945, particularly at high latitudes of Europe and over the northern North Pacific and Canada. This is distinct from the warming pattern since the 1970s (see Sect. 4.5.1). A latitude-time plot of decadally averaged temperature anomalies (Fig. 4.41, right) reveals a rapid change from the 1910s to the 1920s in the entire Northern Hemisphere and tropics, followed by a strong high-latitude warming in the 1930s and 1940s. The latter period was also accompanied by more frequent heatwaves (see Fig. 4.11).

Since the work of Callendar (1938), many hypotheses have been proposed to explain the warming from 1890 to 1945. Solar irradiance increased from 1900 onward (the period around 1890 is sometimes called the "Gleissberg minimum"). However, the increase may have been rather small (see Sect. 3.3.2). Tropical

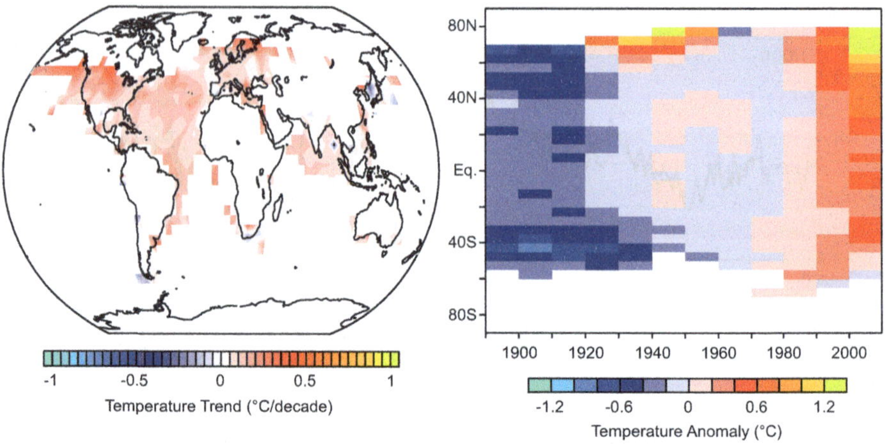

Fig. 4.41 (*left*) Map of annual-mean temperature trends from 1890 to 1945 from HadCRUT4. (*right*) Latitude–time plot of decadal temperature anomalies (with respect to 1961–1990). At least 6 months per year, 5 years per decade, and (*right*) 10 grid boxes (5° × 5°) per latitude circle must be available

volcanic eruptions, which were still frequent in the 1880s (e.g., Krakatau), became increasingly rare after the Santa Maria eruption of 1902 (Fig. 3.44). The reduction of volcanic activity implies a positive forcing—another possible contribution to the warming. In fact, Scherhag (1939b) attributed the warming to a combination of solar and volcanic forcing. In addition, anthropogenic greenhouse gas emissions might have already had an effect, as suggested by Callendar (1938).

Our simple forcing–temperature regression (Eq. 4.1) suggests that half of the observed trend of land surface temperatures (0.1 °C decade^{-1} for 1890–1945) can be explained by forcings, most of which comes from greenhouse gases. The contributions of solar forcing and tropospheric aerosols cancel each other out (+0.02 and −0.02 °C decade^{-1}, respectively). In this calculation, the warming due to the lack of eruptions is negligible.

More sophisticated attribution studies (Hegerl et al. 2007; Schurer et al. 2014) find that anthropogenic forcing likely contributed to—or even dominated—the warming. However, the relative role of solar, volcanic, greenhouse gas, or aerosol forcing is not robust across studies (Bindoff et al. 2013). Similar to our simple approach, the sum of the forced components is still less than the observed trend. A large contribution, therefore, must have come from internal variability of the climate system (see Delworth and Knutson 2000; Zhang et al. 2007)—a major gear shift in the machinery.

4.3.1.1 Changes in Circulation Indices

So, how did the machinery change? For the oceanic part of the machinery, we have seen in Fig. 4.7 that indices of the PDO and AMO were low in the 1890s. The

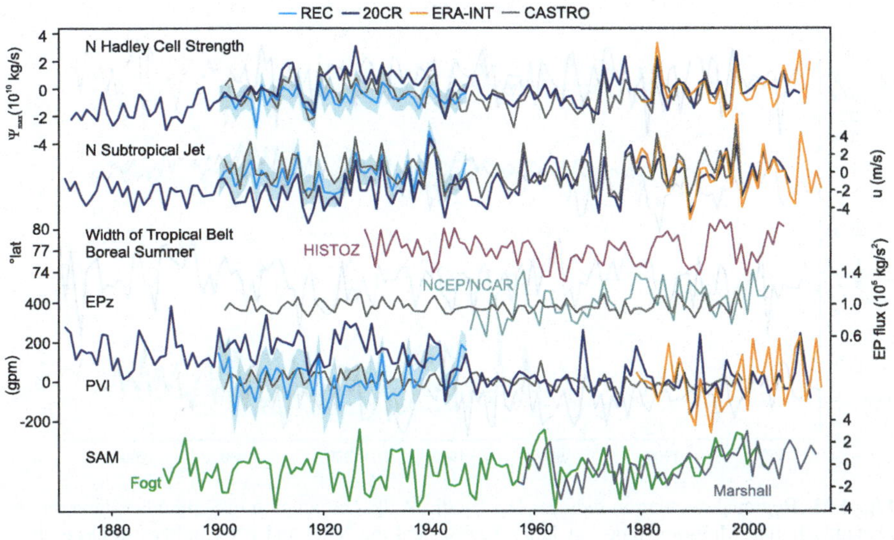

Fig. 4.42 Zonal-mean circulation indices. The strength of the northern Hadley cell and the subtropical jet in December–March (anomalies with respect to 1961–1990), the width of the tropical belt in May–October, the vertical component of the Eliassen-Palm, flux and the boreal Polar Vortex Index in December–March, and the Southern Annular Mode in December–February (Fogt et al. 2009; Jones et al. 2009a; Marshall 2003). *Shading* indicates the 95 % confidence interval of the reconstructions (see Brönnimann et al. 2009b)

PDO index first increased rapidly, then increased slowly, and peaked in the early 1940s. The AMO index first remained low, then increased sharply in the 1920s, and was high in the 1940s. Tropical sea surface temperatures and, in CCC400, the NINO3.4 index (of El Niño) increased similarly to the PDO. In the following, I will take a look at the same atmospheric circulation indices as in Sect. 4.1.4, but now using reanalyses (20CR, NCEP/NCAR Reanalysis, and ERA-Interim; see Table 2.3), statistical reconstructions, and the CASTRO simulations (see Figs. 4.9 and 4.10 for EKF400). I will also show some new stratospheric indices (described below) that can only be derived after 1900 from HISTOZ and CASTRO. Indices are subdivided into zonal-mean circulation indices (Fig. 4.42) and regional circulation indices (Fig. 4.43), and the discussion will give a preview of what follows in the remainder of the book.

A strengthening of the *boreal winter Hadley cell* until aorund 1930 is seen in 20CR but not statistical reconstructions. However, trends in the strength of the Hadley circulation need to be addressed cautiously; even trends over the past 30 years are debated because the quality of reanalysis datasets precludes firm conclusions (see Hartmann et al. 2013). In fact, the signal seen in EKF400, CASTRO, and 20CR might reflect the model response to sea-surface temperatures rather than atmospheric observations. However, interannual variability is consistent across datasets and well reproduced by GCM simulations.

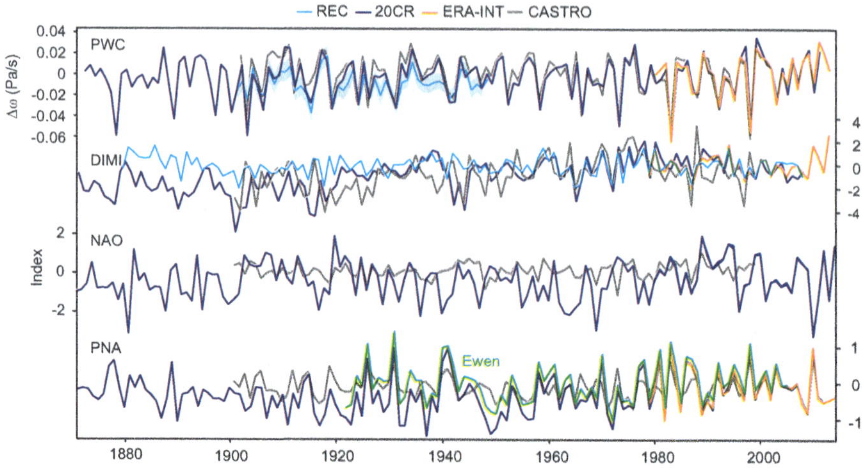

Fig. 4.43 Regional circulation indices. The Pacific Walker circulation in September–January, the Dynamic Indian Monsoon Index in June–August, and the NAO and PNA indices in December–March. *Shading* indicates the 95 % confidence interval of the reconstructions (see Brönnimann et al. 2009b)

The time series of the *boreal winter subtropical jet strength* (see Table 3.2) shows interannual variability but no trend. Agreement is good among datasets and the model simulations reproduce the observations well. Strong El Niño events (such as those in 1918/1919 and 1939–1942 (Sect. 4.3.4)) were accompanied by a strong jet.

Two new indices of the zonal-mean circulation are introduced here: the width of the tropical belt in the boreal warm season (based on the strongest meridional column ozone gradients in HISTOZ)[12]; and an index of the wave-driving of stratospheric circulation, that is, the vertical component of the *Eliassen-Palm flux* at 100 hPa averaged from 40°to 80°N in January–February from NCEP/NCAR Reanalysis and CASTRO simulations.

The *width of the tropical belt* shows no centennial trend. However, a contraction between the 1940s and the 1970s (Allen et al. 2014) is followed by a widening. Both changes will be addressed in more detail in the remainder of the book. The *Eliassen-Palm flux index* from the NCEP/NCAR Reanalysis shows a long-term increase (which, however, is not reflected in the CASTRO simulations), decadal changes, and interannual variability. The interannual variability is partly reflected in the CASTRO simulations, which implies an effect of external forcings and sea-surface temperatures.

[12]The index of the width of the tropical belt is defined as the distance between the two latitudes with the strongest zonal-mean total ozone gradient in each hemisphere (restricted within 32° and 45° in each hemisphere) in monthly data, averaged for the May–October season (see also Hudson et al. 2006).

The boreal winter *Polar Vortex Index* (see Table 3.2) is correlated with the *Eliassen-Palm flux*. A weak vortex is seen in 20CR until the 1940s, much less so in reconstructions. A considerable strengthening from 1980 to the 1990s is seen in the reanalysis (very slightly also in the CASTRO simulations) and in other indices such as the NAO (see below). This strong trend will be discussed in Sect. 4.5.4.

The *Southern Annular Mode index* (Table 3.2) shows a clear increase from the 1960s to 2000 as well as interannual variability. An SAM reconstruction for the first half of the 20th century (Fogt et al. 2009) shows no trend.

Regional circulation indices are shown in Fig. 4.43. While the Hadley circulation might have been strengthening, the *Pacific Walker Circulation* (see Table 3.2) shows a slight weakening up to the 1990s. Based on sea-level pressure, Vecchi et al. (2006) and others found a weakening of the Pacific Walker circulation between the late 19th century and the 1990s (Power and Kociuba 2011). This was particularly apparent during boreal summer (Nicholls 2008). However, trends in boreal winter or in other indices are inconclusive (Hartmann et al. 2013). Over the last two decades, the Pacific Walker circulation has strengthened considerably (Sect. 4.5). Interannual variability in the Pacific Walker circulation is in very good agreement among the datasets and is well reproduced by the model simulations. Individual events such as the 1918/1919 (Giese et al. 2010) or 1939–1942 El Niños (Sect. 4.3.4) appear in all series.

The Indian monsoon, according to the *DIMI* (Table 3.2), was rather strong in the 1880s and then dropped into a pronounced minimum from 1900 to 1915. This was accompanied by other climatic changes in the Australasian and Asian regions (see Sect. 4.1.3). After a recovery in the 1920s, interannual-to-multiannual variability dominated and no strong decadal signal is seen. The interannual variability of DIMI is less consistent among the datasets than other indices. Furthermore, it is not well reproduced in the model simulations.

A strong peak in the early 1940s and during other El Niño events is also found in the *Pacific-North American pattern* (see Table 3.2) from 20CR and reconstructions (Ewen et al. 2008). These peaks are captured in GCM simulations, arguably due to the close relation between the PNA and sea-surface temperatures in the tropical Pacific. Noteworthy is a shift in the level of the PNA in the 1970s, which will be discussed in more detail in Sect. 4.4.5.

The *North Atlantic Oscillation* (see Table 3.2) index shows pronounced variability and a strengthening from 1965 to 1995, concurrent with the strengthening of the Polar Vortex Index. The latter will be discussed in detail in Sect. 4.5.4. A cluster of strong negative NAO winters occurred around 2010.

What is the suspected major gear shift of the machinery in the early 20th century? While large changes appear in the Pacific and Atlantic Oceans in Fig. 4.7, changes in the large-scale atmospheric circulation are less pronounced. The subtropical jet, PNA, DIMI, and Hadley cell were arguably strengthening; the NAO was weakening. However, regional circulation changes over the Arctic occurred in the early 20th century, as discussed in the Sect. 4.3.2. Additionally, there are pronounced deviations on an interannual scale in most series, such as in 1940–1942 (Sect. 4.3.4) or in 1998 (Sect. 4.5.5). Both of these are related to El Niño events.

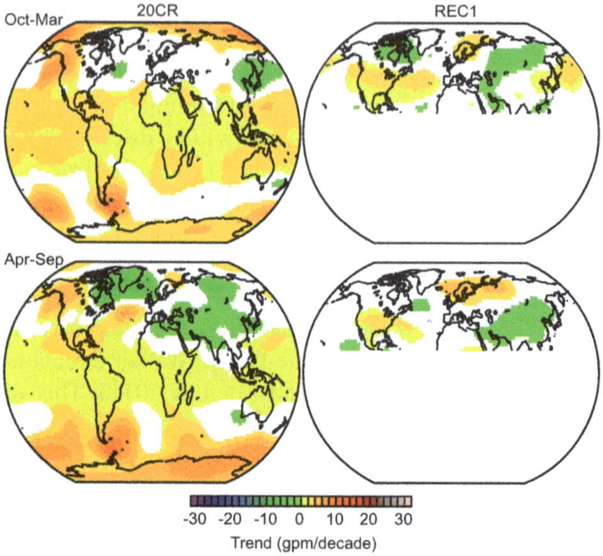

Fig. 4.44 Map of boreal (*top*) winter and (*bottom*) summer trends in 700 hPa geopotential height from 1890 to 1945 in (*left*) 20CR and (*right*) REC1 (only for the northern extratropics). Trends are only shown if significant at the 90 % level after accounting for autocorrelation (as in Hartmann et al. 2013)

Time series give an incomplete view of atmospheric circulation. Upper-level fields are available from several data sources, but their trends are uncertain. Trend maps for 700 hPa geopotential height for summer and winter from 20CR and reconstructions REC1 (Fig. 4.44) are clearly different, even though the datasets are not entirely independent. Obviously, these datasets have not reached trend quality. The features that do agree among the datasets are the increase over northern Russia and western North America, and the decrease over Labrador. In boreal summer, a decrease over large parts of the Eurasian land mass appears.

Apart from the global temperature change during 1890–1945, several prominent climatic events occurred. Before I address the global view of early 20th century warming in Sect. 4.3.5 and before I discuss the role of various possible drivers, I will analyse some of these climatic events—the Arctic warming from the 1910s to the 1940s, the "Dust Bowl" droughts of the 1930s, and the global climatic anomalies related to El Niño from 1939 to 1942.

This is a narrow selection of climatic events in this period. Other possible choices include the "Federation drought" in Australia (1895–1902), the Indian summer monsoon weakening in the early 20th century, the already mentioned 1918/1919 El Niño (Giese et al. 2010), the pluvial period in the western U.S. in the 1910s (Cook et al. 2004), the decrease in European storminess during the first half of the 20th century, and the 1936–1945 "World War Two drought" in Australia (briefly discussed in Sect. 4.5.6). The external forcing from the eruptions

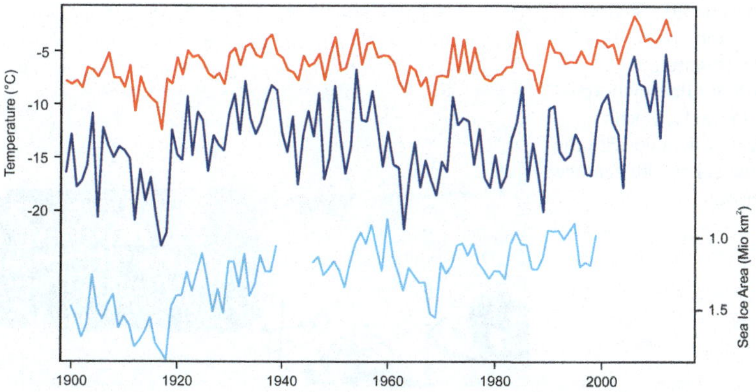

Fig. 4.45 Temperatures at Spitsbergen for winter (December–February) and the annual average (from Nordli et al. 2014), along with sea ice area in the Greenland and Barents Seas (note the reverse scale) from Johannessen et al. (2004)

of Santa Maria (1902) and Katmai (1911) or the famous but inconsequential Tunguska meteorite impact in 1908 could have also been addressed.

4.3.2 The Warming of the European Arctic from the 1910s to the 1940s

Few cases exist where a step change in a temperature time series of several degrees Celsius is accepted as real. This is the case for Spitsbergen (Fig. 4.45). From 1918 to 1919 (note that this coincided with the above-mentioned El Niño), annual-mean temperatures rose by more than 3°C from an average of −9.7°C (1913–1918) to −6.5°C (1919–1924). The step change was even more pronounced—almost 7°C—for winter (Dec.–Feb.) averages. Although contemporary and subsequent scientists noticed and questioned this big step, no changes could be found in the instrument or the station (Birkeland 1930; Scherhag 1939a). The series has been reanalysed many times and was recently extended backwards (Nordli et al. 2014). Another large step in winter-average temperature of 4.2 °C (again comparing 6 years before and after the step) took place in 2005/2006 (Fig. 4.45).

After the step in 1919, temperatures at Spitsbergen remained high throughout the next decade. Then, temperatures increased even further, remaining at that level until the late 1930s (or, after a dip, even to the 1950s). The marked temperature increase in Spitsbergen is arguably the strongest expression of the early 20th century warming. Overall, this warming was pronounced in the Arctic—specifically the European Arctic—in winter. The launch of the first Soviet drifting ice research station in 1937 ("North Pole"; Fig. 4.46) coincided with the peak of the warming.

Fig. 4.46 The first Soviet drifting station "North Pole" during the research expedition of Glavsevmorputi in 1937. In the European Arctic, this coincided with the peak of the early 20th century warm period

The change in Arctic temperatures, which is evident in proxies (see Fig. 4.4) and in records of sea ice in the Greenland and Barents Seas (Fig. 4.45; Johannessen et al. 2004), did not remain unnoticed. Scherhag (1939a,b) and others discussed the warming, including its relation to the global scale, to spatial progression, and to North Atlantic sea surface temperatures. However, the Second World War put an end to many scientific activities.

4.3.2.1 Mechanisms of the Early 20th Century Arctic Warming

Today, the Arctic is seen as an important component of the climate system due to many known or suspected feedback mechanisms, some of which have been suggested to affect European weather and climate (see Sects. 3.1.4 and 3.4.3). Studying the early 20th century Arctic warming in comparison with the present-day warming is therefore seen as relevant for understanding Arctic climate processes.

The early 20th century Arctic warming has often been addressed in terms of circulation changes, that is, a strengthening of the NAO and a southwestward shift of the Icelandic low (e.g., Fu et al. 1999; Rogers 1985). Wood and Overland (2010) found that a meridional circulation pattern known as the Arctic Dipole (Wu et al. 2006) was responsible for the warming. The pattern consisted of a dipole of pressure anomalies over the eastern and western Arctic, particularly between northern Eurasia and Greenland. This led to the advection of warm air into the European sector of the Arctic. The pattern was accompanied by unusual sea surface temperatures, which may have invoked the atmospheric circulation anomalies or, conversely, may have been a consequence of the atmospheric circulation anomalies (Wood and Overland 2010). Thus, the question remains as to whether the early 20th century Arctic warming was a large, random climatic event (Wood and Overland 2010); whether it was forced by changes in sea ice or oceanic circulation within the Arctic (Polyakov et al. 2004) or the North Atlantic inflow region (Semenov and Bengtsson 2003); or whether anthropogenic and decreasing volcanic forcings were the main cause.

4.3.2.2 The Early 20th Century Arctic Warming in CCC400, 20CR, and REC1

How is the early 20th century Arctic warming expressed in our datasets? In addition to instrumental observations from the ground and upper air, 20CR and REC1 are used to address atmospheric circulation and CCC400 is used to address the forced component. The focus is on winter in the European sector of the Arctic, the region of most pronounced warming.

Several factors distinguish the early 20th century Arctic warming from the present Arctic warming. The sea ice area was larger than today and the vertical structure of the warming was different. Sea ice was decreasing during the early 20th century in the European sector (see Fig. 4.45) and possibly other areas of the Arctic (Polyakov et al. 2003) but was still well above present-day levels.

With respect to the vertical temperature structure, few direct measurements of upper-level temperatures are available for the early period. Thus, the uncertainty is large and conclusions are speculative. However, it is interesting to note that the sparse observations show exceptionally high winter temperatures at 700 hPa in the mid-1930s (Fig. 4.47). In fact, the anomaly at 700 hPa was as large as—or even larger than—at the surface. The same result is found in 20CR, which does not include upper-air data (note that 20CR suffers from a bias in Arctic surface air temperatures due to a misspecification of sea ice; see Compo et al. 2011). A different vertical structure is found during the present warm period. According to observations and all reanalysis datasets, current surface temperature anomalies are almost twice the magnitude of 700 hPa anomalies (Brönnimann et al. 2012a). Section 4.5.8 will go into more detail for the recent Arctic warming.

Large amounts of heat are transported into the Arctic at 850–700 hPa (Overland and Turet 1994). A stronger warming at 700 hPa than at the surface is consistent with circulation changes being the main cause. In fact, our reconstructions (REC1) reveal distinct atmospheric circulation patterns during the pre-warming phase (1912–1918), the warm period (1919–1929), and the peak of the warm phase (1930–1939; Fig. 4.47; Grant et al. 2009). The first period shows weak anomalies of 850 hPa geopotential height over the region. The second period shows an east–west dipole between a strong Siberian high and a strong Icelandic low, similar to the Arctic Dipole. The pressure contrast further increases in the third period when the anomaly centre of the Siberian high strengthens and moves westward to Scandinavia. The large gradients in 850 hPa geopotential height over the North Atlantic are consistent with warm air being advected into the Arctic. The patterns seen in this analysis also dominate the trend maps of 700 hPa geopotential height for the 1890–1945 period shown in Fig. 4.44.

20CR shows similar circulation patterns (not shown). Support for these patterns also comes from an unexpected source: ice cores. Ice cores archive aerosols, which originate from lower-latitude continental regions and are transported to the Arctic by atmospheric circulation, particularly in winter. Thus, large amounts of aerosols provide indirect evidence for increased meridional transport. In fact, peculiar spikes in sulphate concentrations are found in an ice core from Svalbard, which are consistent with reconstructed circulation changes (note, however, that ice cores do

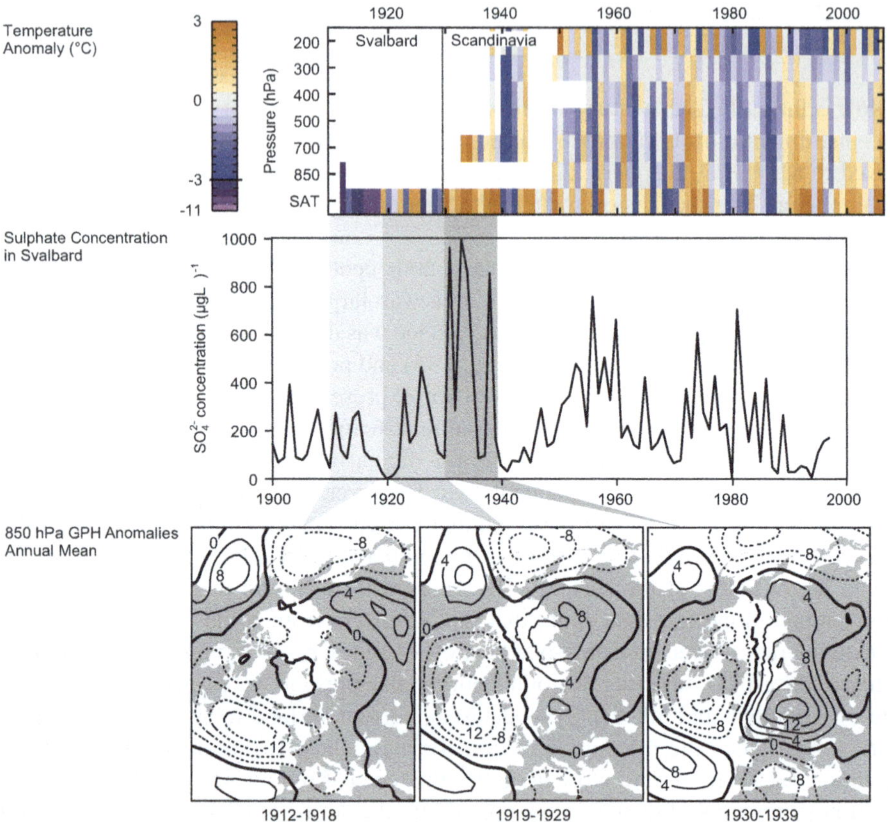

Fig. 4.47 (*top*) Time–height cross section of seasonal-mean temperature anomalies as a function of pressure and year for the European Arctic in winter (Data: CHUAN/IGRA at upper levels from Stickler et al. (2010) and CRUTEM3v at the surface from Brohan et al. (2006)). (*middle*) Sulphate concentrations as a function of time from the Lomonosovfonna ice core (Svalbard). (*bottom*) Reconstructed 850 hPa geopotential height anomalies (relative to 1961–1990) for annual means in (*left*) 1912–1918, (*middle*) 1919–1929, and (*right*) 1930–1939 (From Grant et al. 2009, reproduced with permission, www.schweizerbart.de)

not provide a reliable record of year-to-year variability). The first period was cool and exhibited low aerosol deposition at Svalbard, presumably reflecting clean Arctic air. The first warming phase (1919–1929) had an east–west pressure gradient over the northern North Atlantic that brought more warm (though still rather clean) air from the Atlantic sector. Sulphate in the ice core increased to medium values. During the peak of the warming, the pressure gradient was strengthened and the advection of warm and polluted air from western Europe was more frequent. This led to a large spike in sulphate concentrations (Grant et al. 2009). The machinery was operating at full strength. Over the Pacific, a similar east–west dipole had developed earlier and remained remarkably persistent.

The poleward flow over the eastern North Pacific and especially over the eastern North Atlantic indicates an increased poleward atmospheric heat flux. For the entire

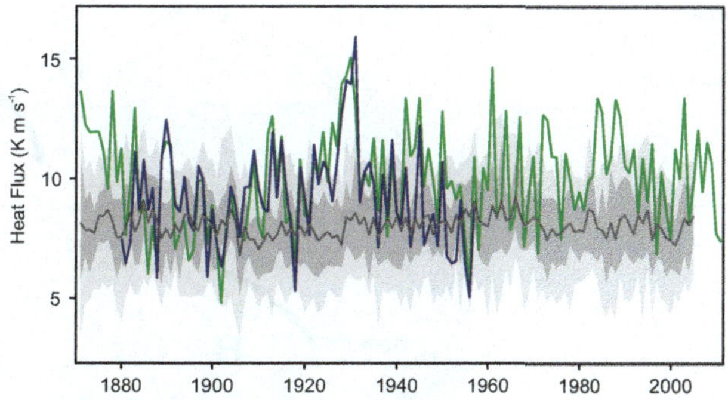

Fig. 4.48 Atmospheric meridional heat flux by stationary eddies at 60° N and 700 hPa, calculated based on monthly data and then averaged for December–February from 20CR (*green*), REC1 (*blue*; here 3-km winds were used), and CCC400 (*grey*; *shading* indicates the 50 % and 90 % ensemble spread)

Arctic, we can quantify the role of the unusual circulation pattern by analysing the contribution of stationary waves (monthly averages) to the meridional heat flux $[\overline{v}^{*}\overline{T}^{*}]$ at 60°N and 700 hPa (see Fig. 3.7) from 20CR, REC1, and CCC400. The results (Fig. 4.48), averaged for the December–February season, show considerable year-to-year variation. They also imply an increase in the poleward heat flux from stationary waves (note that the data do not allow us to quantify the contribution of eddies to the heat flux) from the 1890s to the 1940s, with a peak around 1930. The agreement between 20CR and REC1 is very good. This increase in stationary wave heat flux is another feature that distinguishes the early 20th century warming of the Arctic from the recent warming of the Arctic (see Sect. 4.5.8). Although no conclusions about the total heat flux can be drawn, the trend in the stationary wave component is unique to this period.

Thus, anomalous atmospheric circulation over the North Pacific and North Atlantic sectors arguably played a role for the early 20th century Arctic warm period. A schematic depiction of the processes leading to the Arctic warming is given in Fig. 4.49.

Was the anomalous atmospheric circulation a forced response of the machinery or random variability? A candidate for the latter is sea surface temperatures, particularly in the North Atlantic. Our CCC400 simulations (Fig. 4.50), forced with observed sea-surface temperatures, allow us to address the forced component to an extent. The simulations reproduce the Arctic warming in the early 20th century. They also reproduce a pressure dipole over the eastern North Atlantic, a strengthening of the Siberian high, and a dipole over the eastern North Pacific, particularly in winter. CCC400 simulations also show a slight increase in the stationary wave heat flux from 1906 to 1960 (Fig. 4.48). However, the magnitude is strongly underestimated. Thus, sea surface temperatures might have played a role in forcing and maintaining the anomaly, but they are insufficient to explain the warming.

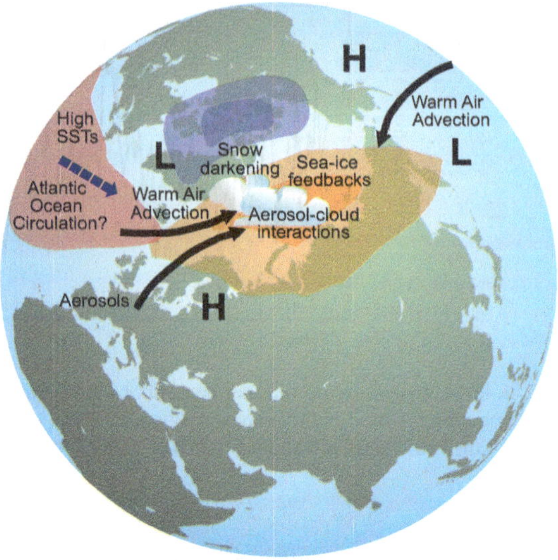

Fig. 4.49 Schematic depiction of the early 20th century Arctic warming

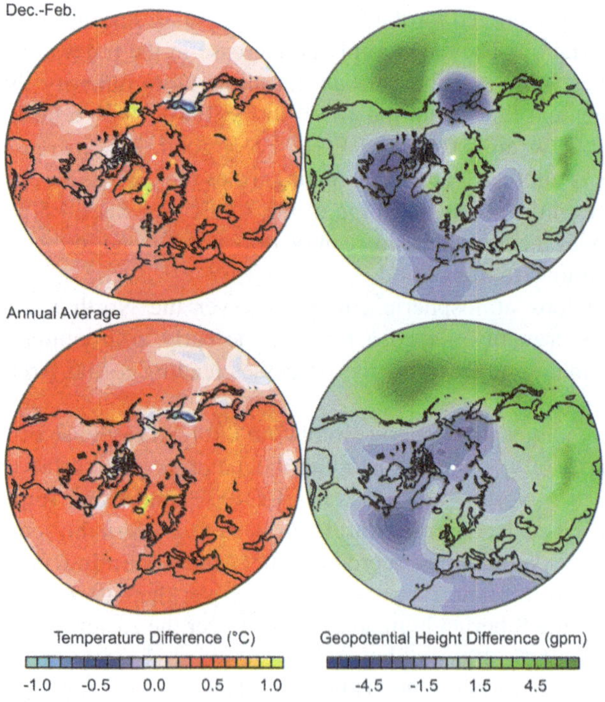

Fig. 4.50 Difference fields of (*left*) surface air temperature and (*right*) 850 hPa geopotential height in CCC400 between the periods 1920–1939 and 1900–1919 for the (*top*) boreal winter (Dec.–Feb.) and (*bottom*) annual average

Once the anomalous circulation was established, Arctic amplification processes may have set in through changes in snow, ice, radiation, clouds, and water vapour. These powerful feedback mechanisms can amplify an initial warming (see Sect. 3.4). In addition, the transport of aerosols into the Arctic might have further enhanced the warming. This may sound counterintuitive at first, since aerosols are normally expected to have a cooling effect. In the Arctic, however, the aerosol-induced reduction of surface solar shortwave radiation is less important than on a global scale because of the bright Arctic surface. During the polar night, indirect effects such as an aerosol-induced increase in cloud cover—and, thus, in downward longwave radiation—may become relevant (e.g., Garrett and Zhao 2006). Moreover, aerosol-induced snow darkening may have a converse effect on the energy balance. However, both hypotheses remain to be confirmed.

The early 20th century is a key period for our understanding of global climate. It seems that increased atmospheric heat flux from the midlatitudes was mainly responsible for the warming. Anomalous sea-surface temperature in the Atlantic may have played a role in setting up the atmospheric circulation anomaly. The magnitude of the event might point to additional feedback mechanisms, some of which remain to be discovered. This case demonstrates that the machinery may produce large variations.

4.3.3 The "Dust Bowl" Droughts of the 1930s

In the 1930s, concurrent with the Arctic warming, the United States experienced a decadal-scale climatic anomaly: the "Dust Bowl" droughts. This event was particularly devastating because it concurred with the Great Depression. The drought, which primarily affected the U.S. Midwest, was accompanied by intense dust storms, sometimes lasting for days, which blew away the topsoil and turned farmland into desert (see Fig. 4.51). Thousands of farmers in Oklahoma and Kansas were forced to leave their homes. The social history and economic consequences of the droughts are described in the works of Worster (1979) and Hurt (1981). This drought, more than any other, persists in the collective memory of America through Steinbeck's "The Grapes of Wrath", the songs of Woodie Guthrie, or the photos of Dorothea Lange.[13,14,15] It is often seen as a manmade catastrophe because unsustainable land use techniques enhanced soil erosion. However, climate was exceptional during that period. What was the cause of this decade of drought? Could

[13]John Steinbeck, 1902–1968, was an American writer. Also of interest to climatologists is his 1945 novel "Cannery Row". It describes the boom of sardine fishery in California in the early 1940s, which was followed by rapid demise in the 1950s, both of which have been attributed to the Pacific Decadal Oscillation.

[14]Woodie Guthrie, 1912–1967, was an influential singer and songwriter of folk music.

[15]Dorothea Lange, 1895–1965, was an American documentary photographer and photojournalist.

Fig. 4.51 Aerial view of the beginning of a dust storm over the prairies east of Denver in 1935. Northerly winds removed topsoil and clouds of dust were raised to the free troposphere. Prevailing west winds carried some of the dust as far east as the Atlantic coast (Photo from NOAA's NWS Collection)

it have been predicted? In this section, I will summarise recent studies on this climate event and trace its development in reconstructions and historical upper-air data.

4.3.3.1 Development and Causes of the Dust Bowl

Tree ring-based drought reconstructions for the U.S. (Cook et al. 2004) show that the Midwest and western U.S. are frequently affected by droughts. From a long-term perspective, it was not the drought but rather the pluvial period preceding the drought and peaking in the 1910s (see Cook et al. 2004) that was unusual and facilitated the intensification of agriculture in the region (see Fig. 4.6). In fact, Brückner witnessed and reported this pluvial period in 1912 and forecasted it to peak in 1920.[16] The study of past droughts, some of which have lasted for several decades, helps to clarify the processes responsible for droughts in the Midwest and western U.S., including the Dust Bowl droughts of the 1930s.

[16]Brückner participated in an expedition across the U.S. in 1912. He reported on agricultural activities in the semi-arid Great Basin and noted that settlement in the western U.S. during 1870–1890 was facilitated by pluvial conditions, while land was abandoned in the 1890s due to drought.

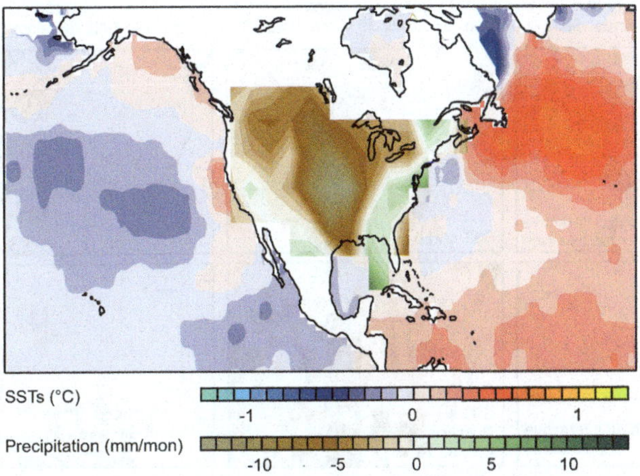

Fig. 4.52 Sea-surface temperature and precipitation anomalies during the Dust Bowl (1931–1939) relative to 1921–50 (From Brönnimann et al. 2009a, John Wiley & Sons)

Considerable progress was made during the last decade to isolate the mechanisms behind this event. Based on model simulations, oceanic forcing has been suggested as a trigger (e.g., Cook et al. 2008; Schubert et al. 2004a; Seager 2007). This is consistent with studies based on climate archives (McCabe et al. 2004; Woodhouse and Overpeck 1998). The oceanic setting that is conducive to droughts in the U.S. Midwest consists of a cold tropical Pacific and a warm tropical Atlantic—exactly the pattern that occurred in the 1930s. The sea-surface temperature anomalies during the Dust Bowl (Fig. 4.52) show a cool Pacific (both tropical and extratropical) and a warm Atlantic (also, both tropical and extratropical). The negative precipitation anomaly covered the entire U.S. Midwest and extended to the Canadian Plains and to the Pacific Northwest.

Prescribing these sea-surface temperatures in climate model simulations, many authors find a precipitation deficit in the U.S. Midwest (see Fig. 4.53) and an increase in the Gulf of Mexico (the latter is also seen in the observations; see Fig. 4.52) in the 1930s. However, the precipitation anomalies are neither pronounced nor in the right place in many studies (Fig. 4.53). For instance, in CCC400 the anomaly is too far east and in CASTRO too far south. While there is a lot of evidence that the oceans have provided the large-scale setting for drought, they alone are not sufficient to explain the Dust Bowl droughts.

The precipitation deficit induced by anomalous sea surface temperatures was probably amplified by land–atmosphere interactions (e.g., Schubert et al. 2004a; also see Sect. 3.4). Additional contributions may have come from atmospheric dust and human-induced land degradation (see Cook et al. 2008, 2009). When Cook et al. (2009) included these factors in their model simulations, the drought strengthened and became increasingly realistic (see Fig. 4.53).

To understand how sea surface temperatures may have triggered the Dust Bowl drought, we have to look at the machinery on a continental scale (see

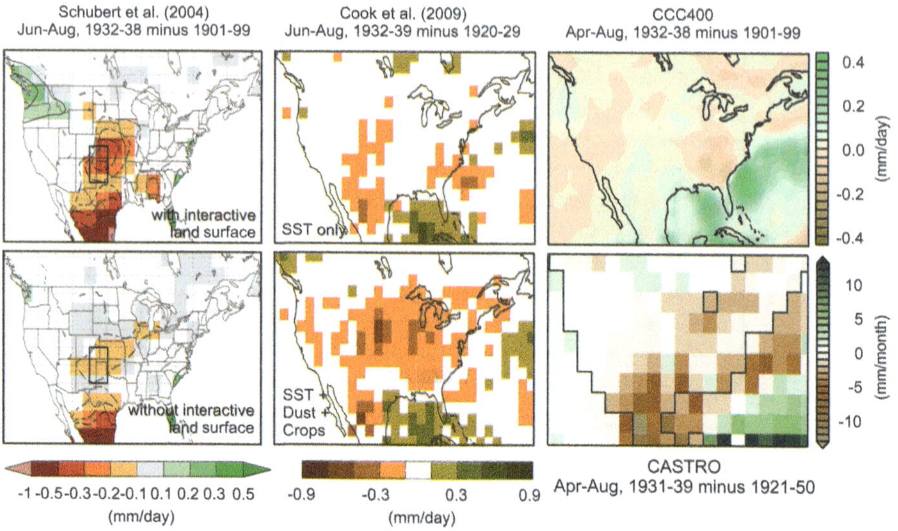

Fig. 4.53 Modelled precipitation deficit during summer for the "Dust Bowl" drought using different models and setups. Results from (*left*) Schubert et al. (2004b, reprinted with permission from AAAS), (*middle*) Cook et al. (2009, reproduced with permission), and (*right*) from (*top*) CCC400 and (*bottom*) CASTRO. Note the different definitions of drought years, reference period, and season. The *bottom left* and *middle panels* show experiments in which the land surface was not interactive (*left*) or in which additionally atmospheric dust and crops were considered (see Schubert et al. (2004b) and Cook et al. (2009) for details)

Brönnimann et al. (2009a) for the following). Specifically, we need to consider the typical pathways of moisture advection into the region during summer. Apart from occasional storms from the Pacific, most of the moisture originates from the Gulf of Mexico and the Caribbean Sea. It is drawn into the region by a nocturnal circulation system over the Great Plains known as the "Great Plains low-level jet" (Helfand and Schubert 1995; Higgins et al. 1997; Rasmusson 1967). Apart from transporting moisture, the Great Plains low-level jet also affects low-level convergence and nocturnal convective systems (Higgins et al. 1997).

Interannual changes in the Great Plains low-level jet can be induced by changes of the subtropical highs: the North Pacific high and especially the Azores high. In turn, these changes can be initiated by sea-surface temperature anomalies in the tropical Pacific and the Atlantic (e.g., Rodwell and Hoskins 2001).

A second factor affecting drought in the Great Plains is persistent midtropospheric ridging (Chang and Wallace 1987; Namias 1982). The accompanying subsidence suppresses convection to reduce precipitation (note that precipitation is predominantly convective in the U.S. Midwest during the warm season). The flow in the upper troposphere must be conducive to ridge formation, with the jet shifted poleward. Various factors such as large-scale oceanic forcing, regional land–atmosphere interaction, and atmospheric dust have been suggested to be important in maintaining the ridge (Namias 1982).

Another factor for drought is low antecedent precipitation. Soil moisture, and hence precipitation in spring, has been suggested to affect summertime precipitation in the Great Plains (Oglesby 1991; van der Schrier and Barkmeijer 2007). In winter and spring, depending on the upper-level flow, rain-bearing disturbances from the Pacific can propagate onto the continent. This is also the season when the flow over the North Pacific responds to tropical Pacific sea-surface temperatures via the Hadley circulation (see Sect. 3.2.3).

Thus, decisive factors for drought in the U.S. Midwest are a decreased strength of the low-level jet, a buildup of the Great Plains ridge, and an anomalous upper-level flow from late winter to summer, which includes a poleward shift of the jet and positive geopotential height anomalies over the contiguous United States.

4.3.3.2 The Dust Bowl in CCC400, 20CR, Reconstruction and Upper-Level Data

Were these features present during the Dust Bowl? The available gridded datasets as well as historical upper-air data can help to diagnose atmospheric circulation during the Dust Bowl. In the following, we complement our previous studies on this topic (Brönnimann et al. 2009a) with analyses of new datasets such as 20CR.

Historical upper-air wind observations allow us to focus on nocturnal conditions over the Great Plains. When we used these data in 2009 to compare the low-level jet in the 1931–1939 period with that in the wet years 1941–1944 (Brönnimann et al. 2009a), we were unable to assess their homogeneity. Only with the emergence of 20CR, could the data be assessed. Ramella Pralungo and Haimberger (2014) found a major network-wide inhomogeneity in wind direction around 1940. This shows how important it is to work with homogeneous data (see Sect. 2.5.2).

The new homogenised data (Fig. 4.54) show a change in the Great Plains low-level jet relative to the surrounding years that is similar to (although less pronounced than) the original results. The jet was weakened above the maximum and east of the core during the Dust Bowl. However, the 1941–1994 period was also atypical. Relative to 1901–1999, the 700 hPa wind in reconstructions and 20CR show a different zonal wind anomaly, but the weakening of the southerly component remains (Fig. 4.55). Thus, the weakening of the low-level jet is confirmed by various analyses.

The Great Plains ridge is also prominent during the Dust Bowl droughts in 20CR and in reconstructions. Subsidence associated with the ridge may have led to a more stable atmosphere. Sparse aircraft ascents (Fig. 4.54) indicate that this might have been the case in the lowest 1.5 km.

With respect to the upper-level flow, reconstructions and 20CR show that during the Dust Bowl drought in spring and summer, positive 200 hPa geopotential height anomalies extended from the North Pacific across North America to the Atlantic. Thus, the expected features can be confirmed from the historical data products and a good mutual agreement is found among the products.

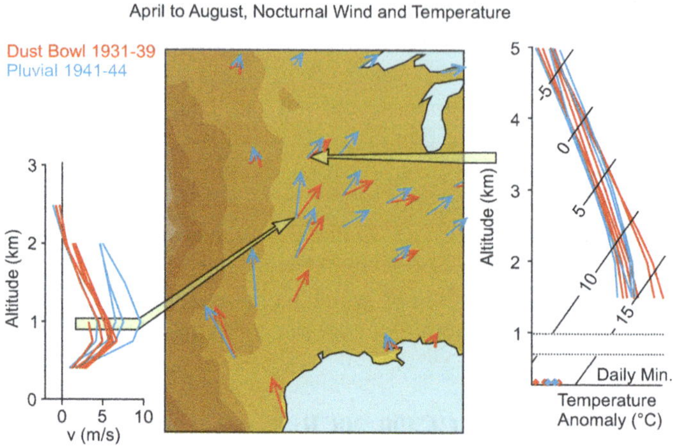

Fig. 4.54 Observed nocturnal winds over the Great Plains during the "Dust Bowl" droughts 1931–1939 and the subsequent wet years 1941–1944. (*left*) The vertical profile of meridional wind at Wichita, Kansas, averaged for each summer. (*middle*) Map of the nocturnal wind at 1 km altitude averaged for the two periods. (*right*) Temperature profiles from Omaha, Nebraska, averaged for each summer (Adapted from Brönnimann et al. (2009a), using homogenised winds from Ramella Pralungo and Haimberger (2014))

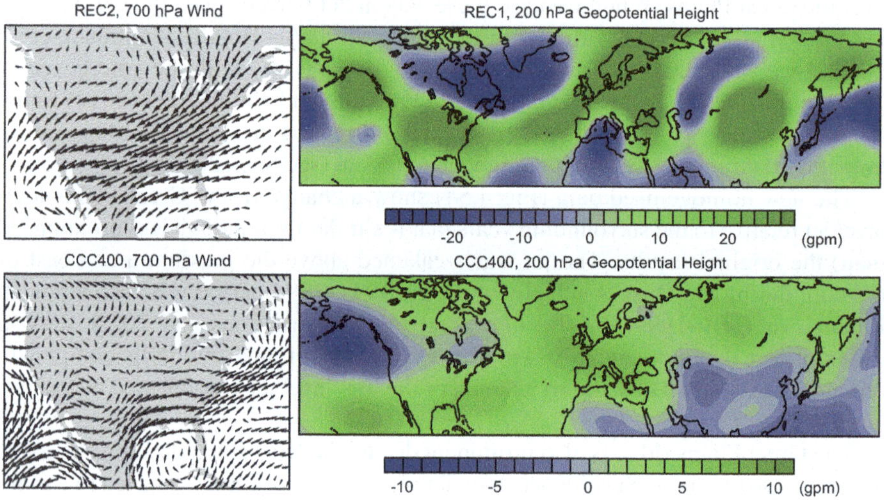

Fig. 4.55 Anomalies of (*left*) 700 hPa wind and (*right*) 200 hPa geopotential height in (*top*) reconstructions and (*bottom*) CCC400 for April to August, 1931–1939 relative to 1902–1999

To what extent were these features caused by sea-surface temperature anomalies? It turns out that CCC400 and CASTRO reproduce the three features to some extent, although the drought is too far east. The magnitudes of the anomalies are much smaller in the ensemble mean of CCC400 than in the observation-based data, but individual members capture the drought. The flow pattern at 700 hPa (Fig. 4.55) shows a decreased southerly flow—similar to reconstructions and 20CR—pointing to a weakening of the Great Plains low-level jet. The Great Plains ridge is less well reproduced in CCC400. Geopotential height anomalies at 200 hPa in CCC400 over the Great Plains roughly correspond to those in REC1. Because sea surface temperature anomalies are arguably the most important boundary condition in CCC400, the flow structure at 200 hPa might be a response to sea surface temperature anomalies.

4.3.3.3 Puzzle Solved?

From these results and the numerous recent papers (e.g., Cook et al. 2009; Schubert et al. 2004a; Seager 2007), a conceptual view of the Dust Bowl droughts emerges that exposes some characteristics of the machinery (Fig. 4.56). The air flowing into the Gulf of Mexico was arguably moister than normal due to the warmer Caribbean Sea. However, it rained out more quickly (leading to increased precipitation over the Gulf of Mexico) so that less precipitation fell over land. The tropical Atlantic (by affecting the low-level flow from the tropical Atlantic to the Great Plains) and the tropical Pacific (by inducing a planetary wave pattern that reduces the Pacific influence and enhances the Great Plains ridge) may have led to the establishment of a circulation pattern that caused a precipitation deficit in the Great Plains. The pattern might have been further amplified by land–atmosphere interaction (which reduced evaporation) and dust (which heated upper layers and cooled the surface, thus stabilising the atmosphere), both of which also depend on land management.

The suspected oceanic trigger is relevant because sea-surface temperatures are partly predictable a year ahead. In this sense, the Dust Bowl remains an interesting

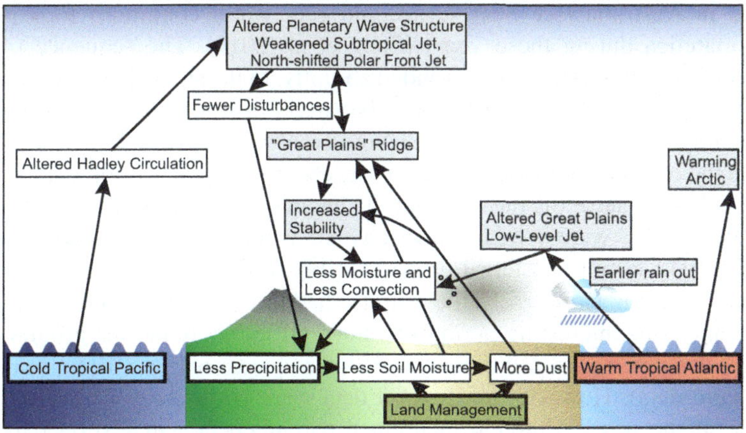

Fig. 4.56 Schematic depiction of the mechanisms causing the "Dust Bowl" droughts (Adapted from Brönnimann et al. 2009a)

Fig. 4.57 German soldiers in Russia in January 1942. Temperatures dropped to −56 °C (Provided by Barry L. Geipel, www.geipelnet.com/war_albums/otto/)

test case for seasonal forecasting. Although not all U.S. Midwest droughts are of this type, the Dust Bowl serves as an example for other droughts, including those of 2010–2013. Thus, the machinery seems to fall into this mode repeatedly.

4.3.4 Global Climate Anomalies During the 1939–1942 El Niño

In 1939, the world plunged into the Second World War. Concurrently, the global climate entered a peculiar phase with various climatic anomalies during the 1940–1942 period. Winters were bitter cold in eastern Europe (Fig. 4.57), but very warm in Alaska, resulting in an increased forest fire area (though non-climatic factors also contributed; see Lutz 1956). Springs were rather wet in central Europe, drought occurred in Australia (see Sect. 4.5.6), and very warm conditions took place in West Antarctica during these years (Steig et al. 2013). The sequence of warm winters in Spitsbergen (Sect. 4.3.2) ended abruptly. In the stratosphere, total column ozone was extremely high over Europe, the Arctic, North America, and East Asia. Concurrently, a long-lasting date line El Niño event was ruling in the tropical Pacific. So, were the climate anomalies different aspects of the same phenomenon?

Despite (or because of) the fact that the anomalies occurred during the Second World War, some were noted and discussed, while others were not. The cold winter of 1942 in northeastern Europe (see Fig. 4.57) even had an effect on the outcome of the war. Conversely, the war also had large effects on observations and, consequently, this interesting period was not well monitored at the ground (where data were destroyed or measurements were interrupted; see Fig. 2.19 for sea-surface temperature data). However, upper-level data were particularly well documented (because of the war). Although mentioned parenthetically in several papers, the 1940–1942 climate anomalies were only later on revisited in more detail (Lejenäs 1989).

4.3.4.1 The 1940–1942 Climate Anomaly in REC1, CCC400, and HISTOZ

This period is analysed in more detail for several reasons. First, I think it is one of the most pronounced and interesting climatic anomalies of the 20th century. Second, for me personally this climatic event stood at the beginning of a long enterprise that started in 1999 and eventually led to this book. In an effort to explain total column ozone anomalies over Switzerland during 1940–1942, I started to digitise long-forgotten historical upper-air data and produced three-dimensional hemispheric climate reconstructions that reach into the stratosphere. Finding El Niño as a major player was then a complete surprise. This not only convinced me to consider climatic changes as variations in a global three-dimensional, physical-chemical system, but also showed me the potential of sparse historical weather information. Recovering historical upper-air data remained a key activity. Fifteen years later, I am still devoting much of my energy towards that aim.

Let us first focus on the three (boreal) late winter–early spring periods (Jan. to Apr.) of 1940, 1941, and 1942 (Fig. 4.58) relative to 1961–1990 (CCC400 data are contrasted with La Niña winters 1939, 1943, and 1944). The three El Niño winters were characterised at the surface by extreme negative temperature anomalies over northeastern Europe but positive anomalies in Alaska. The sea-level pressure field shows a negative NAO and positive PNA pattern. Statistical reconstructions indicate a weak polar vortex and a lower stratosphere that was warm over northern Eurasia but cold over North America.

The cold Eurasian winters can be related to frequent cold-air outbreaks that can be directly studied in historical radiosonde ascents (Fig. 4.59, left) and 20CR. They were often related to blocking situations over the Atlantic and upper-level troughs or cutoff lows over central Europe. Two examples of omega-shaped blocks are shown in Fig. 4.59. The frequency of green stripes (as opposed to brown ones) in the lower left part of the figure indicates that such situations were particularly frequent during the 1940–1942 period. These situations are accompanied by a low tropopause and, thus, a high stratospheric mass, with locally increased total column ozone due to convergence in the lower stratosphere (Fig. 4.59, upper left). This explains part of the ozone increase over Switzerland.

However, this is not the whole story. If redistribution alone was responsible for the increase, we would expect it to cancel out when averaging in space. Although our sampling is sparse (there are six series from the northern extratropics), it is noteworthy that positive anomalies occurred at all locations, including central Europe, North America, China, and the Arctic (Fig. 4.60, left). Our HISTOZ data (Fig. 4.60, right) show an ozone increase in the mid- to high latitudes in the years 1940–1942 relative to the neighbouring years 1939, 1943, and 1944; and a decrease in the tropics (where no observations are available). This pattern is typical for a change in the strength of the Brewer-Dobson circulation (also see Sect. 3.1.3), transporting ozone from its tropical source regions to the extratropics.

How do these pieces fit together? The large-scale circulation anomaly featured a negative NAO and positive PNA pattern. This anomalous wave pattern expressed itself in the surface climate as more frequent troughs and cutoff lows over central

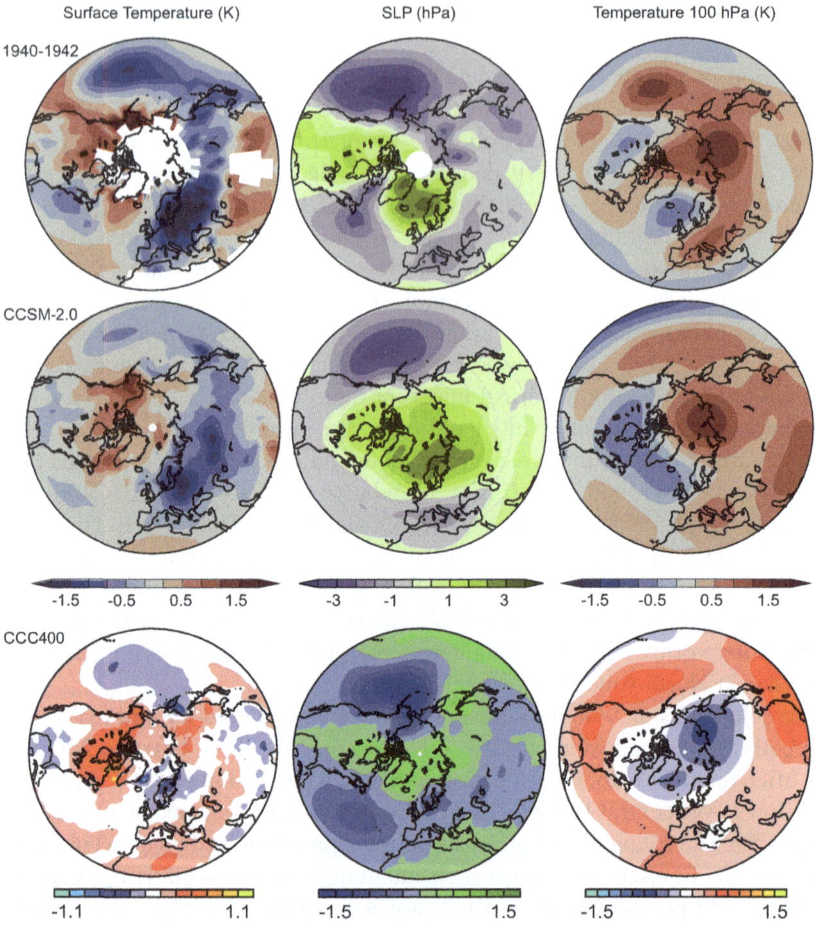

Fig. 4.58 Boreal late winter (Jan.–Apr.) anomalies of surface air temperature, 100 hPa geopotential height, and 100 hPa temperature for 1940–1942 (relative to 1961–1990) from (*top*) observations and reconstructions, and (*middle*) strong El Niño events in the control simulation of the Community Climate System Model CCSM 2.0 (from Brönnimann et al. 2004). (*bottom*) The corresponding anomalies (relative to La Niña years 1939, 1933, and 1944) from the CCC400 ensemble mean

Europe. In the troposphere, these were accompanied by frequent cold-air outbreaks. In the stratosphere, these affected the local total column ozone through redistribution. Easterly weather types in 20CR (see Sect. 2.9) were more frequent over Switzerland than normal; westerly to northerly types were rare. Weather patterns also changed in the North Pacific region. In the Southern Hemisphere, a change in the planetary wave pattern led to high temperatures over West Antarctica. The combination of a positive PNA and a negative NAO was conducive to increased upward propagation of planetary wave activity to the stratosphere (see Sect. 3.2.5

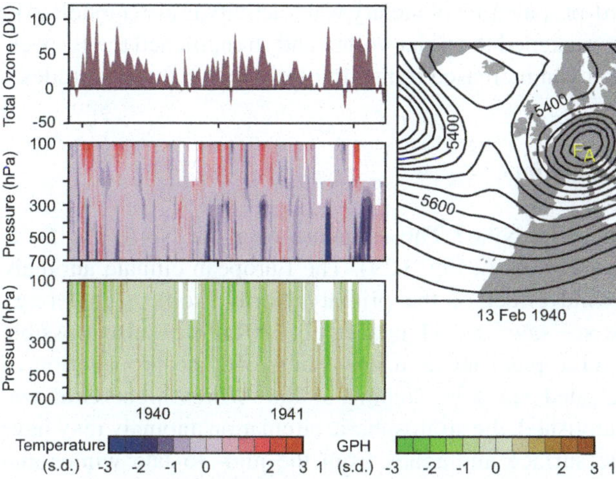

Fig. 4.59 (*left*) Daily anomalies of total ozone at Arosa, Switzerland, and standardised daily anomalies (with respect to the 1948–2001 mean seasonal cycle) of geopotential height and temperature at Freiburg, Germany, from December 1939 to March 1942, smoothed with a Gaussian filter ($\sigma = 3$ days) (Brönnimann et al. 2004, John Wiley & Sons). (*middle*), (*right*) Two examples of omega-shaped blockings over the Atlantic, with cutoff lows over central Europe from 20CR. The letters "F" and "A" mark Freiburg and Arosa (Adapted from Brönnimann and Compo 2012)

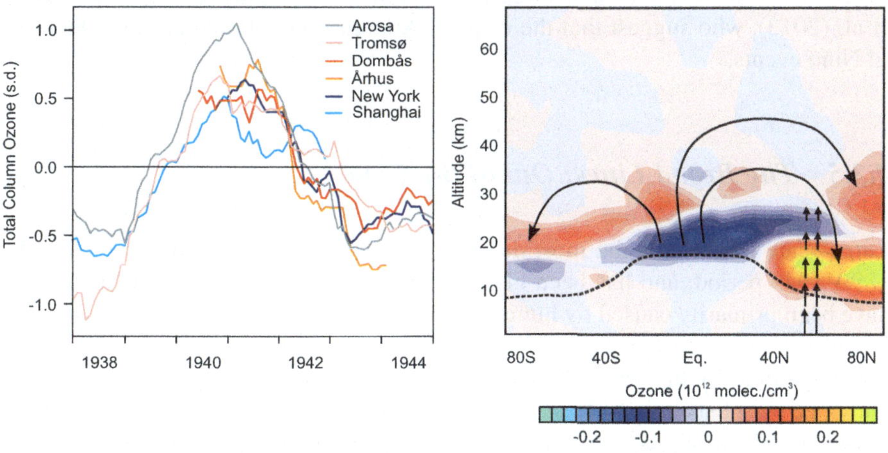

Fig. 4.60 (*left*) The 1940–1942 total column ozone anomaly. Standardised monthly anomalies (with respect to the 1938–1944 mean annual cycle) of total ozone at Arosa (Switzerland), Tromsø and Dombås (Norway), Århus (Denmark), New York (United States), and Shanghai (China) smoothed with a 12-month moving average. (*right*) Zonally averaged ozone concentrations from HISTOZ in 1940–1942 (annual means of the ensemble mean) relative to the annual averages of 1939, 1943, and 1944. The *short arrows* denote the upward-propagating planetary wave activity. The *long arrows* indicate the Brewer-Dobson circulation, which was strengthened in this case (Adapted from Brönnimann et al. 2004, 2013a)

and Fig. 3.26). More upward-propagating planetary wave activity is also modelled in CASTRO (Fig. 4.42). This weakened the polar vortex and strengthened the Brewer-Dobson circulation, leading to an increased transport of ozone to the midlatitudes.

4.3.4.2 Wider Effects of El Niño

So, was all of this related to El Niño? The signature over the North Pacific is typical for El Niño events (Sect. 3.2.6; Fig. 3.29). The European climate anomaly could have been a downwind effect of the primary North Pacific signal (e.g., via the Aleutian-Icelandic see-saw; see Honda et al. 2001). It is also possible that ENSO activated the subtropical jet as a waveguide and the propagation of the signal to Europe proceeded via a weakening of the Azores high (Graf and Zanchettin 2012). Once established, the stratospheric circulation anomaly may have also helped to prolong the surface anomalies from the mid- to late winter and early spring (Ineson and Scaife 2009). The change in the stratospheric ozone field may have further weakened the stratospheric flow by changing the equator-to-pole temperature gradient. In any case, the climatic anomaly pattern of the 1940–1942 El Niño, though much larger in amplitude, fits well with statistical results for date line El Niño teleconnections.

We do not know why the machinery produced a date line El Niño event in the first place. Internal variability of circulation over the tropical Pacific is the most likely trigger. Nevertheless, it is interesting in this context to read the recent work of Ham et al. (2013), who suggest that the tropical Atlantic can possibly activate date line El Niño events.

4.3.5 The Broad Lines: Out of the Cold

Large-scale warming is the most distinctive feature of climatic change during the 1890–1945 period, and has been given as an example of a warming that may not have been primarily caused by human activity. In a pioneering study, Delworth and Knutson (2000) showed that climate models were in principle able to reproduce the magnitude of the early 20th century warming. However, it is a rare event. Only one out of six ensemble members showed a prominent warming. More recent model studies mostly confirm this view (see Bindoff et al. 2013). In Sect. 4.3.1, we argued that part of the warming might have been forced, while another fraction can likely be explained by internal variability such as changes in oceanic or atmospheric circulation—an extreme mode of operation of the machinery.

The cases discussed in this chapter elucidate this mode of operation. The Arctic warming was related to circulation anomalies, which may have been triggered by a warm North Atlantic. However, regional amplification factors such as aerosols, clouds, and sea ice may have also played a role. Likewise, the Dust Bowl was triggered by sea-surface temperature anomalies—the same warm North Atlantic and the cool Pacific. Again, there is a possible role for regional feedback mechanisms such as dust emissions or land–atmosphere interaction. The global climatic anomalies of 1940–1942 were due to high tropical sea surface temperatures with a possible role of stratospheric processes.

The early 20th century warming allows an appreciation for the connectedness of the earth's climate system. The climatic changes can be seen in the ocean, atmosphere, land surface, and sea ice, and reached into the stratosphere. Atmospheric constituents such as mineral dust, sulphur, and ozone may have also played a role.

Together, these three cases make up a large fraction of the early 20th century warming. Excluding the Arctic winters and the North American summers from the global land surface temperature record, the warming appears weaker and smoother than the global average (Brönnimann 2009). This suggests that the warming tropical oceans and the abruptly warming North Atlantic shaped the early 20th century warming, with regional feedbacks possibly amplifying the responses.

A key role seems to have been played by the Atlantic Ocean. In fact, many studies on the early 20th century warming in the 1990s and 2000s focused on the Atlantic (Schlesinger and Ramankutty 1994). The early 20th century warming thus might be part of an oscillatory mode of the Atlantic meridional overturning circulation. On the other hand, Fyfe et al. (2013) found that climate models are able to reproduce the early 20th century Arctic warming without reproducing changes in the Atlantic circulation.

Recently, Thompson et al. (2015) suggested that an increase in westerly winds over the central tropical Pacific (i.e., more frequent El Niño situations) might have caused the initial phase of the early 20th century warming by releasing heat to the atmosphere. They show that understanding the accelerated global warming phase of the early 20th century also has implications for understanding the current slowdown of global warming (Sect. 4.5.9).

While speculating about oceanic warming, it should be kept in mind that sea-surface temperatures are also uncertain. Thompson et al. (2008) and Brohan et al. (2009) showed that the well-known (but inadequately corrected) change in the measurement technique (see Sect. 2.6.1) and the poor data coverage during the Second World War (see Fig. 2.19) led to large errors in records of global sea surface temperatures. The latest version of sea surface temperatures has corrected some of these errors (Kennedy et al. 2011). However, uncertainties remain.

If the Atlantic played a key role, what was driving its sea-surface temperatures? Obviously, forcings such as greenhouse gases may have played a role. Booth et al. (2012), in a much-discussed paper, point to aerosols as possible drivers. Häkkinen et al. (2011) suggest that atmospheric blocking warmed the subpolar Atlantic. A recent paper by Estrada et al. (2013) also detected anthropogenic influences down to an interannual scale. Perhaps the atmosphere was driving the oceanic anomalies (Wood and Overland 2010). Or, perhaps the causes are to be found within the Atlantic itself, for instance, a change in deep water formation.

An understanding of processes in the Atlantic Ocean and its interaction with the atmosphere is considered important for seasonal-to-decadal climate predictions in Europe and North America. This is not new. In fact, the relevance of a connection between the tropical Atlantic and Europe for climate prediction was already recognised during Brückner's time. In 1905, Napier Shaw (Shaw 1905) wondered

> "whether there is any definite and (...) useful connection between the pulsations of the south-east trade wind [over the tropical Atlantic] and the rainfall in north-western Europe"

If only we had a better understanding of the machinery!

Box 4.2 Early greenhouse effect theories: A bright future

The theory of temperature changes being caused by greenhouse gases goes back to the work of Tyndall and, in the 1890s, De Marchi (1895b) and Arrhenius (1896).[17]

Arrhenius' famous 1896 paper was the first to quantify the global effects of a doubling of CO_2 levels (see Fig. 4.61). His work sought an explanation for the ice ages—the major unsolved problem of that time. His theory was rapidly dismissed and only resurfaced in the 1930s (Callendar 1938) and in the 1950s (Plass 1956). Several books have been written about the history of the greenhouse gas theory (Fleming 2007; Weart 2008). From a present-day perspective, the value judgements inherent in some of these early greenhouse gas theories are interesting. This is exemplified by the quote from Arrhenius (1906; translated from p. 57):

> "As a consequence of increased carbonic acid in the air, we may look forward to times of more stable and better climatic conditions, particularly in the colder regions of the world; times in which the earth will be able to bear a multiple of today's harvest yields, for the benefit of the growing human population."

Likewise, Guy Stewart Callendar,[18] in his seminal 1938 paper (the first paper relating a global temperature increase to CO_2; see Fig. 4.62; also see Fleming (2007) and Weart (2008)), concluded that

> "the combustion of fossil fuel (...) is likely to prove beneficial to mankind (...)."

In fact, global warming (including the anthropogenic greenhouse effect) had for a long time been considered beneficial from the dominating temperate latitude point of view. Even worse, plans were discussed (but mostly not taken seriously) to make the climate warmer (e.g., Wexler 1958). Weather modification was a major military driver of climate research at that time

(continued)

[17]John Tyndall, 1820–1893, was a British physicist known for his work on the interaction between radiation and air.

Luigi De Marchi, 1857–1936, was an Italian geophysicist.

Svante Arrhenius, 1859–1927, was a Swedish physicist and physical chemist. He received the Nobel Price in chemistry in 1903.

[18]Guy Stewart Callendar, 1898–1964, was a British steam engineer. He wrote several papers on atmospheric carbon dioxide and global temperature.

Box 4.2 (continued)
(Edwards 2010; Weart 2008). A different angle was taken by Revelle and Suess (1957) for the problem of carbon dioxide in the air. They wrote that

> "Human beings are now carrying out a large-scale geophysical experiment of a kind that could not have happened in the past nor be reproduced in the future. (...) This experiment, if adequately documented, may yield a far-reaching insight into the processes determining weather and climate."

Although the anthropogenic greenhouse effect is not depicted as negative or as a threat, this often-quoted perspective puts an emphasis on the scale, irreversibility, and responsibility of the carbon dioxide problem. The positive impression slowly gave rise to a predominantly negative one, but this process took a long time (also see Kellogg (1987) and von Storch and Stehr (2000); also see Box 4.3, p. 280). Well into the 1970s, many media reports on the carbon dioxide greenhouse effect emphasised increased food production and a hope for palms trees in England.

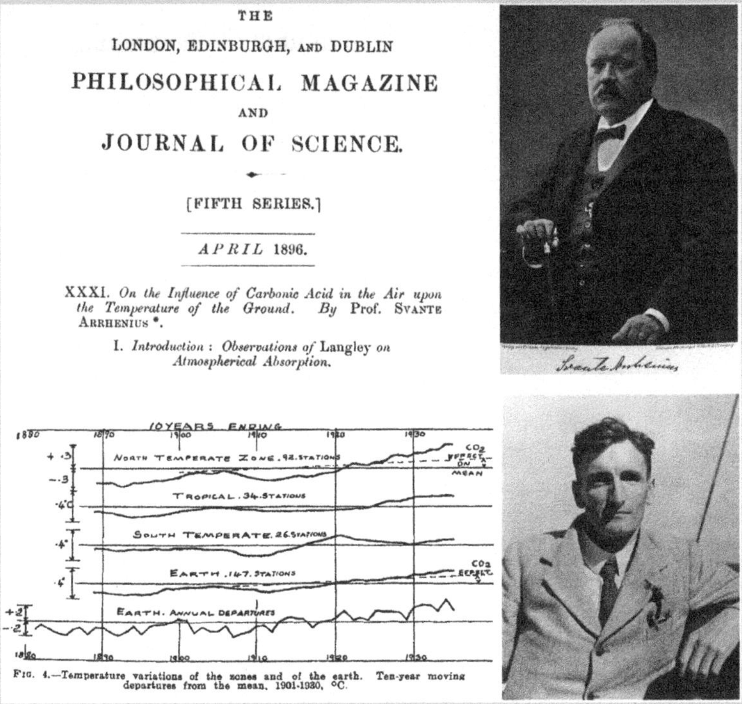
THE
LONDON, EDINBURGH, AND DUBLIN
PHILOSOPHICAL MAGAZINE
AND
JOURNAL OF SCIENCE.
[FIFTH SERIES.]
APRIL 1896.
XXXI. *On the Influence of Carbonic Acid in the Air upon the Temperature of the Ground.* *By* Prof. SVANTE ARRHENIUS *.
I. *Introduction: Observations of* Langley *on Atmospherical Absorption.*

Fig. 4.61 Two pioneers of the science of the greenhouse effect. (*top*) Svante Arrhenius with his 1896 paper "On the Influence of Carbonic Acid in the Air on Temperature of the ground" and (*bottom*) Guy Stewart Callendar (Callendar Archive, University of East

(continued)

Box 4.2 (continued)

Anglia) with his 1938 paper "The Artificial Production of Carbon Dioxide and Its Influence on temperature" (Copyright RMetSI). Note that Callendar, in the tradition of Köppen and Brückner, continued a global-mean temperature series

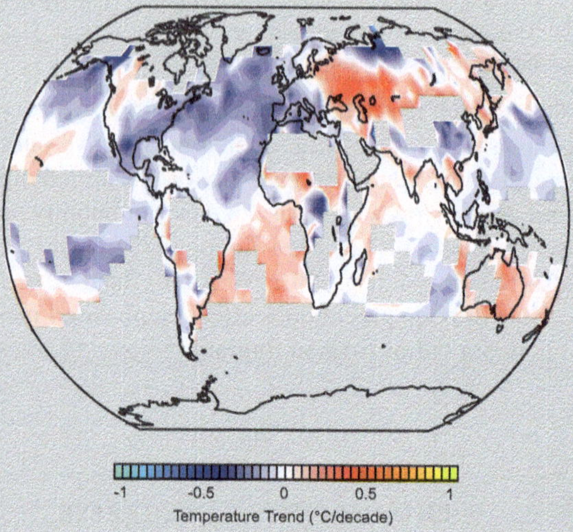

Fig. 4.62 Map of annual-mean temperature trends from 1945 to 1985 (based on anomalies from 1961 to 1990). At least 9 months per year and 75 % of all years must have data (Data source: HadCRUT4)

4.4 The Period from 1945 to 1985: Delusive Stability

A new era started after the Second World War. Geopolitics, societies, and economics changed. It was also a new era with respect to institutions, international organisations, and technology. Furthermore, the earth's climate experienced a turning point around 1945, entering a new phase. In terms of global-mean temperatures, the next three decades were a period of stagnation. This is often attributed to an increase of cooling anthropogenic aerosols in the atmosphere after the Second World War.[20]

Although the 1950s were a decade of increased drought frequency (Sheffield et al. 2009), many people perceived it to be the start of a period of climatic stability. The technological futurism of the 1950s and 1960s included climate and its consequences. Climate control appeared feasible, and even if climate cannot be fully controlled, then humankind would easily master the consequences. However, the climate stability was delusive in several ways. In Switzerland, for instance, the rather stable climate during this period (which coincided with a period of economic growth and planning of large infrastructure projects) and the absence of climate-related natural disasters (Pfister 2009) led to a loss of traditional disaster memory. In the Sahel region, steady rainy seasons facilitated the expansion of agriculture and growth of the population, which could not be sustained during the 1970s and 1980s. The stable global-mean temperature was also delusive because greenhouse gas forcing continued to increase. However, it was masked by other factors. When greenhouse gas forcing resurfaced in the late 1970s, it led to an even faster rise of temperatures.

The period 1945–1985 also witnessed considerable changes to the atmospheric constituents. With the post-war recovery of economies, there was an increase in production and consumption. In the United States and Europe, the automobile provided mobility for many, and a new lifestyle of mass production and consumption began—the "1950s syndrome" (Pfister 1995). The increase of CO_2 accelerated and followed an exponential growth curve. Air pollution became a large-scale (even global) issue. In the 1960s, acid rain started to damage sensitive lake ecosystems in Sweden, and transboundary pollution effects became a political issue. In short, atmospheric constituents became a new challenge and component of the machinery.

For contemporary climate scientists analysing past climate variations, a new era started in the years following the Second World War due to changes in the availability of data. The Conference of Directors in 1947 established the World Meteorological Organization. Within the United Nations, this finally achieved the goals of its precursor organisation, the International Meteorological Organisation (Edwards 2010). Scientific collaboration resumed after the Second World War (although now in a Cold War setting), especially with the "International Geophysical Year" in 1957/1958, a programme that has affected the state of climate sciences up to the present (see Sect. 2.2.2). At the same time, electronic processing of geophysical

[20]The stagnation ended in the 1970s. Nevertheless, I chose another year for delimitating the last phase of the book—1985, when the ozone hole was discovered.

data began. This enabled (at least) portions of weather data to be electronically archived, copied, analysed, aggregated, supplemented, and repeatedly analysed up to the present. Due to their repeated analysis, these data are better understood than those that were more recently digitised. For a good reason, many of today's global climate datasets (including some reanalyses; see Table 2.3) reach back to 1957.

In the following, we will take a closer look at the climate from 1945 to 1985 using the same datasets and models as in the previous section. This section starts with the global temperature record and some other characteristics. A subsection looks at changes in atmospheric constituents during this period. Then, I go into more details with respect to individual climate anomalies. These include the droughts and heatwaves in Europe in the late 1940s, the Sahel pluvial in the 1950s and 1960s and subsequent droughts of the 1970s and 1980s, and the 1976/1977 climate shift in the Pacific. Again, there were many other possible choices. These include the snow-rich and cold Eurasian winters of the 1950s, the Mt. Agung eruption of 1963, or the "Great Salinity Anomaly" in the North Atlantic (Häkkinen 1999) with a (possibly associated) shift in sea-surface temperatures in the late 1960s and early 1970s (Thompson et al. 2010).

4.4.1 Global Drivers and Global Changes

Even though global-mean temperatures do not show a trend over the 1945–1985 period, it is still interesting to analyse a map of temperature trends (Fig. 4.62). While some continental areas (large parts of Eurasia and Australia) and the South Atlantic continued to warm; cooling occurred over the Pacific, Indian, and North Atlantic Oceans. Cooling also occurred in northern Siberia and the United States. In the tropical Atlantic, the temperature trend resembles the Atlantic Meridional Mode (see Sect. 3.2.7). Thus, the spatial pattern of this temperature trend is quite different from the 1890–1945 trend (note, however, that the periods are of different lengths).

Trends in precipitation are extremely uncertain (see Hartmann et al. 2013). Those shown in Fig. 4.63 are taken from 20CR over the oceans and GPCC (Schneider et al. 2014b) over land. Results imply a southward shift of the ITCZ (which is clearly seen in EKF400; see Fig. 4.9). Regions of increasing precipitation are found in South America, Western Eurasia and North America, decreasing rainfall occurred in China and in the Sahel. In fact, the change from a pluvial to a pronounced drought in the Sahel is one of the most characteristic features of global climate trends in this period. This will be analysed in this section.

Trends in sea-level pressure show an increase of atmospheric mass in the tropics and a decrease at polar latitudes of the winter hemisphere, particularly at high southern latitudes. The hemispheric asymmetry in the temperature trend is mirrored in the 500 hPa geopotential height, particularly in boreal winter, with a decrease in the northern extratropics and an increase in the tropics and Southern Hemisphere. Note, however, that uncertainties in the trends of these variables are large in 20CR.

The boreal winter Hadley cell might have been weaker than before or after this period (Fig. 4.42). Furthermore, the northern edge of the boreal winter Hadley

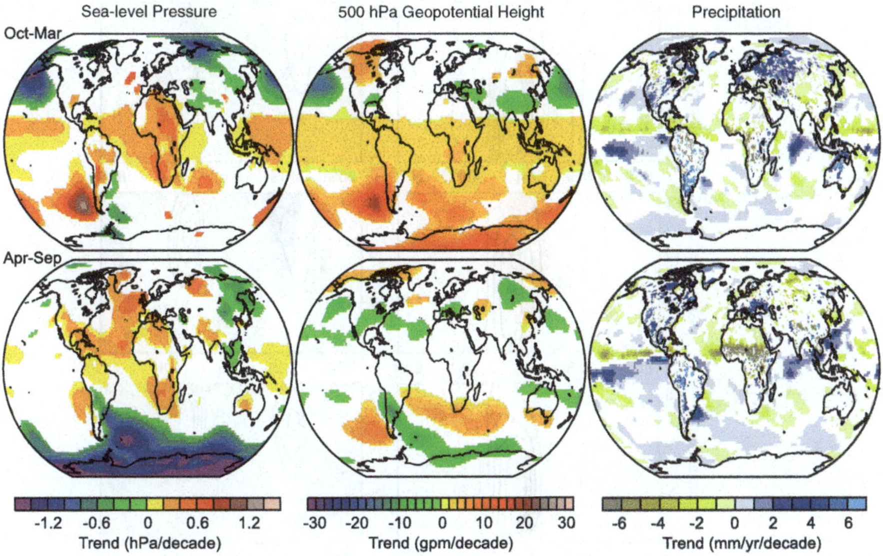

Fig. 4.63 Map of boreal (*top*) winter and (*bottom*) summer trends from 1945 to 1985 of (*left*) sea-level pressure, (*middle*) 500 hPa geopotential height, and (*right*) precipitation from 20CR (over the oceans) and GPCC (over land). Trends are only shown if significant at the 90 % level after accounting for autocorrelation (as in Hartmann et al. 2013)

circulation seems to have moved equatorward during this period (a tropical contraction is seen in Fig. 4.42 for the May–October period). Interestingly, Lamb (1969) diagnosed an equatorward shift of circulation and arid zones. Figure 4.64 shows latitude–height cross sections of trends in the zonal-mean meridional mass streamfunction and in the zonal-mean zonal wind for boreal winter from 20CR, the NCEP/NCAR Reanalysis (starting in 1948), and the CASTRO simulations. 20CR and (to some extent) the NCEP/NCAR Reanalysis suggest a strengthening of the ascending branch of the Hadley cell and a southward shift of the subtropical jet. Overall, they suggest a shift of tropical circulation towards the warming hemisphere (similar to precipitation; Fig. 4.63). The same features, although weaker, are also seen in the CASTRO simulations. Hence, the contraction of the tropical circulation can (at least partly) be reproduced from sea-surface temperatures and forcings.

Greenhouse gases started to increase more rapidly after the 1950s (see Fig. 3.53), which implies a stronger greenhouse gas forcing. So, what caused the stagnation of warming in the 1945–1985 period? The early 20th century warming was partly an expression of internal climate variability such as due to changes in the oceans. If the sign of this internal contribution changes—and eventually it must—this would affect the trend. In fact, the trends in the two main oceanic modes of variability, the AMO and PDO (see Table 3.2), reversed. Furthermore, from the 1950s onward, aerosol concentrations increased. This had a cooling effect on climate (Sect. 3.3.4), temporarily offsetting greenhouse gas warming.

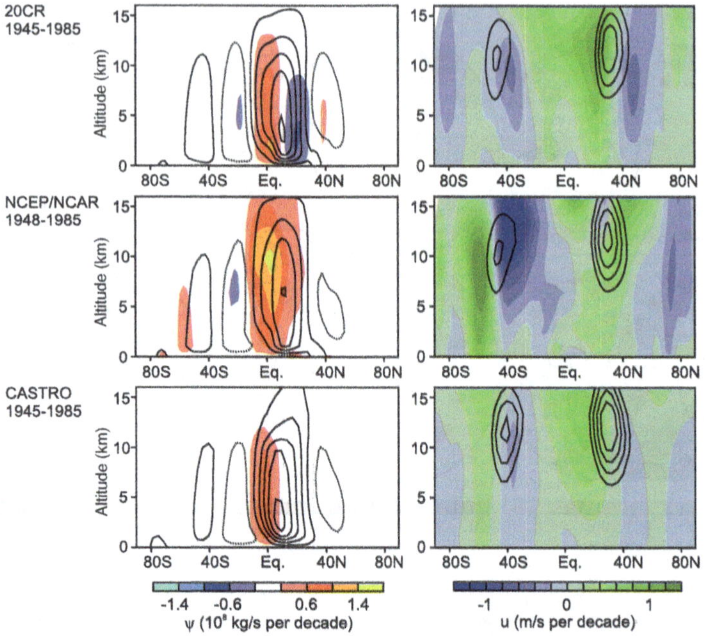

Fig. 4.64 Zonal-mean trends (colours) and mean values (contour spacing is $0.04 \cdot 10^{10}\,\mathrm{kg\,s^{-1}}$ centered around zero) from boreal winter (Dec.–Mar.) 1945–1985 (*left*) meridional streamfunction and (*right*) wind from (*top*) 20CR (ensemble mean), (*middle*) NCEP/NCAR reanalysis (1949–1985), and (*bottom*) CASTRO (ensemble mean). Wind contours start at $25\,\mathrm{m\,s^{-1}}$; spacing is $5\,\mathrm{m\,s^{-1}}$

In our simple forcing–temperature regression (Eq. 4.1), forcings alone yield an increase in global land surface temperature of 0.068 °C decade^{-1} from 1945 to 1986. However, the observed trend was only 0.036 °C decade^{-1}. Thus, natural variability might have compensated part of the forced signal. The main forced contributions are from greenhouse gases (+0.133 °C decade^{-1}) and tropospheric aerosols (−0.055 °C decade^{-1}). The next section goes into more details with respect to the latter.

4.4.2 *Changes in Atmospheric Constituents*

After the Second World War, a new component of the machinery became noticeable: atmospheric constituents. I will briefly present examples of trends in atmospheric constituents, based on our own datasets and on model simulations. I will address the increase in tropospheric aerosol concentrations, the change in tropospheric ozone and the oxidising capacity of the atmosphere, and changes in the stratospheric ozone distribution and their relation to climate.

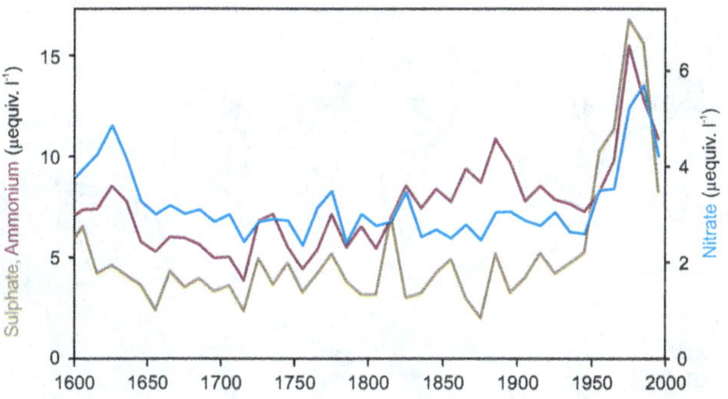

Fig. 4.65 10-yr average of major ion concentrations from an ice core from the Belukha glacier, Siberian Altai. Dating uncertainties are <3 (1815–2000) and <5 yr (earlier) (Data from Eichler et al. 2009a, 2011)

4.4.2.1 Tropospheric Aerosols

The increase in tropospheric aerosols is arguably one of the most characteristic features of the 1945–1985 period. It can indirectly be inferred "bottom up" from emissions data or "top down" from measured optical properties. It is also well documented in ice cores from the Alps, Greenland, and remote continental sites. Figure 4.65, for example, shows ion concentrations in an ice core from the Altai in Siberia (Eichler et al. 2009a). Ammonium, nitrate, and (particularly) sulphate increased steeply from the 1950s to the 1980s, demonstrating the global reach of anthropogenic aerosols. After the 1980s, aerosol concentrations decreased.

The increase in aerosols led to a decrease in atmospheric transmissions from the 1950s to the 1980s (see Sect. 3.3.4). This is known as the "global dimming" period (Ohmura and Lang 1989) and was accompanied by a decrease in surface solar radiation. In Sect. 4.5.1, I will go into further details with respect to atmospheric transmission changes since the 1950s.

A spatial view of aerosol trends (both emission and deposition) is given in Fig. 4.66 based on our ECHAM5-HAM nudged simulations (see Sect. 2.9). During the period 1960–1985, aerosol emissions changed dramatically. Emissions of sulphate and black carbon aerosols increased in most regions of the globe. For sulphate, this was strongest in industrialised regions of the northern extratropics (sulphate), and over the oceans (due to increasing ship traffic), for black carbon the increase was strongest in tropical-subtropical regions (where population growth might have been an important driver). A decrease, particularly of black carbon emissions, is found in some traditional industrial regions. For instance, the typical "London smog" of the early 1950s subsided in the 1970s and 1980s.

Trends in the modelled atmospheric deposition (Fig. 4.66, bottom) reflect both emissions and transport of aerosols. Increasing emissions from tropical continents are transported westward with the trade winds, leading to increasing aerosol deposition over the tropical North Atlantic. Conversely, the aerosols from the midlatitude

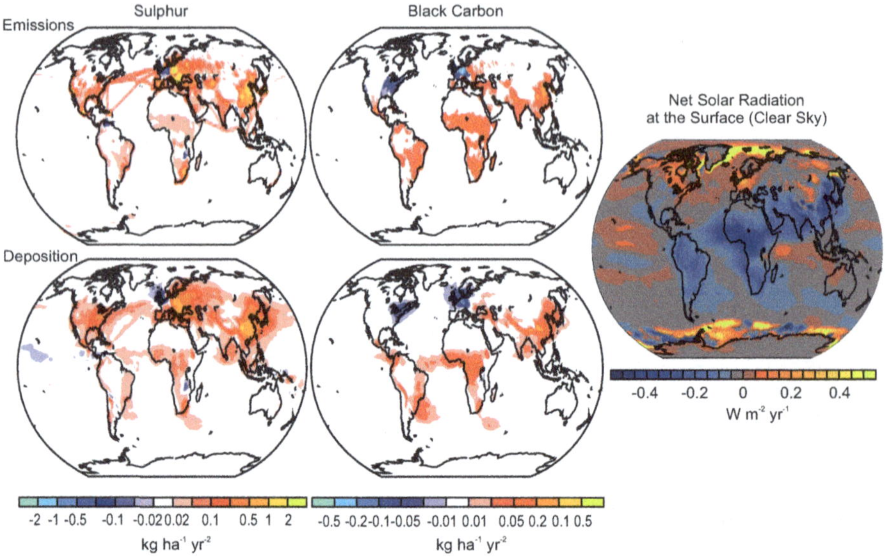

Fig. 4.66 Trends of (*top*) emissions and (*bottom*) deposition of (*left*) sulphate and (*middle*) black carbon aerosols from 1960 to 1985. (*right*) The trend in net solar clear-sky radiation (note that this includes volcanic aerosol effect, but not indirect effect due to clouds). Atmospheric aerosol concentrations and radiation are from ECHAM5-HAM simulations, nudged with ERA-40 (From Florian Arfeuille)

continents are transported eastward (e.g., in East Asia, South America, or North America) and deposited over the oceans. The trend in aerosol concentrations is reflected in the trend in net surface clear-sky solar radiation, whose decrease in our nudged model simulation was particularly apparent in tropical and subtropical continents.

There are similarities among the spatial distribution of aerosol trends, radiation trends, and temperature trends (Fig. 4.62). Most importantly, the Northern Hemisphere cooled during this period, whereas the Southern Hemisphere did not. Aerosols had become part of the machinery.

4.4.2.2 Ozone, Methane and Earth's Changing Oxidation Capacity

In addition to aerosols, gas-phase pollution increased during these years. With a growing fleet of automobiles, this change became relevant. In the 1950s, the first episodes of photosmog occurred in Los Angeles (Fig. 4.67) and ozone was found to be one of its constituents. Although photosmog was first considered to be a peculiar local phenomenon, emissions of NO*x* and hydrocarbons increased over industrialised countries.

Changes in air pollution after the 1950s, particularly in the form of increasing tropospheric ozone, affected the atmospheric oxidising capacity on a global level.

Cycle Messengers Wear Gas Masks in L.A. Smog

LOS ANGELES (AP)—No, they're not men from Mars. They are motorcycle messengers with a unique solution to the Los Angeles smog problem.

The 40 motorcycle riders said yesterday the smog was so thick it hurt their eyes. They refused to deliver for a blueprint firm.

So General Manager Henry Davis went to a war surplus store, bought 40 gas masks, with goggles, and told the riders to carry on.

Fig. 4.67 Messengers of the "Rapid Blueprint Company" in Los Angeles were equipped with gas masks in the fall of 1955. Ellensburg Daily Record, Washington, 15 September 1955

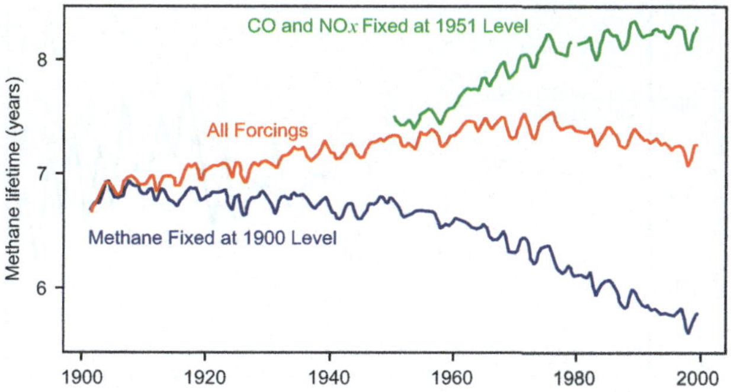

Fig. 4.68 The global methane lifetime in CASTRO chemistry-climate simulations (see Fischer et al. 2008), smoothed with a 12-month moving average. Results are from a simulation with all forcings (*red*), with methane concentrations held constant at their 1900 values (*blue*), and with emissions of CO and NO*x* held constant at their 1951 levels. Note that methane is not emitted in the model, but concentrations are prescribed (Figure by Dogushan Kilic)

This lowered the lifetime of many trace constituents, including the greenhouse gas methane. To illustrate this, I will use CASTRO simulations over the 20th century (see Sect. 2.9), which include basic tropospheric chemistry (Fischer et al. 2008; Schraner et al. 2008). In the "all forcings" setup (Fig. 4.69), the global methane lifetime increases until the 1970s and then decreases (note that, in general, the methane lifetime is underestimated in this model). Sensitivity simulations were performed with either methane concentration fixed at its 1901 starting value, or with emissions of NO*x* and CO (the main precursor gases of tropospheric ozone formation) fixed at their 1950s levels (see Fig. 4.68). If the methane concentration is kept constant, methane lifetime decreases. This is most notable after the 1950s. If NO*x* and CO concentrations are fixed, methane lifetime continues to increase.

How can we interpret this result? Changes in the lifetime of methane strongly depend on the concentration of the hydroxyl radical OH. Conversely, according to our results, up to the 1960s the concentration of OH was determined by methane (also see Sect. 3.4.4). As the amount of methane increased, there was a higher consumption of OH. In other words, methane was increasing its own lifetime.

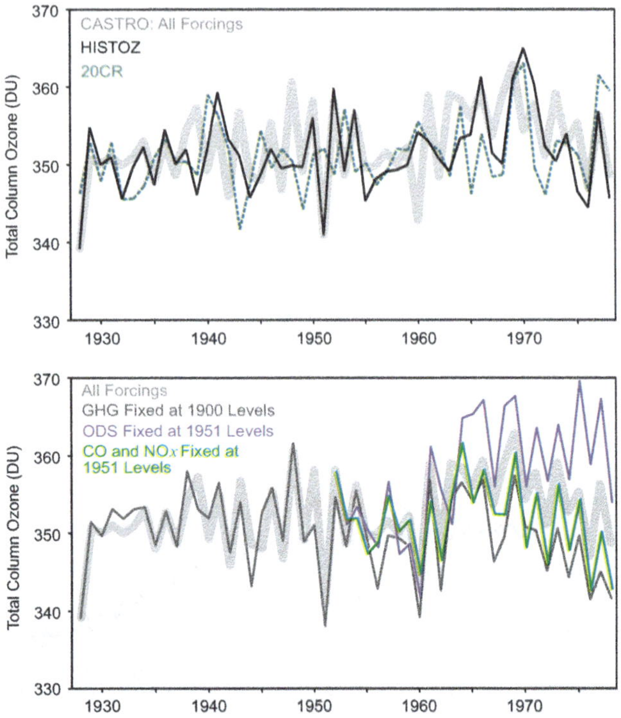

Fig. 4.69 November–April averages of zonal-mean total column ozone near 48°N in (*top*) HISTOZ (ensemble mean), 20CR, and CASTRO (*thick line*); and (*bottom*) four ensembles of CASTRO chemistry-climate model simulations (all-forcings simulations (*thick line*) as well as scenarios with fixed greenhouse gas (GHG) concentrations, ODS concentrations, or emissions of CO and NO*x*) (Brönnimann et al. 2013a)

Around the 1960s, the increase in OH due to ozone formation became more important than the decrease due to methane emissions. Thus, the methane lifetime decreased. This result is also interesting because of the observed decline in the methane growth rate since the 1980s (see Hartmann et al. 2013). This phenomenon is not fully understood. Possible causes are changes in (anthropogenic or natural) emissions or changes in the atmospheric lifetime of methane.

In CASTRO, the prescribed methane concentrations only change on a decadal scale. Nevertheless, the simulations exhibit interannual variability in the methane lifetime. This must be related to climatic factors. An important contributor to the variability is ENSO. One way that ENSO is able to affect methane lifetime is through changes in water vapour and the formation of OH. However, these variations are rather small. Montzka et al. (2011) find interannual variations in global OH concentrations of only a few percent and conclude that the system is well buffered. Nevertheless, the relation between methane lifetime and El Niño demonstrates the close link between chemistry and climate—another cog in the machinery.

4.4.2.3 A Thickening Ozone Layer?

Although little was known about atmospheric chemistry in the 1950s, total column ozone was considered an important geophysical variable. Therefore, since the 1950s total column ozone has been observed with a global network. These observations show an increase in total column ozone over the northern midlatitudes from 1950 to 1970 (Goldsmith et al. 1973; Johnston et al. 1973; Komhyr et al. 1971). The HISTOZ total column ozone data at 48°N in boreal winter clearly reflect this trend (Fig. 4.69). Was the ozone layer getting thicker?

CASTRO "all forcing" simulations and 20CR (which does not have atmospheric chemistry) reproduce the trend in total column ozone over the northern midlatitudes (Fig. 4.69, top). This agreement allows us to explore contributors to the trend. That 20CR reproduces the trend suggests that there were dynamical contributions. As expected, an opposite trend is found in 200 hPa geopotential height at 48°N during this period (not shown). According to Shindell and Faluvegi (2002), an increase in tropospheric ozone contributed to the trend in total column ozone, which is consistent with the CASTRO sensitivity simulations with CO and NO*x* emissions fixed at their 1951 levels (note, however, that tropospheric chemistry in this model is very basic). Further simulations were made with greenhouse gases fixed at their 1900 levels and ozone-depleting substances (ODS) fixed at their 1950s levels. These simulations indicated that greenhouse gas-induced climate change enhanced the total ozone trend, while ozone-depleting substances diminished the trend (also see Sect. 4.5.2).

The dynamical contribution to the midlatitude ozone trend is consistent with an equatorward shift of the poleward edge of the northern Hadley cell. This was mentioned earlier in this section for the boreal summer season (Fig. 4.42) and was seen in trends of the zonal-mean meridional streamfunction (Fig. 4.64). As a result, the midlatitudes were more often in the range of the subpolar, high ozone column. Allen et al. (2014) found a contraction of tropical circulation from 1950 to 1979 using observations and models. They ascribe this to a decrease in the Pacific Decadal Oscillation, possibly induced by increasing amounts of aerosols.

The examples in this section show that atmospheric composition changed on a large scale, interacting with climatic changes. Even larger changes were yet to come, as will be addressed in Sect. 4.5. For the remainder of this section, I will turn to changes of climate near the ground.

4.4.3 European Summers of 1945–1949

Between 1945 and the mid-1950s, a sequence of years with droughts occurred in central Europe. When they struck, the economy had not yet fully recovered from the Second World War. Because some of the winters were cold in central Europe, several scientists and the media spread fear of an increase of continentality and desertification (Groissmayr 1949; Paret 1947; Schmauss 1948).

Particularly noteworthy was the summer of 1947, whose temperature records in many regions were not broken until the summer of 2003. In Basel, for instance, 38.7 °C was measured on 29 July 1947 (Grütter et al. 2013). During the same year, Switzerland experienced the largest reduction of crop yield (Calanca 2007). In Germany, heatwave and droughts affected transportation, energy production, and forests (e.g., through pests and fires; Baumgartner 1950).

The droughts from the mid-1940s to mid-1950s coincided with a high phase of the Atlantic Multidecadal Oscillation (see Sect. 3.2.7 and Fig. 4.6). The AMO index had increased until the 1940s but remained high into the 1950s. At the same time, the Pacific Decadal Oscillation dropped to a negative state. Was this another consistent global mode of the machinery, or regional (seemingly random) processes?

4.4.3.1 Causes and Mechanisms

Central Europe is a region that is regularly—but not permanently—affected by droughts. Consequently, droughts are a major economical and ecological hazard in this area (e.g., Ciais et al. 2005). Atmospheric precursors to droughts and heatwaves in western Europe are well understood. Typically, atmospheric blocking or sustained high pressure influence accompanies heatwaves (e.g., Carril et al. 2008; Della-Marta et al. 2007; Fischer et al. 2007b; Schär et al. 2004). The anomalous circulation pattern persists for about a week or longer, causing high temperatures. A sequence of such situations can cause a drought. On a seasonal scale (the scale of a drought), high pressure anomalies are found over west-central (Fischer et al. 2007b), northwestern (Cassou et al. 2005; Folland et al. 2009), or northern Europe (Della-Marta et al. 2007). This depends on the spatial pattern of the drought. Similar to the "Dust Bowl" case, we can ask what triggered the drought and what affected its persistence?

For the triggers of European droughts, various authors have emphasised the role of sea-surface temperatures with a focus on the Atlantic Ocean (e.g., Sutton and Hodson 2005), but also the Indian and Pacific Oceans (see Black and Sutton 2007). The Mediterranean region is influenced by Asian summer monsoon systems (Gill 1980; Rodwell and Hoskins 1996; Tyrlis et al. 2013). The monsoon heating can trigger a Rossby wave response, which propagates westward and induces descending motion over southern Europe. In addition, the African monsoon has been suggested to affect precipitation in southern Europe via changes in the Hadley circulation (Gaetani et al. 2011). In terms of persistence, as for the "Dust Bowl" droughts (see Sect. 4.2.3), a precipitation deficit that is set in place by a circulation anomaly can be amplified and maintained through land–atmosphere interactions.

Which mechanisms were at work during the droughts of the 1940s? Sutton and Hodson (2005) found that the AMO may have facilitated the occurrence of these droughts. In the following section, we will revisit these droughts using our new datasets and model simulations and focusing on Switzerland.

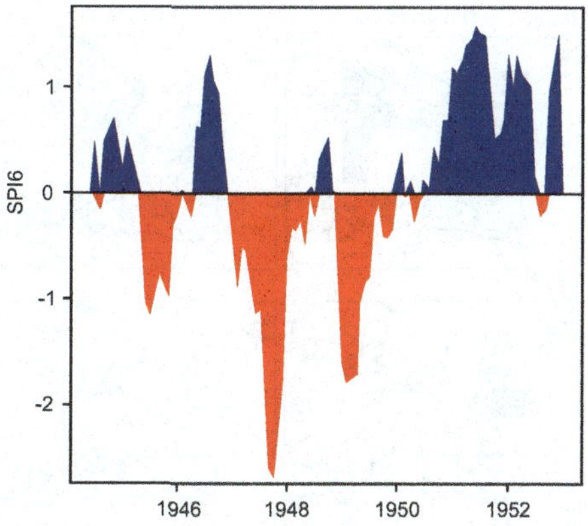

Fig. 4.70 Time series of the 6-month Standardized Precipitation Index (with respect to 1961–1990) at Bern from observations (Hirschi et al. 2013)

4.4.3.2 The Droughts in CCC400, 20CR, and REC2

An analysis of the droughts of the late 1940s was made using the 6-month standardized precipitation index (McKee et al. 1993) with data from a meteorological station in Bern, Switzerland (Fig. 4.70, Hirschi et al. 2013). The index was almost always negative from 1945 to 1950. The most pronounced drought years were 1947, followed by 1949 and 1945. Analysing precipitation data from these three summers reveals distinct spatial differences (Fig. 4.71). The 1945 drought affected mostly the Mediterranean area, the 1947 drought was in central Europe, and the 1949 case concerned western Europe. Interestingly, the drought of 1947 was preceded by a dry winter in Switzerland (Hirschi et al. 2013).

All drought summers were accompanied by above-normal 500 hPa geopotential height over south-central Europe and negative anomalies over North Africa. However, the position of the strongest anomaly is not the same for all summers and the patterns differ over the North Atlantic.

Sea-surface temperatures were anomalously high over the Atlantic. To assess their effect, we analysed the CCC400 simulations (Grütter 2014). It turns out that CCC400 reproduces a considerable part of the precipitation signal (Fig. 4.71) and, in some cases, even the spatial pattern. In the model, this must be due to sea-surface temperature anomalies. Similarly, some of the anomalies in 500 hPa geopotential height are reproduced. Thus, anomalously high sea-surface temperatures in the Atlantic presumably contributed to the drought.

How did the influence work? Droughts are intimately related to anticyclonic conditions over central Europe. In fact, analysing daily weather types in 20CR (GWT18; see Sect. 2.9), we find an increase in the frequency of anticyclonic weather types during the late 1940s (Fig. 4.72). Interestingly, the filtered time series closely resembles the AMO index, as does the heatwave area (Fig. 4.11). Thus, the positive AMO is related to more frequent anticyclonic situations and heatwaves over central Europe. A schematic view of the processes responsible for the drought summers is given in Fig. 4.73.

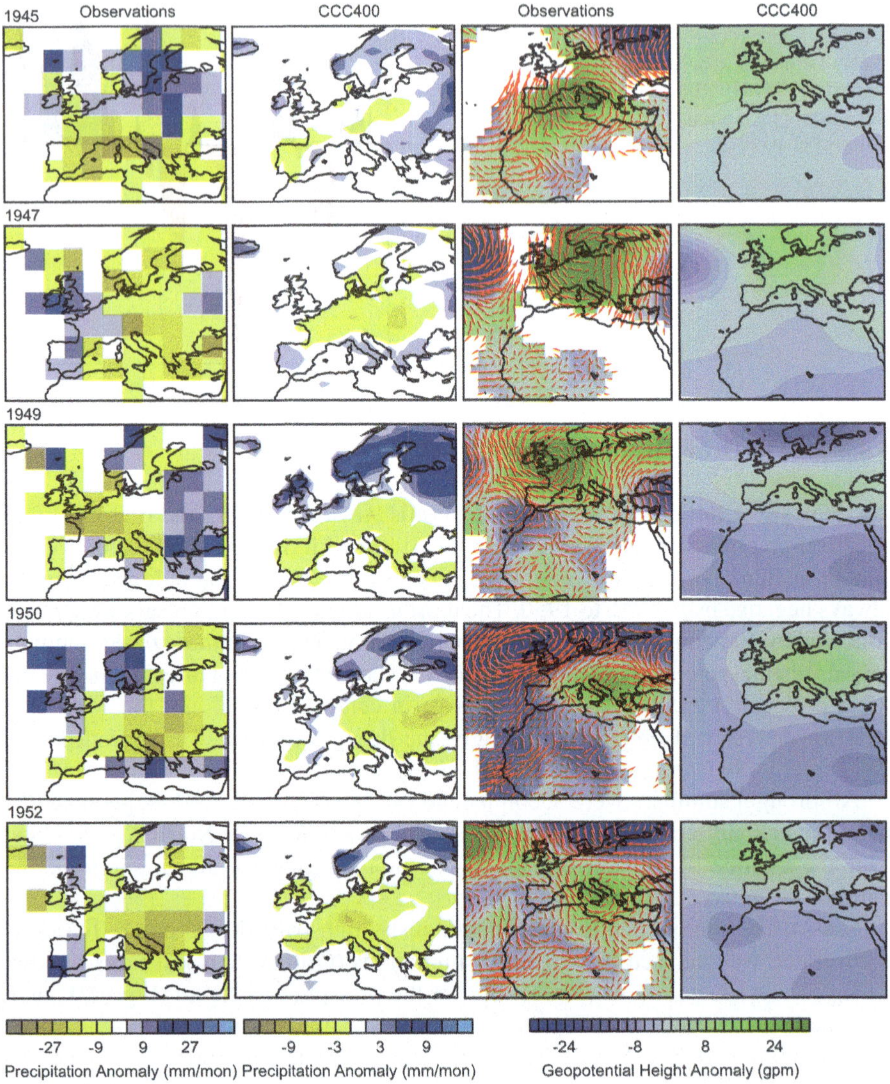

Fig. 4.71 Anomalies (with respect to 1961–1990) of precipitation, 500 hPa geopotential height, and 850 hPa wind for the drought summers (Apr.–Sep.) of 1945, 1947, 1949, 1950, and 1952 and in observations (GHCN for precipitation; REC2 for geopotential height and wind) and CCC400

In some respects, this situation is similar to that around 1800 (Sect. 4.2.3). However, the AMO for that period is not well known. Circulation indices suggest that the tropical belt may have been relatively wide in the 1940s (Fig. 4.42), the ITCZ (in EKF400) was shifted northward (Fig. 4.9), and the Hadley circulation was possibly stronger than in the following years. However, the available datasets are still uncertain and more evidence is needed.

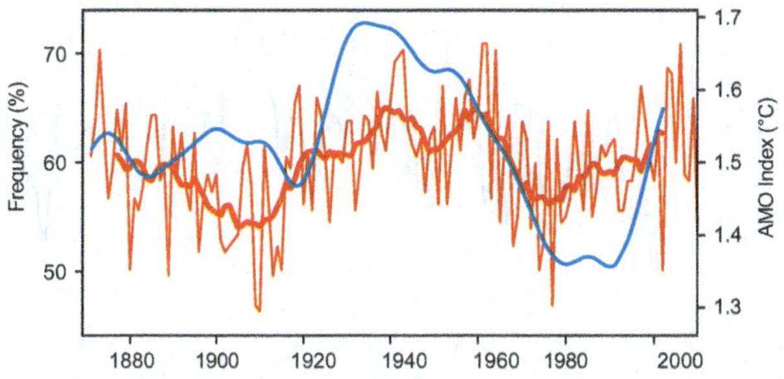

Fig. 4.72 Frequency of anticyclonic weather types in the GWT18 weather type classification based on sea-level pressure in April–September (Weusthoff 2011) in 20CR (*thin red line*) and smoothed with an 11-point moving average (*thick red line*). The *blue line* shows the AMO index (Data by Jenny Grütter and Marco Rohrer; Grütter 2014)

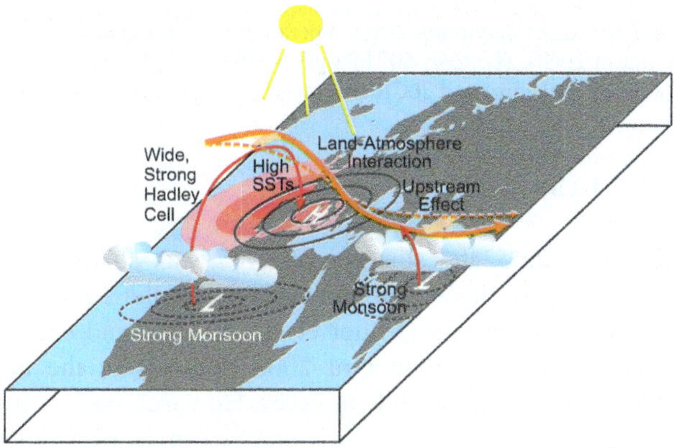

Fig. 4.73 Schematic of the suggested processes contributing to European droughts in the late 1940s

Remote influences might have played a role in the European droughts. Some of the drought years coincided with wet years in the African and Asian monsoon regions (variations in Sahel precipitation are discussed in Sect. 4.4.4) and strong monsoonal circulation as indicated by converging wind anomalies over the Sahel-Sudanese region at 850 hPa in REC2. Both might have affected the high pressure anomalies over south-central Europe. More detailed studies are needed to separate the different influences. Most likely, there were different causes for the European drought years of the 1940s and early 1950s.

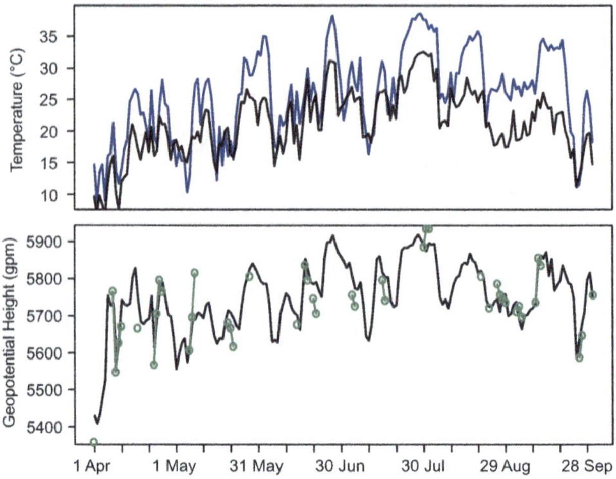

Fig. 4.74 Time series of (*top*) maximum 2-m temperature from Basel station data (*blue line*) and the closest grid point in 20CR. (*bottom*) 500 hPa geopotential height from the station of Payerne (*green line*; increased by 200 gpm) and 20CR from April to September 1947 (Grütter et al. 2013)

4.4.3.3 The 1947 Heatwaves

On the weather time scale, the summer of 1947 was characterised by intense heatwaves. In many places, these were the strongest of the 20th century. Focusing on the heatwaves in Switzerland, Grütter et al. (2013) found that the summer of 1947 was comparable to the summer of 2003 in terms of the number of heat days and the number and duration of heatwaves. However, peak values were lower than in 2003. In Basel (Fig. 4.74), there were five main heat episodes between May and September 1947 during which the daily maximum temperatures exceeded 30 °C over consecutive days. 20CR represents these episodes well (the offset can be explained by the difference in altitude between the station of Basel and the grid cell in 20CR). Each of these heat periods coincided with elevated 500 hPa geopotential height (in radiosonde data from Switzerland and 20CR; Grütter et al. 2013). However, the absolute values in this case remain uncertain because the observations and 20CR could have biases.

In central Europe, the heatwave of 1947 marked the peak of the early 20th century warming. An increase in heatwave area over Europe is also found in CCC400 during these years (see Fig. 4.11). Thus, apart from internal atmospheric variability, an ocean-forced contribution to the 1940s heatwaves is plausible. Brückner did not live to experience these drought years, which were associated with increased continentality that was much discussed in the public. These years would have nicely fit his concept of a warm and dry period.

4.4.4 *Sahel Pluvial and Drought*

Among the most prominent features of global, decadal-scale hydroclimatic variability during the 20th century are the changes in the Sahel region between the 1950s and the 1980s. From the 1920s to the early 1960s, but particularly in the 1950s, precipitation was abundant (Paeth 2007). Then, precipitation decreased, which led to droughts after the late 1960s with two main peak phases (in the early 1970s and in the mid-1980s). Droughts in the Sahel region have a large effect on the local population, which depends on agriculture and pastoralism. The famine accompanying the 1970s and 1980s droughts triggered worldwide attention and, starting in 1973, international disaster relief activities led by the United Nations and the United States. The first drought period (1968–1973) caused some 250,000 fatalities. It also destroyed the livelihood for two million pastoral people and severely impacted societies in the region (Sheets 1974). The second drought period in the early 1980s in Sudan and Ethiopia coincided with conflicts and caused an excess mortality of approximately one million people (Devereux 2000). Overall, the droughts caused considerable economic losses, triggered large-scale migration processes, and led to irreversible land degradation (Paeth 2007).

It is also important to consider the influence of the pluvial period preceding the drought. In fact, similar to the "Dust Bowl" case, severe and long droughts are a recurring phenomenon in the region (Shanahan et al. 2009) and the pluvial period of the 1950s also was not a common occurrence. However, the pluvial came at a critical moment in the regional history. Rapid population growth in the 1950s due to decreased mortality (Norton-Griffiths and Rydén 1989) was accompanied by agricultural expansion. Between 1956 and 1960, many Sahel countries obtained their independence, just before climate turned from pluvial to drought conditions.

One interesting fact is that the Sahel has pronounced climatic persistence. This is also apparent in Fig. 3.18, where observations of temperature and precipitation exhibit a red spectrum. Therefore, a mechanism explaining the drought must be able to explain a high persistence.

4.4.4.1 Causes and Mechanisms of the Drought

Summer rainfall in the Sahel is related to the West African monsoon (see Sect. 3.1.4). The heated Sahara desert draws the ITCZ to the north, leading to monsoon flow from the tropical Atlantic to the Sahel-Sudanese region. Was the monsoon flow weakened in the 1970s and 1980s and, if so, why persistently? Or, were local factors such as land degradation and feedback to the atmosphere responsible for the drought?

The Sahel droughts were attributed by some scientists to unsustainable land use practices, which led to desertification (Hammer 2000). Other scientists put climatic factors in the foreground. A schematic of suggested mechanisms is shown in Fig. 4.75. In the simplest form, Sahel rainfall depends on the interhemispheric

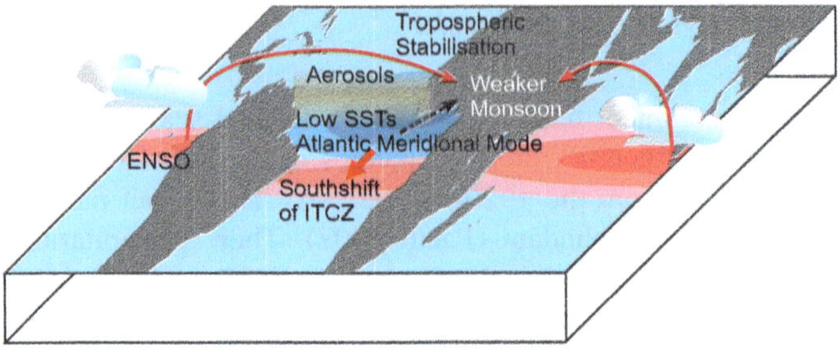

Fig. 4.75 Schematic of the suggested processes leading to Sahel drought

temperature gradient. If the African continent to the north of the equator is heated, the monsoon is intensified and the Sahel rain belt is drawn toward the warmer region (Held et al. 2005). Such interhemispheric temperature gradients are often ascribed to changes in sea-surface temperatures, as is outlined in more detail below (also see the temperature trend map in Fig. 4.63). Additionally, teleconnections to other ocean basins alter the monsoon circulation and lead to a shift of the rain belts (e.g., Diatta and Fink 2014; Giannini et al. 2003; Janicot et al. 1998; Nicholson 2001).

A second mechanism is related to tropospheric stabilisation. A warming of tropical sea-surface temperatures (e.g., related to ENSO) leads (via zonal circulation cells) to areas of increased subsidence and a large-scale warming of the free troposphere. Over regions with sea-surface temperatures that are not as warm, this mechanism tends to stabilise the atmosphere and suppress convection (see, e.g., Herceg et al. 2007).

Some model-based studies also suggest a role of aerosols through changing sea-surface temperatures (Frierson et al. 2013; Rotstayn and Lohmann 2002), the vertical temperature profile, or cloud microphysics (e.g., Konare et al. 2008). Aerosols increased during the period of drought, but mostly north of the equator (see Fig. 4.66). They could have affected the cross-equatorial temperature gradient.

Because approximately half of the moisture in the Sahel region is recycled, land surface processes might also play an important role (Nicholson 2000). Vegetation reacts slowly (Wang and Eltahir 2000). Therefore, such a mechanism could produce decadal changes. Giannini et al. (2003) found that their model is only able to reproduce the Sahel drought if the land surface is allowed to react to the atmosphere. Therefore, Giannini et al. (2008) conclude that the ocean acts as a trigger for drought, which is then amplified through land–atmosphere interactions. This opens the possibility for an effect of land management on climate. However, Lauwaet et al. (2009) and Seaquist et al. (2009) find a negligible influence of land use changes on Sahelian droughts and total precipitation. As for the "Dust Bowl", climate simulations with prescribed sea surface temperatures generally reproduce the multidecadal rainfall variability qualitatively. However, they may miss the amplitude (Scaife et al. 2009). These studies indicate a possible role of the ocean, which is relevant for predictability.

4.4.4.2 The Sahel Pluvial and Droughts in CCC400, REC2, and ECHAM-HAM Nudged Simulations

In the following section, we will analyse the changes in Sahel rainfall encompassing the pluvial and the subsequent drought. In addition to 20CR, we analyse REC2 (which is based on upper-level wind observations) because the African monsoon is not well reproduced in reanalyses (see Stickler and Brönnimann 2011; note that the wind field in REC2 is not necessarily physically consistent) and model simulations (CCC400, CASTRO, and the nudged aerosol simulations performed with ECHAM5-HAM).

Consistent with other studies (Diatta and Fink 2014), we find that Sahel rainfall correlates with an index of the Atlantic Meridional Mode prior to monsoon onset (see Sect. 3.2.7) and with the AMO (Fig. 4.76). In addition, we analysed the interannual variability of aerosols over the tropical Atlantic in our nudged aerosol simulations. A gradient between the areas from north to south of the equator in aerosol deposition and radiation appears in Fig. 4.66. The time series of the corresponding difference prior to the monsoon onset (Fig. 4.76) shows a significant correlation with Sahel rainfall in summer, in line with the hypothesis that tropospheric aerosols contributed to the drying by changing the meridional gradient. However, a change in the temperature gradient could have other reasons including ocean-induced changes in the Atlantic Meridional Mode.

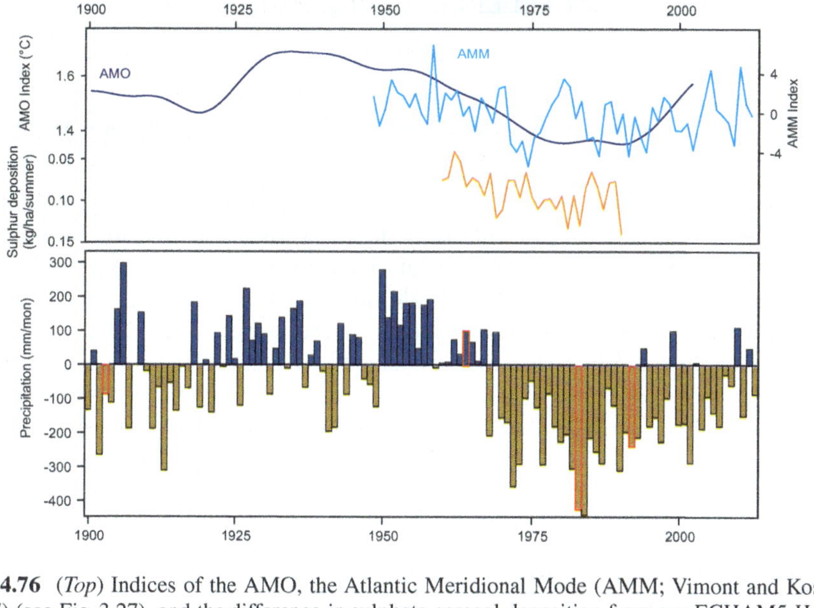

Fig. 4.76 (*Top*) Indices of the AMO, the Atlantic Meridional Mode (AMM; Vimont and Kossin 2007) (see Fig. 3.27), and the difference in sulphate aerosol deposition from our ECHAM5-HAM nudged simulation between the areas (0°–20° N, 60°–0° W) and (20° S–0° N, 40°–0° W), both during April–July (note reverse scale). (*Bottom*) July–October averaged precipitation over the region (10°–20° N, 20° W–10° E). *Red-outlined bars* denote summers following major tropical volcanic eruptions (Data source: GHCN/JISAO, Univ. Washington, doi:10.6069/H5MW2F2Q)

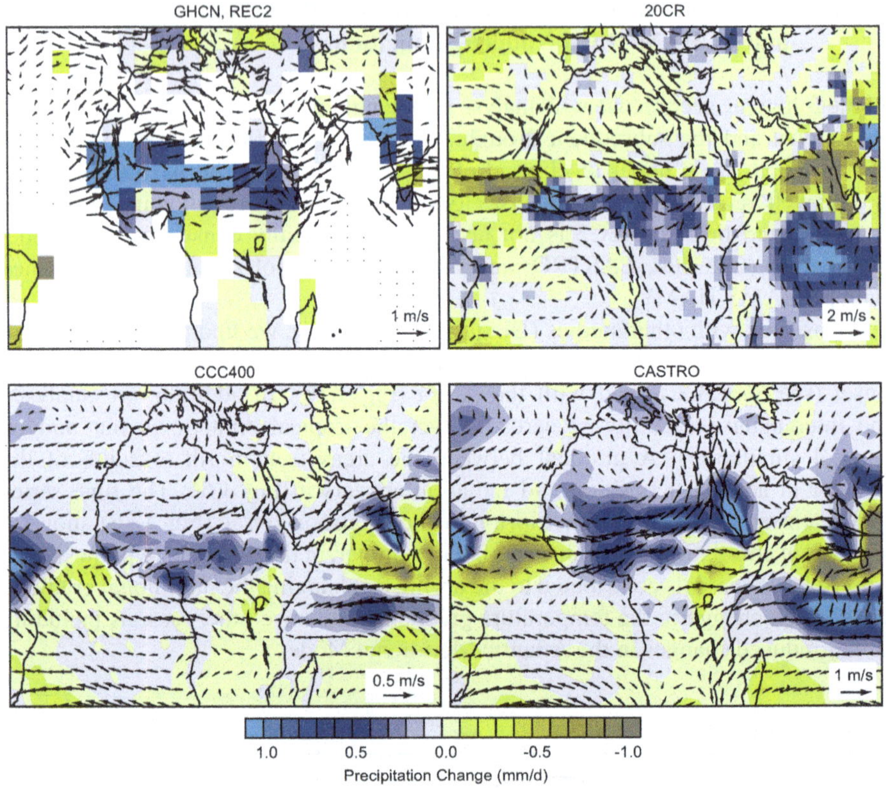

Fig. 4.77 Difference in precipitation and 950 hPa winds in July–October between 1950–1970 and 1970–1990 in GHCN (precipitation), REC2 (850 hPa winds), 20CR, CCC400, and CASTRO

Most climate models reproduce the variations in Sahel rainfall over the 20th century (Held et al. 2005). Similarly, both CASTRO and CCC400 simulations (Fig. 4.77) reproduce the decrease in rainfall from the 1950–1970 period to the 1970–1990 period, meaning that a sizable fraction of the decrease is reproducible from forcings and sea-surface temperatures. In the model, as in REC2 and 20CR, the decrease in rainfall is due to a weakening of the African monsoon system. Although a minor influence on a decadal scale, tropical volcanic eruptions (marked in Fig. 4.76) also contribute to a precipitation decrease in the following summer (see Sect. 3.3.1).

4.4.5 The Climate Shift of 1976/1977

We end our characterisation of the 1945–1985 period with a brief description of a truly global climate change that originated in the Pacific Ocean: the climate shift of 1976/1977. This was not a single event in the sense that it had severe impacts in a certain region. Rather, it was a change in the entire system—a gear shift of the "machinery".

Around 1976, the Pacific Decadal Oscillation (see Sect. 3.2.7 and Table 3.2) shifted from predominantly negative to positive values (Mantua and Hare 2002; also see Fig. 4.7). A strong eastern Pacific La Niña gave rise to a weak El Niño (see Fig. 3.27), and a period began in which the base state of the equatorial Pacific was more El Niño-like. Consistently, many ENSO time series (Fig. 4.7) indicate a shift towards El Niño conditions. Between 1978 and the 1990s, El Niños became more frequent, although of varying strengths. Accompanied by this shift was a strengthening of the Aleutian low and a change in the boreal winter subtropical jet and the PNA index (see Figs. 4.42 and 4.43). These are reproduced by atmospheric models forced with sea-surface temperatures. Variations in Alaskan temperatures are reported (Hartmann and Wendler 2005), as well as changes in storms, hydroclimate, and other climate variables over North America (see Miller et al. 1994, for a review). Changes are also found in more distant regions. The South Pacific high weakened, an anticyclonic anomaly appeared at the tip of South America, temperatures in the Andes increased (Jacques-Coper and Garreaud 2015; Vuille et al. 2008), and Sabeerali et al. (2012) found a change in the Indian rainy season.

In Fig. 4.78, I show the upper-ocean heat content in the North Pacific from January to March and the equatorial SOI (see Table 3.2) calculated from 20CR data

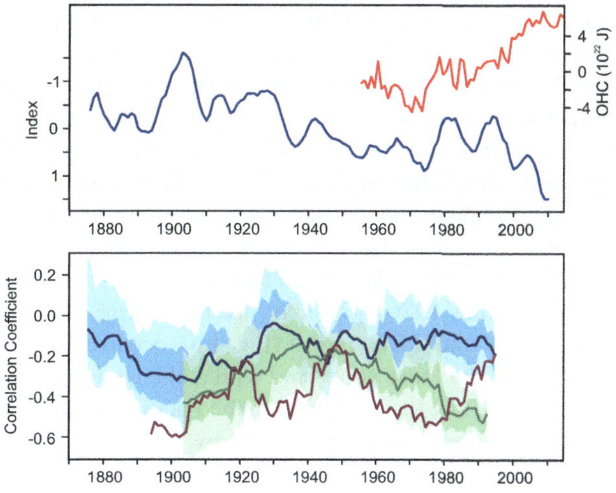

Fig. 4.78 (*top*) September–January averages of the equatorial SOI (see Table 3.2) in 20CR (*blue*, note the inverse scale), filtered with a 9-point triangular filter, and annual values of ocean heat content in the uppermost 700 m of the North Pacific Ocean (*red*, data source: NOAA/NODC). (*bottom*) 29-point moving correlation between the DIMI and NINO3.4 from observations and reconstructions (*purple*), CCC400 (*blue*), and CASTRO (*green*). *Light* and *dark shading* denotes the 50 % and 90 % ensemble spread of the correlations

for the September–January season (the latter series is smoothed to focus on decadal variability). The shift in 1976 appears as a sudden jump in ocean heat content and as a drop in the equatorial SOI. Similar changes are found in other indicators from the Pacific region.

What happened? How could the machinery produce such a massive change in a large ocean body? Some authors attribute the change in the North Pacific to a remote forcing from the tropical Pacific; others attribute it to oceanic changes in the North Pacific. In fact, the concept of the PDO partly grew out of the analyses of this climate shift. However, extraordinary atmospheric circulation anomalies might have also driven the 1976/1977 shift, which is often called a "regime shift" (Miller et al. 1994).

Above all, the 1976/1977 climate shift was a change in the base state. This is reflected in different climate variables in different places. It was also a change in variability, which primarily occurred in the tropical Pacific. El Niño events were not only more frequent after the shift, but more occurred in the eastern Pacific and fewer occurred along the date line. The change in variability might have had additional effects. As a consequence of the fewer date line El Niños or for other reasons, teleconnection patterns between the tropical Pacific and other areas of the globe might have changed.

One of the changes in teleconnections was the disappearance of the long-standing relation between El Niño and the Indian monsoon (Kumar et al. 1999). In Fig. 4.78, this is reflected in the 29-yr moving-window correlation between NINO3.4 and DIMI. The correlation is currently around zero; earlier it was −0.6. Whether this is related to the 1976/1977 climate shift is unclear. It has been suspected that the warming of the Eurasian land mass interfered with the ENSO-monsoon relationship (Zhou et al. 2010). If this is the case, the observed result may be more than just a temporary lack of correlation. Rather, it could be a change in the structure of the teleconnection (because continents warm faster than the oceans in the future). Do climate models reproduce this result? The corresponding moving-window correlations for CCC400 and CASTRO are also shown in Fig. 4.78. Both models show decadal changes in the moving-window correlations and those of CASTRO loosely resemble the observed changes. However, it cannot be claimed that the models have skill in reproducing the changes.

Reverting to the Pacific, 1976/1977 may not have been the only shift in the Pacific. Past changes in the PDO index (Fig. 4.6) occurred in the 1950s and (arguably) in the 1900s. The latter concurred with a change in the Indian monsoon system (see Fig. 4.10). A more recent change in the Pacific Ocean occurred after the 1997/1998 El Niño. This period was dominated by a La Niña-like base state and weak El Niños. The consequences of this change for the global energy balance and global-mean temperature are further discussed in Sect. 4.5.9.

4.4.6 The Broad Lines: Delusive Stability

Let us briefly pause to summarise climatic changes during the 1945–1985 period; a period of true or perceived climatic stability. Stability means stagnating global temperatures. The Northern Hemisphere even cooled up to 1970. This stagnation is often attributed to aerosol cooling balancing greenhouse warming, which illustrates that atmospheric composition became an important player on a global scale. Aerosols may have also changed the mode of operation of the machinery. They are suspected to have caused changes in the North Atlantic (Booth et al. 2012) and Pacific (Allen et al. 2014) sea-surface temperatures, which may have contributed to anomalies in atmospheric circulation. Through forcing an interhemispheric gradient in sea-surface temperatures, aerosols have also been considered for the Sahel drought.

The lack of warming was also due to changes in the oceans. First the PDO, then the AMO changed from positive to negative states during the period. Changes in PDO and AMO are suspected to affect ocean heat storage (Chen and Tung 2014). Hence, the Pacific and Atlantic Oceans may have taken up heat from the atmosphere and stored it below the surface.

As found in previous sections—and, in fact, throughout this book—many of these changes in atmospheric circulation and climate are reproducible by models from external forcings and sea-surface temperatures. This raises the question of whether oceanic influences are the key to these anomalies. Because imputed sea-surface temperatures carry the signature of atmospheric circulation anomalies, a clear separation is not possible. However, the importance of tropical sea-surface temperatures and their effects on the extratropics is well established (see Sect. 3.2.6). In any case, possible effects of the tropical or subtropical ocean temperatures on atmospheric circulation appear for the three climatic events analysed in this section: the European droughts, the Sahel pluvial, and the climate shift of 1976/1977. To some extent, all three were reproduced in climate model simulations forced by observed sea-surface temperatures.

With respect to large-scale atmospheric circulation, there is evidence that the tropical belt contracted (the northern tropical edge retracted) and the ITCZ shifted to the south during the 1945–1970 period. The European droughts at the beginning and the Sahel droughts towards the end of the period may indicate an equatorward shift of subtropical anticyclones and arid zones in boreal summer. This is the characteristic of a different mode of operation of the machinery than in the preceding period. Figure 4.79 summarises the suspected main features and mechanisms of global climatic change during the 1945–1985 period.

The period of climate stability may have allowed some sectors or regions to prosper. However, the stability was delusive. Soon, fear emerged of a climate system that is out of balance (see Box 4.3, p. 280).

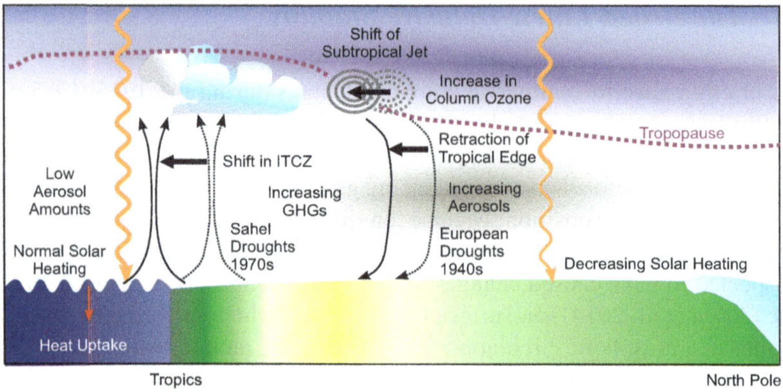

Fig. 4.79 Schematic showing some of the mechanisms affecting large-scale climatic variations in the 1945–1985 period

Box 4.3 The Great Fear

The scientific optimism and technological futurism with respect to the 1950s and 1960s climate slowly gave rise to fear in the 1970s. Political, social, and cultural changes, as well as new attitudes towards technological solutions contributed to this fear. Environment-related problems called for political solutions: nuclear threat, energy crisis, technology disasters (such as Three-Mile Island in 1979 and Bhopal in 1984), hunger catastrophes (such as that in the Sahel), Malthusian fears (also reflected in the Club of Rome Report "Limits to Growth" (Meadows et al. 1972)) and omnipresent pollution worried Europeans and Americans. The environmental concerns were expressions of a new age. Although many scientists in the 1960s and 1970s discussed the possibility of future greenhouse gas-induced climate change, aerosol-induced cooling was also considered a possibility (Rasool and Schneider 1971), particularly in view of the downward temperature trend in the Northern Hemisphere (Fig. 4.80). Among the public, this triggered fears of an impending ice age, as visualised by an artist in Fig. 4.80.

(continued)

Box 4.3 (continued)

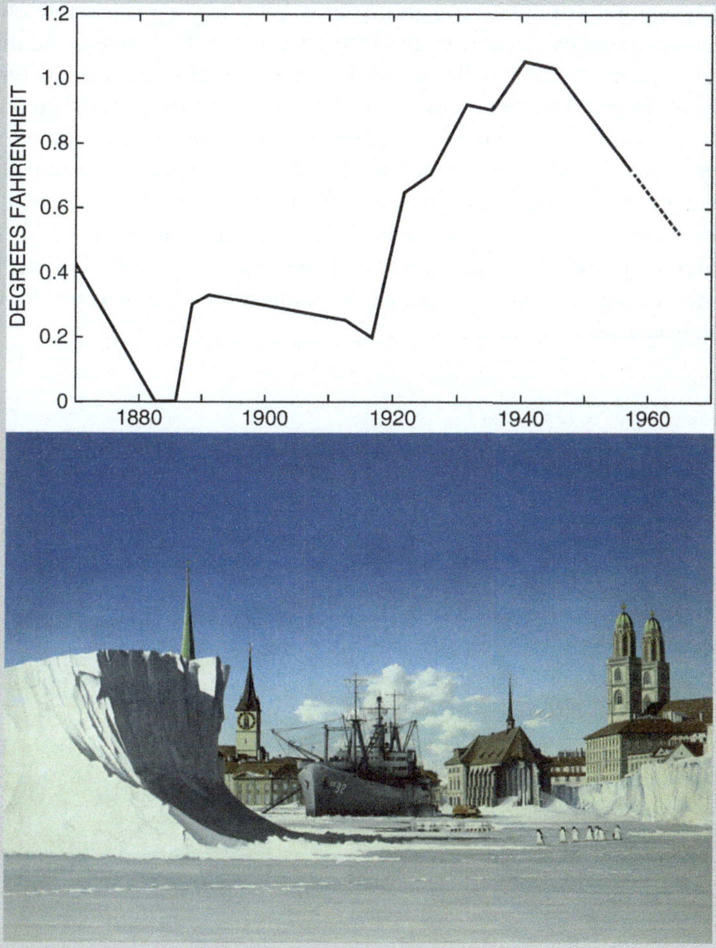

Fig. 4.80 (*Top*) Temperature from 10° to 80° N after Bryson and Kutzbach (1968). (*Bottom*) On the verge of a new ice age? "Zürich-Eiszeit" (Zurich-Ice Age) gouache on pavatex, 90 × 120 cm, by Giuseppe Reichmuth, 1975, reprinted with permission

(continued)

Box 4.3 (continued)

The possibility of solar-terrestrial catastrophes leading to climate change was demonstrated by Alvarez et al. (1980), who identified an asteroid impact and subsequent climatic collapse as the cause for the Cretaceous-Tertiary extinction. Inspired by this work, Crutzen and Birks (1982) and Turco et al. (1983) analysed the effect of a nuclear war on climate, later known as a "nuclear winter" (Edwards 2010). Interestingly, the Tambora eruption and its climatic aftermath, the "Year Without a Summer" of 1816 (see Sect. 4.1.4), was revisited in the context of these discussions (also see Box 3.2, p. 124). Apocalyptic forest dying scenarios and the discovery of the ozone hole, discussed in the next section, seemed to confirm the biggest fears. In 1985, the end was near.

4.5 The Period Since 1985: Accelerated and Slowed Warming

In the late 1970s, scientists such as Bert Bolin[21] and Hans Oeschger[22] (Siegenthaler and Oeschger 1978) began to warn their peers and the public about the potential impact of the carbon dioxide greenhouse effect on global climate. Around the same time, temperatures began to rise. In 1988, with a record heatwave in the United States, the topic of greenhouse gas-induced climate change attained global awareness. Rapidly increasing atmospheric temperatures (into the 2000s), losses of sea ice, rising sea levels, megaheatwaves, and epochal droughts characterise this ongoing period.

In addition to a warming climate, this period was characterised by atmospheric composition change. Pollutants had begun to alter the atmosphere in the 1950s and were becoming a major global environmental concern. The term "global atmospheric change" was coined and encompasses global warming, ozone depletion, and air pollution. I will start the discussion of this time period with the discovery of the ozone hole (including its implications for climate). However, this is just one example. Photochemical pollution and aerosols—and their effects on climate—became important topics (see Sect. 4.4.1). The eruption of Pinatubo in 1991 highlighted the importance of atmospheric chemistry and stratospheric processes. The "Atmospheric Brown Cloud" (ABC; masses of aerosols and haze over the Indian subcontinent, the Indian Ocean, and Southeast Asia during the dry season) affects temperature and possibly monsoon systems. The following sections will give more details.

The changes in atmospheric constituents were just one element in a system that was about to change fundamentally. In addition, strong decadal variability continued in the North Atlantic and the Pacific. Frequent winter storms hit Europe, rain belts shifted, and droughts occurred in Australia, North America, and elsewhere. It is daunting to list, let alone analyse, the major climatic events of this period that is closest to our memories and, perhaps, too recent to be assessed.

Nevertheless, this section will briefly take a look at the most recent 30 years of climate history. From the perspective of a present-day scientist, this is the satellite era. Spaceborne observations of the earth provide more detail than previous data sources. Satellite data have now reached the length (though maybe not yet the quality) necessary for climate trend calculations. Satellite information has also been used to create numerous reanalysis datasets for this period. This section will make use of these datasets—in particular, ERA-Interim (also see Sect. 2.9, Table 2.3).

---

[21] Bert Bolin, 1925–2007, was a Swedish meteorologist. He was one of the founders and first chair of the IPCC.

[22] Hans Oeschger, 1927–1998, was a Swiss physicist and climate scientist. Apart from his work on carbon dioxide in the atmosphere, Oeschger pioneered the measurements of trace gas concentrations and climatic changes in polar ice cores.

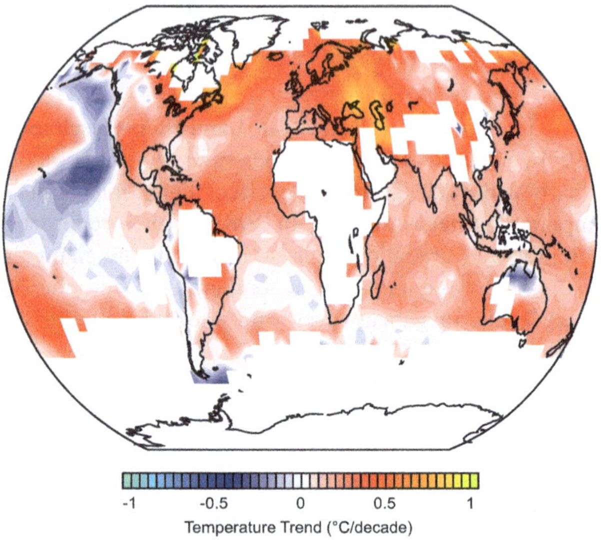

Fig. 4.81 Map of annual-mean temperature trends from 1985 to 2013 from HadCRUT4

4.5.1 Global Drivers and Global Changes

4.5.1.1 Global Warming and More

From 1985 to 2013, global temperatures increased strongly, but at a variable rate. After the 1997/1998 El Niño, a rapid temperature increase was followed by a period of slower change. This is sometimes referred to as the "global warming hiatus" (further discussed in Sect. 4.5.9). The annual-mean temperature trend during the 1985–2013 period (Fig. 4.81) is almost globally positive. The strongest temperature increase was over northern midlatitudes and certain subtropical continental regions. Satellite data indicate that the warming of the Arctic was particularly strong. According to Bromwich et al. (2013), the post-1950s warming of West Antarctica was dramatic and possibly at a similar rate to the Arctic. Conversely, parts of the North Pacific cooled over this period. Pronounced warming is also found over the western Pacific, the Indian Ocean, and the tropical Atlantic; the latter projecting onto the Atlantic Meridional Mode (see Sect. 3.2.7 and Table 3.2).

Figure 4.81 reveals large white areas that indicate a lack of surface observations. Other data products try to fill these gaps through interpolation (e.g., making use of satellite data or other information). Interpolated data suggest a stronger global-mean warming than purely instrumental records (Cowtan and Way 2014, see Sect. 4.5.9).

The lower troposphere warmed at a similar rate to near-surface air (Fig. 4.82, right). With respect to the vertical structure of the trend (Fig. 4.82, left), warming since 1985 encompasses the entire troposphere (except for in the southern high latitudes) and cooling is observed in the stratosphere. In the tropics and subtropics,

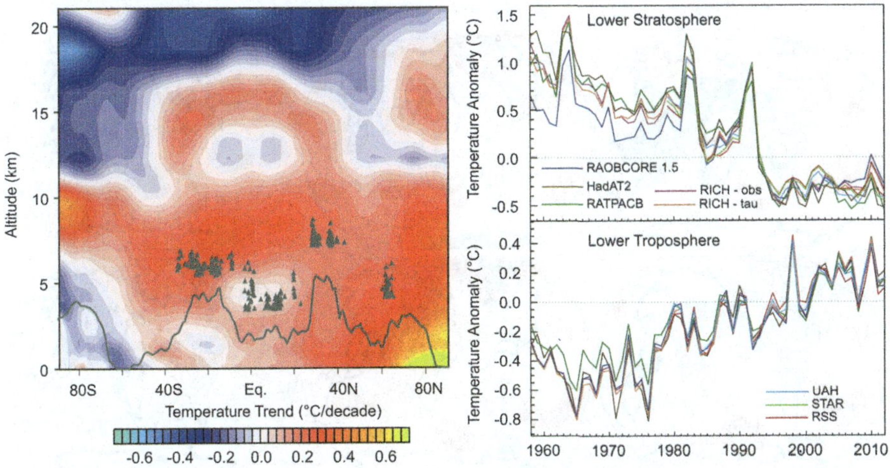

Fig. 4.82 (*left*) Latitude–height cross section of the zonal-mean temperature trend from 1985 to 2013 from ERA-Interim. The *solid line* represents the highest orography in the model and *triangles* show the altitude of selected peaks. (*right*) Global temperatures of the lower stratosphere (From different radiosonde datasets) and lower troposphere (From different satellite datasets) (Source: Hartmann et al. 2013)

rates of warming peak in the upper troposphere because of latent heat release. This feature is expected to continue into the future. Tropical mountains reach this altitude (see triangles in Fig. 4.82), highlighting the role of vertical trend structure for water resources for regions such as the tropical Andes (Bradley et al. 2006; Pepin et al. 2015). The stratospheric cooling trend, which is expected from the increase in greenhouse gases and stratospheric ozone depletion, is punctuated with temperature increases due to volcanic eruptions. Since 2000, stratospheric cooling has been levelling off. However, this is dataset dependent and not yet fully understood (Thompson et al. 2012).

Along with this warming, the water cycle has sped up during the last 30 years. Specific humidity has increased at approximately the rate expected from the Clausius-Clapeyron relation. Relative humidity has stayed relatively constant. Precipitation trends are more difficult to measure. Trends in Fig. 4.83 are from ERA-Interim. They show an increase over tropical regions such as the Amazon-Orinoco basin, East Africa, India, and the western tropical Pacific. Decreasing trends are indicated over subtropical regions such as North America and the Sahel. There are differences across datasets. However, the spatial patterns of observed trends in precipitation and evaporation are mirrored in those of oceanic surface salinity (Stocker et al. 2013). Other aspects of the water cycle have also changed. Precipitation events seem to have become more extreme (Hartmann et al. 2013; also see Fischer and Knutti 2014). Combined satellite and in situ observations show a reduction in Northern Hemispheric snow cover since the 1920s (but an increase over Siberia in fall, see Cohen et al. 2012). Most of the reduction occurred in the 1980s (Stocker et al. 2013); a phenomenon that will be revisited in Sect. 4.5.4.

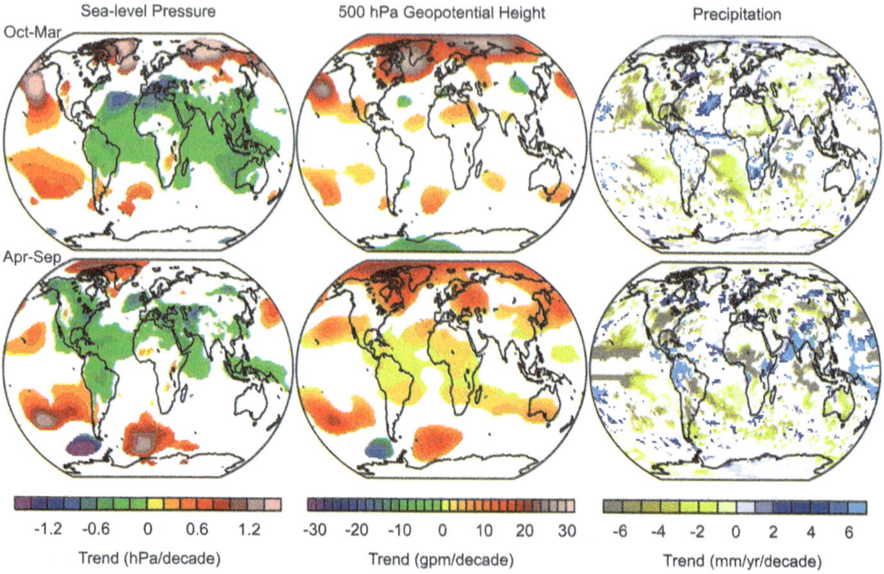

Fig. 4.83 Maps of boreal (*top*) October–March and (*bottom*) April–September trends from 1985 to 2013 of (*left*) sea-level pressure, (*middle*) 500 hPa geopotential height, and (*right*) precipitation from ERA-Interim. Trends are only shown if significant at the 90 % level after accounting for autocorrelation

Changes in the water cycle are influenced by atmospheric temperature, sea-surface temperatures, the land surface, and atmospheric circulation. So, how has the machinery changed over the period 1985–2013? According to reanalysis datasets, whose trends may not be reliable, the Hadley circulation has strengthened (also see Hartmann et al. 2013). Also, since the late 1990s (and following a long period of gentle decrease), the Pacific Walker circulation has strengthened (Kosaka and Xie 2013). This is puzzling because climate models usually respond to increases in greenhouse gases with a weakening of tropical circulation (Collins et al. 2013).[23] It is, however, consistent with other changes observed in the Pacific Ocean (see Sect. 4.5.9).

With respect to the zonally symmetric circulation, a widening of the tropical belt since the 1970s is found (Forster et al. 2011; Hu et al. 2011; Seidel et al. 2008). Although widening rates are uncertain (they depend on the datasets and methods used, see Birner 2010), the widening itself is considered robust because it can be derived from independent observing systems (e.g., outgoing longwave radiation, total column ozone, and reanalyses). For example, Fig. 4.84 shows the trend in

[23]This is because, for a given temperature change, the latent energy response according to the Clausius-Clapeyron relation is very large. Atmospheric circulation becomes overly efficient in transporting energy, which cannot be radiated equally efficiently. Slowing down the atmospheric circulation is the only way to thermodynamically compensate for this discrepancy.

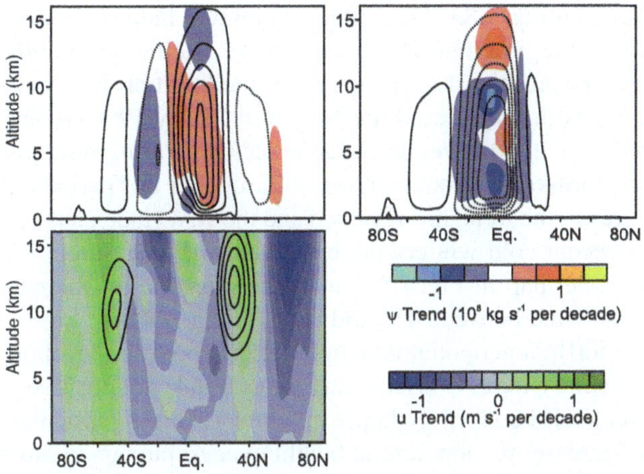

Fig. 4.84 Trends (colours) and mean values (contour spacing is 0.04×10^{10} kg s^{-1}, centered around zero) from 1985 to 2013 in the zonal-mean (*top*) meridional mass streamfunction and (*bottom*) zonal wind during boreal (*left*) winter (Dec.–Mar.) and (*right*) summer (Jun.–Sep.) (Data source: ERA-Interim)

the meridional mass streamfunction from ERA-Interim. The poleward edge of the southern Hadley cell seems to have shifted (negative trends) in both seasons and a similar (though less clear) trend is also found in the Northern Hemisphere.

Consistent with a widening of the Hadley cell, a poleward shift is also found in the maxima of the zonal-mean zonal wind in both hemispheres (Fig. 4.84), a measure of the subtropical jet. More sophisticated definitions of jet position indeed show that the subtropical jets have moved poleward since 1979 (Hartmann et al. 2013). A poleward shift after 1979 is also seen in the various indices displayed in Fig. 4.42.

The widening of the tropical belt is expected from greenhouse gas forcing, according to recent climate model studies (Hu et al. 2013), and responds to warming tropics and changes in meridional sea-surface temperature gradients. Allen et al. (2014) find that the increase in the PDO contributed to the widening trend. Ozone and aerosols have also been suggested to affect the width of the tropical belt (Allen et al. 2012).

Not only did the tropical edge move poleward, but some circulation features of the extratropics (the position of the midlatitude jet stream or storm tracks) showed a poleward migration in reanalysis data along with variables such as cloud cover (Hartmann et al. 2013).

Over the southern extratropics, consistent with the poleward expansion of circulation, the Southern Annular Mode (Table 3.2; Sect. 3.2.4) strengthened since the 1970s. This trend has been partly attributed to the ozone hole over Antarctica (Gillett et al. 2003; Thompson and Solomon 2002) and is strongest in austral spring and summer. This will be discussed in the next section.

Furthermore, zonally asymmetric circulation features changed over this period. Maps of trends in sea-level pressure and 500 hPa geopotential height from ERA-

Interim are shown in Fig. 4.83. The sea-level pressure field shows a decrease in the tropical-subtropical region, but an increase in the Pacific, the South Atlantic, and the Arctic. Over the Atlantic-European sector, the trend projects onto the NAO. In fact, the NAO index first increased, peaking in the early 1990s (contributing to the poleward shift of circulation over the Atlantic sector). Warm, snow-poor winters and frequent winter storms were characteristic in Europe for this period (see Sect. 4.5.4). Since then, however, the trend has reversed and the NAO index has become negative as reflected by recent cold winters in central and western Europe. The downward trend dominates the map in Fig. 4.83, and is responsible for the negative trend in zonal-mean zonal wind between 45° and 80°N in Fig. 4.84.

The trend in 500 hPa geopotential height reflects both sea-level pressure and the warming of the atmosphere. Consequently, most trends are positive (although with an NAO imprint over the North Atlantic in winter). This is particularly apparent in the tropics. A decrease in geopotential height over Antarctica in austral summer is noteworthy and mirrors the SAM trend (see Sect. 4.5.2).

4.5.1.2 Greenhouse Gases and Global Brightening

What drove temperature changes since 1985? Evidently, greenhouse gas forcing was a major contributor. According to IPCC's Fifth Assessment Report (Stocker et al. 2013), anthropogenic radiative forcing increased by ca. 1 W m^{-2} between 1980 and 2011. Global-mean land surface temperature increased by 0.17 °C $decade^{-1}$ between 1986 and 2013. In our simple forcing-temperature regression (extrapolating the forcing time series, most of which end in 2005, to the present), the total forced trend was 0.32 °C $decade^{-1}$, most of which is attributable to greenhouses gases (0.29 °C $decade^{-1}$).

An additional contribution could have come from the reversal of the aerosol trend. The decrease of surface shortwave radiation due to the increase in aerosols—the "global dimming" (Ohmura and Lang 1989)—came to an end in the 1980s. Environmental concern led to the implementation of air quality measures in Europe and North America. Cleaning of the air was further helped by the collapse of economies after the dissolution of the Soviet Union. The global dimming trend reversed and gave way to a period of "global brightening" (Wild et al. 2005). This change is apparent in measurements of atmospheric transmission (Fig. 4.85). There was a steady decrease from the 1930s to the 1980s, followed by a strong increase since 1990. Since the 2000s, however, increasing aerosol emissions in East Asia have led to a regionally varying picture. Some regions show renewed diming and some show brightening (Wild 2009). Section 4.5.3 will go into more detail with respect to the decrease of transmission following the Pinatubo eruption as well as the suggestion of injecting sulphur into the stratosphere (see Box 4.5, p. 299).

Decadal changes continued in the oceans. After the 1976/1977 shift, eastern Pacific ENSO events became more frequent and the PDO reached high values, but decreased after 1999. The AMO was low at the beginning of the period and increased to the mid 2000s.

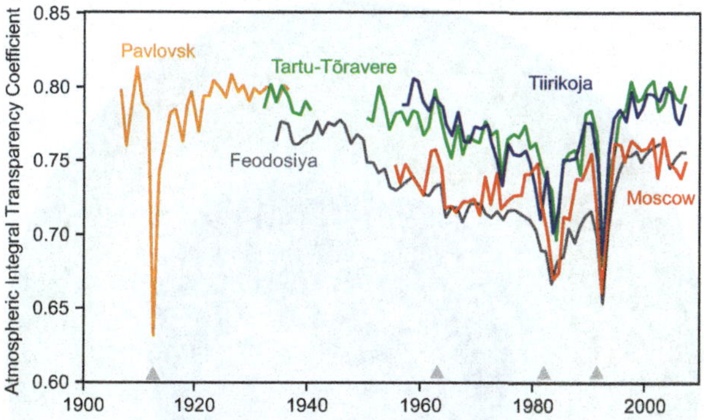

Fig. 4.85 Annual means of the atmospheric integral transparency coefficient at several stations in Eastern Europe. *Grey triangles* denote major volcanic eruptions (Redrawn from Ohvril et al. 2009)

Apart from global warming, changes to the water cycle, and changes to atmospheric circulation, a number of climatic changes on interannual and multiannual scales deserve attention. A small (unaviodably subjective) selection of topics was chosen for this last section, which features changes in atmospheric composition, hydroclimate, storminess, heatwaves, and sea ice. We will look at climatic changes on the Indian subcontinent, the Arctic, Australia, and Europe. I will end this section with the most recent decade of global climate history—the global warming hiatus.

4.5.2 *The Ozone Hole*

It was a wakeup call for the global public. In May 1985, scientists from the British Antarctic Survey reported the discovery of rapid stratospheric ozone loss over Antarctica. They attributed the loss to the emissions of chlorofluorocarbons(CFCs). The phenomenon became known as the "ozone hole". Hardly any other environmental event was as effective at portraying how delicate and vulnerable the global atmosphere is to human disturbance. Images of the ozone hole became an icon of environmental change (see Fig. 4.86). At the same time, the ozone hole is an oft-told success story. The Montreal Protocol in 1987 and its successors were successful in banning CFCs.[24] Although some CFC substitutes are also damaging (to the ozone layer and climate), the atmospheric concentrations of CFCs and hydrochlorofluorocarbons (HCFCs) are decreasing. The discovery of the ozone hole also serves as a reminder of the difficulties of environmental monitoring from space. Satellites were in place and measuring total column ozone. However, the ozone hole was discovered from old-fashioned but carefully checked ground-based equipment (Farman et al. 1985). Meanwhile, the UV radiances measured by the satellite were

[24]The Montreal Protocol was also one of the most effective climate protection measures (Estrada et al. 2013).

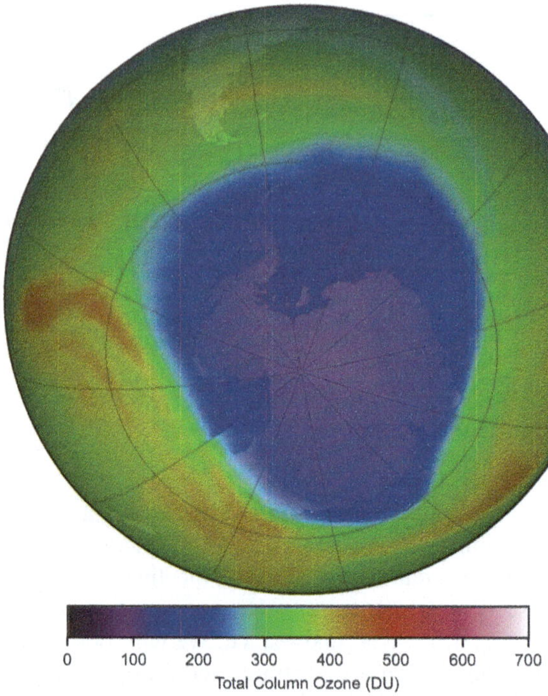

Fig. 4.86 The ozone hole over the Antarctic on 30 September 2014 as seen in total column ozone data from the Ozone Monitoring Instrument (OMI) onboard the Aura satellite

incompatible with the necessary prior assumptions (Sect. 2.3; retrieving column ozone from satellite radiances requires assuming a vertical ozone distribution) and were not published (Bhartia 2009).

Much has been written about the ozone hole. In this section, I will not discuss the complex chemistry of heterogeneous ozone depletion. Interested readers are directed to the overview by Solomon (1999). Furthermore, I will not write about the most discussed effects of the Antarctic ozone hole—the increase of UV radiation. Rather, I will briefly analyse the ozone layer in the CASTRO simulations and discuss the relation between ozone depletion and climate—an important part of the machinery.

4.5.2.1 Ozone Depletion and the Ozone Hole in the CASTRO Simulations

Three-dimensional chemistry-climate models (Sect. 2.7.2) were developed in the 1990s to explore ozone depletion and ozone–climate interactions. The models have been evaluated over the past decade in a coordinated activity (CCMVal SPARC et al. 2010) and proven useful in many respects. Our CASTRO simulations are part of these activities. In the following I focus on annual-mean total column ozone

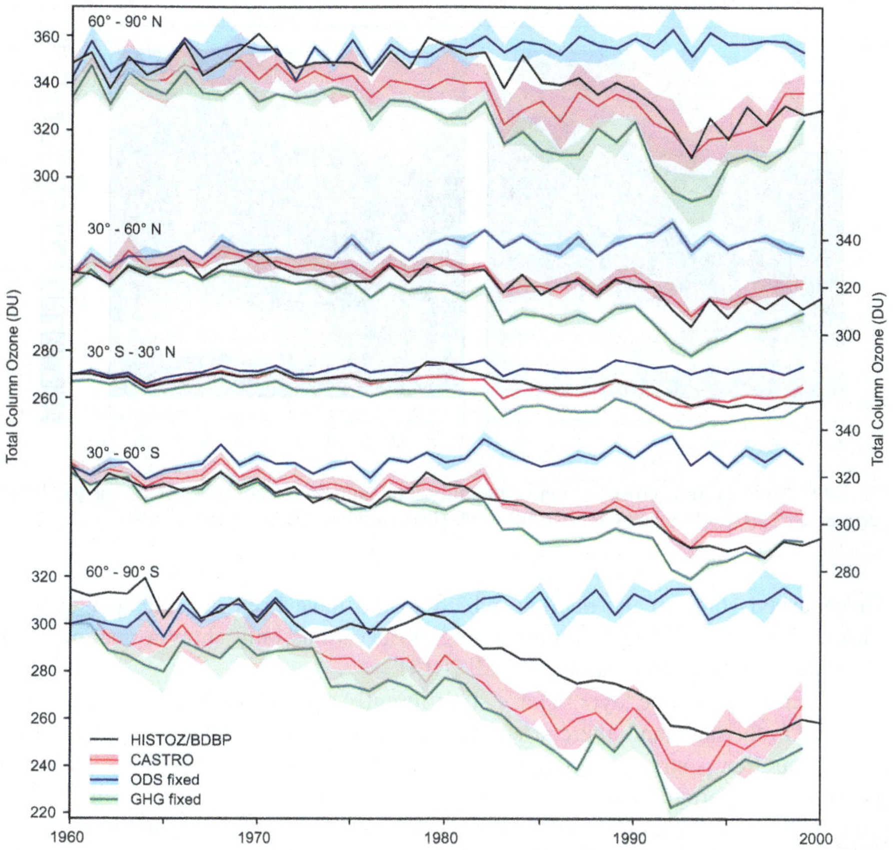

Fig. 4.87 Annual-mean total column ozone during 1960–1999 from the CASTRO simulations (all forcings; an ensemble of nine) and sensitivity simulations (ensembles of three members each) with concentrations of ozone depleting substances (ODS) or greenhouse gases (GHG) held constant at their 1951 or 1901 levels, respectively, for different zonal bands. *Shading* indicates the ensemble range. The *black line* shows observations (HISTOZ/BDBP data). Note that observations only cover 75°S–75°N

for wide zonal bands during the period 1960–2000 (Fig. 4.87; also see Fig. 4.69 for a similar comparison for an earlier period). The observations (HISTOZ/BDBP; Sect. 2.9) show a substantial decline in total column ozone in all latitude bands. However, there are large differences among the latitude bands. The decline in the tropics is much smaller than that in the extratropics.

The trend is strongest in the southern high latitudes; the region of the ozone hole. The CASTRO simulations (*red*) reproduce the observed ozone depletion over the tropics as well as the ozone hole over Antarctica in austral spring. The ozone decline is caused by emissions of ozone-depleting substances such as CFCs. If these substances are kept at their 1950s level in the simulations (*blue*), there is an increase in ozone from the 1960s to the 1980s, mostly due to the effects of greenhouse gases.

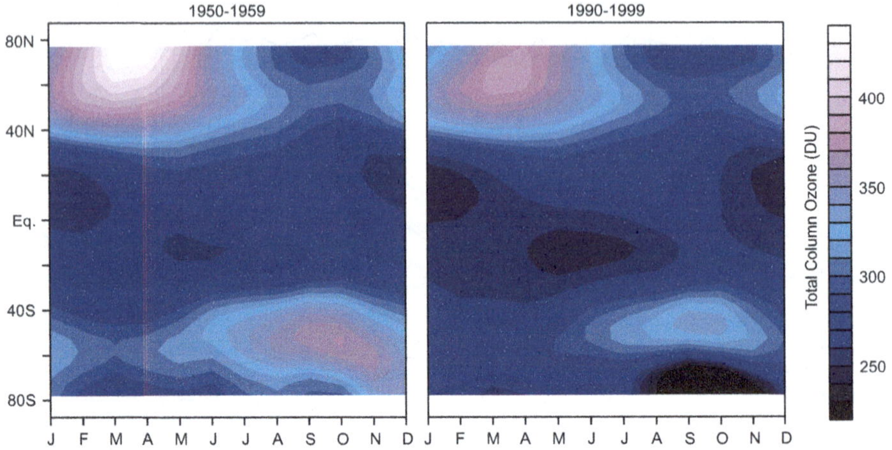

Fig. 4.88 Time–latitude cross section of monthly total column ozone values (DU) for the 10-yr periods (*left*) 1950–1959 and (*right*) 1990–1999 (Data sources: HISTOZ and BDBP)

This masked some of the decline, which would have been even steeper (*green*) without the effects of greenhouse gases. Observations only show a decline after the mid-1970s (see also Shepherd et al. 2014).

Why does the ozone layer behave so differently over the Antarctic than over the Arctic? On the one hand, there are dynamical reasons. The wintertime polar vortex over Antarctica is stronger, more stable, and breaks down later in the season than its Arctic counterpart. This difference in the dynamics explains why in the 1950s, when CFC concentrations were still very low, the ozone column showed a clear springtime maximum over the Arctic, but a weak springtime maximum over the Antarctic (Fig. 4.88, left). The difference in the dynamics also explains the response of the ozone layer over the Arctic and Antarctic to the addition of CFCs (e.g., in the 1990s; Fig. 4.88, right). In addition to sunlight, ozone depletion requires the activation of chlorine by heterogeneous reactions on the surface of cloud particles. Clouds are rare in the Arctic stratosphere because it is typically too warm. Over the Antarctic, the vortex is much colder. Clouds form during the winter, activating chlorine and promoting ozone depletion. Every spring, an ozone hole forms over the Antarctic. Recently, during a dynamically unusual winter, an ozone hole was observed for the first time over the Arctic (Manney et al. 2011).

Because greenhouse gases cool the stratosphere, they are expected to increase the frequency of cold conditions and heighten ozone loss (Austin et al. 1992). In fact, this is the reason that ozone recovery is expected to proceed much slower over the Antarctic. However, with respect to gas-phase chemistry, the equilibrium amount of ozone in the stratosphere also depends on temperature and, in this case, lower temperatures favour more ozone. Therefore, once stratospheric chlorine loading has returned to preindustrial levels, global ozone will be higher than before the onset of ozone depletion; an effect that is sometimes called "super-recovery" (see WMO 2011).

4.5.2.2 Effect of the Ozone Hole on Climate

Between ozone and climate, there is a two-way interaction. Climate affects the ozone layer, but the stratospheric ozone distribution also affects climate. The ozone loss during springtime in the Antarctic reduces heating and leads to a cooler vortex core. The resulting meridional temperature gradient accelerates zonal flow. Radiative and dynamical processes link the troposphere and the stratosphere. Consequently, the change in the stratospheric zonal flow during springtime and summertime in the Antarctic propagates to the troposphere and leads to a poleward migration of jets and strengthened westerlies over the Southern Ocean. These results have been found in observations (Thompson and Solomon 2002) and model simulations (Gillett and Thompson 2003), and explain the puzzling cooling of Antarctica during the 1980s and 1990s. CASTRO simulations reproduce the strong observed trends in lower tropospheric winds, 500 hPa geopotential height, and lower

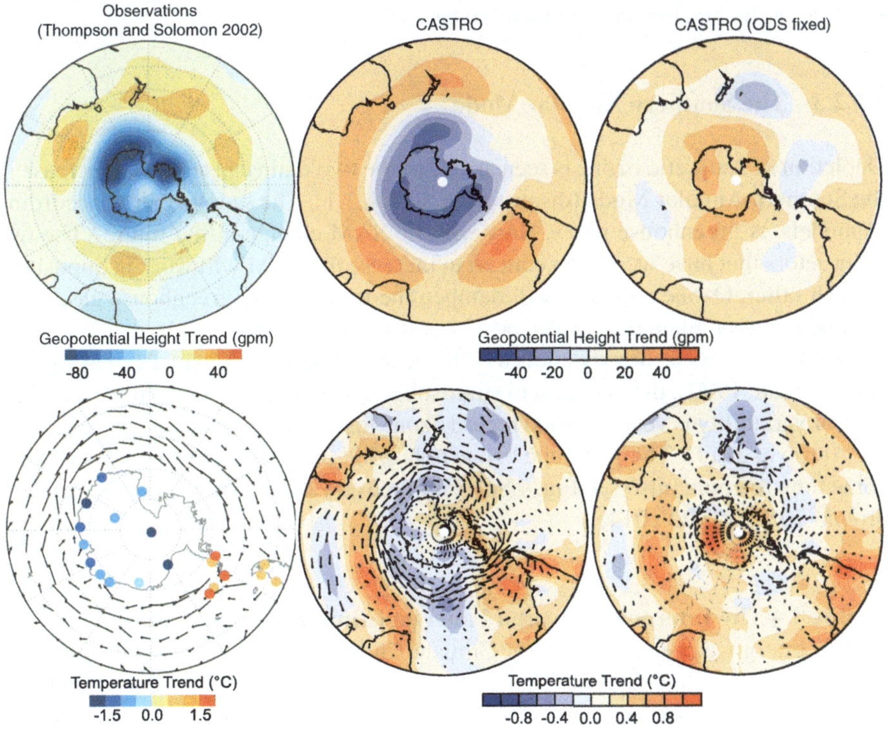

Fig. 4.89 Changes in (*top*) 500 hPa geopotential height and (*bottom*) 925 hPa winds and surface temperatures during December–February 1969–1999 from (*left*) observations and reanalysis (Thompson and Solomon (2002), reprinted with permission from AAAS; geopotential height and wind trends refer to Dec.–Apr. 1979–1999) and (*middle*) CASTRO simulations (ensemble mean; winds and temperature are at 950 hPa). (*right*) CASTRO simulations with ozone-depleting substances held constant after 1951 (mean of ensemble of three)

tropospheric temperatures in austral summer from 1969 to 1999 (Fig. 4.89). This is almost entirely due to the trend in ozone-depleting substances. If the latter are set to constant 1951 values (Fig. 4.89, right), the trend vanishes completely and Antarctica warms rapidly.

The altered circulation further affects summertime temperature and precipitation in the southern extratropics, such as Australia (Thompson et al. 2011). Kang et al. (2011) found increasing in precipitation in the southern subtropics during austral spring. Manatsa et al. (2013) reported an increase in temperatures in South Africa, which they related to the ozone hole. The change in circulation by the Antarctic ozone can also affect ocean–atmosphere interactions in the Southern Ocean. A strengthened circulation (positive SAM) can reduce carbon uptake and enhance ocean acidification (Lenton et al. (2009), see Thompson et al. (2011) for a review).

In retrospect, the relevance of the ozone hole is not just that it indicates the destruction of Earth's protection from harmful radiation. It has farther-reaching effects that may be more subtle but are nevertheless important. The change in atmospheric circulation associated with the ozone hole demonstrates the complex reaction of the machinery to a disturbance.

4.5.2.3 The Southern Annular Mode

Depletion of Antarctic ozone is seen as one of the two main causes for an increase in the Southern Annular Mode (discussed in Sect. 4.3.1). The second cause, according to models, is greenhouse gases. How will the SAM change in the future? The two main actors that have worked in concert to increase the SAM will work in opposition to each other. Ozone recovery will dampen the trend; while greenhouse gases will continue to strengthen the SAM (see Thompson et al. 2011).

Ozone recovery is already underway. Concentrations of CFCs are decreasing and, since the late 1990s, the decrease in stratospheric ozone has slowed. However, year-to-year variability is very high and record ozone losses occasionally occur (e.g., over the Arctic in 2011). The Antarctic ozone hole will continue to appear every year for some time to come (also see Fig. 4.86). Full recovery of the ozone layer is expected around 2050 (WMO 2014).

Box 4.4 The Intergovernmental Panel on Climate Change (IPCC)
The Intergovernmental Panel on Climate Change (IPCC) was established in 1988 by the United Nations Environment Programme (UNEP) and the WMO. Its task is to assess the most recent scientific, technical, and socio-economic information relevant to the understanding of climate change. In 1990, under the leadership of Bert Bolin, the IPCC produced its first assessment report. Since then, assessment reports have been produced every 5–6 years. In addition, the IPCC publishes special reports such as "Managing the Risks

(continued)

Box 4.4 (continued)
of Extreme Events and Disasters to Advance Climate Change Adaptation" (SREX; IPCC 2012).

The IPCC is an intergovernmental body. Therefore, governments are involved in the scoping process and the review of the reports. In turn, the reports are endorsed by governments, giving them political weight. The objective of the IPCC is to produce materials that are "policy-relevant and yet policy-neutral, never policy-prescriptive" (www.ipcc.ch). However, the assessment reports are also a valuable scientific resource; they provide a comprehensive summary of the state of research. In fact, there is hardly any other area of science in which the state of research is so well defined.

Assessments are carried out by three working groups "The Physical Science Basis" (WG I), "Impacts, Adaptation, and Vulnerability" (WG II), and "Mitigation of Climate Change" (WG III). Each working group is led by two co-chairs. Writing an assessment report takes several years of work and involves the voluntary contributions of hundreds of experts from around the world. Then, the assessment report undergoes several rounds of review. The contribution of WG I to the Fifth Assessment Report, for instance, was jointly written by 259 authors. These authors had to reply to 54,677 reviewer comments.

The Fourth Assessment Report (IPCC 2007) was so successful that the IPCC and Al Gore received the Nobel Peace Prize in 2007. At the same time, enormous public scrutiny and tremendous expectations revealed small errors in the report and pointed to several weaknesses in the procedures that were widely debated in blogs. In the interim, the IPCC has reviewed its procedures. The public debate also raised the question of whether assessment reports (as opposed to targeted or regional reviews) are the best format.

The Fifth Assessment Report was published while this book was being written. Overall, the contribution of Working Group I confirms the points made in the Fourth Assessment Report; some are strengthened and some are amended. Some of the key statements of the Fifth Assessment Report (IPCC 2013) are:

> "Warming of the climate system is unequivocal, and since the 1950s, many of the observed changes are unprecedented over decades to millennia."

> "Human influence on the climate system is clear [...] It is extremely likely that human influence has been the dominant cause of the observed warming since the mid-20th century."

The Fifth Assessment Report received less media attention than the Fourth Assessment Report and the "blogosphere" remained calm. Is this a sign that the facts are established?

It is too early to judge the impacts of the Fifth Assessment Report and its usefulness to decision makers. However, it is clear that the Fifth Assessment

(continued)

Box 4.4 (continued)
Report provides another comprehensive summary of the state-of-the art of climate change research. The value of full assessment reports for decision makers—as opposed to more targeted, problem-oriented, regional or sectorial assessments—is open for discussion.

4.5.3 Eruption of Pinatubo

After the 1912 eruption of Novarupta (Katmai), Alaska, volcanic activity decreased and remained low for a half century. The next major eruptions were Mount Agung, Indonesia (1963); El Fuego, Guatemala (1974); and El Chichón, Mexico (1982). The biggest (and most studied) eruption of the century was Pinatubo, Philippines (1991; Fig. 3.33).

Pinatubo was a typical eruption in many aspects and has provided atmospheric sciences a wealth of information. Its effects are well documented in surface and satellite data, can be reproduced in atmospheric models, and are scientifically well understood. Well—are they? There are still uncertainties, model deficiencies, surprises, and opportunities to learn.

In Chap. 3, I discussed many aspects of the Pinatubo eruption and used it as an example of volcanic eruptions in general (e.g., the spreading of aerosols; Figs. 3.35 and 3.36). In this section, I would like to explore the effects on the biosphere and the hydrosphere. To start, however, a look at the stratospheric effects of the eruption reveals interesting insights (also see Fig. 3.37).

4.5.3.1 Stratospheric Ozone and Temperatures

A common model deficiency is the overestimation of stratospheric warming after a volcanic eruption. This does not seem to be caused by aerosol properties (e.g., their size distributions), which are also uncertain (Arfeuille et al. 2013). In Fig. 4.90, the 50 hPa temperature evolution in the tropics during the years surrounding an eruption is given from CASTRO and ERA-Interim. Even when showing absolute temperatures (i.e., without removing the annual cycle), the excessive warming in CASTRO during 1992 is obvious.

Three years after the eruption, temperatures in the lower stratosphere are below pre-eruption values (see Fig. 4.90). Since then, the lower stratosphere does not cool markedly. Thus, Pinatubo introduced a step change in stratospheric temperature trends that is not yet fully understood.

In addition, Pinatubo demonstrated another effect of volcanic aerosols. They can activate chlorine, leading to a decrease of stratospheric ozone (Solomon 1999). Figure 4.90 shows total column ozone as a function of time and latitude for the

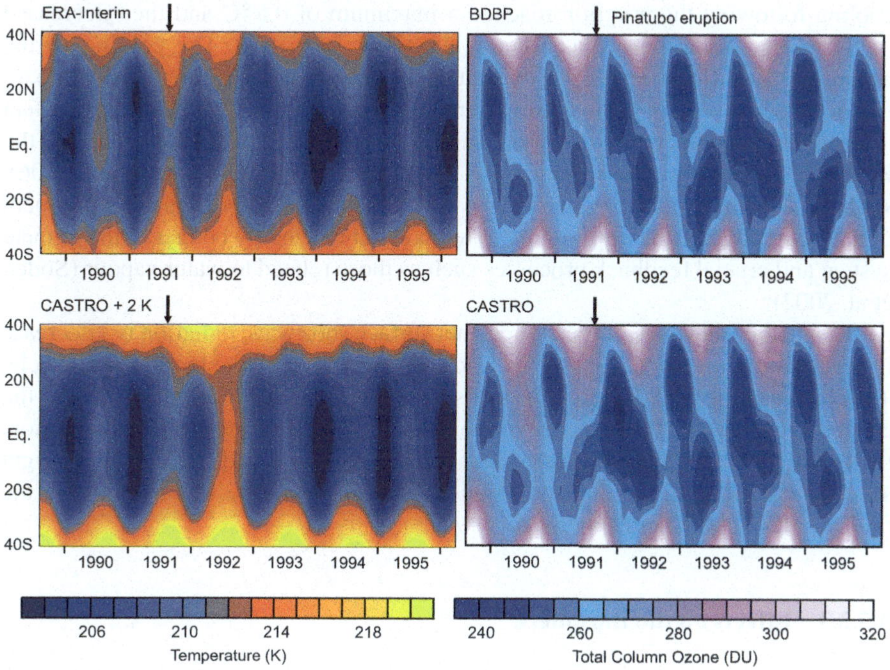

Fig. 4.90 Temperature at 50 hPa and total column ozone as a function of time and latitude from 1989 to 1996 from observation-based data (*top*, ERA-Interim and BDBP) and the CASTRO simulations

period 1989–1996. A clear decrease is seen in the tropics following the Pinatubo eruption. Because the eruption roughly coincided with the peak in atmospheric CFCs, global ozone reached a minimum in models and observations (see Fig. 4.87). The Pinatubo aerosols also changed stratospheric transport, and parts of the ozone changes are related to that (Muthers et al. 2014a). In CASTRO, ozone recovered faster than in the observations, where values remained low for several years.

The change in ozone might have had additional implications for atmospheric chemistry. The methane growth rate (Sect. 4.4.2) showed variations following the Pinatubo eruption, According to Bândă et al. (2013), this might have been due to the stratospheric ozone depletion following the eruption.

4.5.3.2 Effects on the Energy Budget and the Water Cycle

Chapter 3 discusses the impacts of volcanic eruptions on temperatures, the energy budget, and the water cycle because of the decrease in surface solar radiation. Much of the work on these topics is based on the Pinatubo eruption, which (because the observing systems were relatively well developed in the 1990s) provided a quantitative estimate of the cooling effect of stratospheric aerosols. The global cooling following the eruption reached a maximum of 0.5 °C and then decreased (the cooling was also forecasted, see Hansen et al. 1992). The cooling of the ocean surface affects upper-ocean heat content, and cooler waters penetrate to the deep ocean. Through negative thermal expansion, volcanic eruptions affect the sea level. In fact, model-based studies suggest that even the effect of the Krakatau eruption in 1883 persists in the oceans today (Gleckler et al. 2006), slightly offsetting anthropogenic sea-level rise. Thus, Pinatubo—as an experiment of nature—provided the opportunity to learn about the energy budget of the climate system and crucial feedback processes such as those related to water vapour (Soden et al. 2002).

Another consequence of the reduced net surface shortwave radiation is a decrease in evaporation (see Sect. 3.3.1). This leads to a slowdown of the hydrological cycle. The change in the land–sea contrast leads to a weakening of the monsoons. After the eruption of Pinatubo, the response of the water cycle is seen in data of precipitation, the Palmer Drought Severity Index, and runoff (Trenberth and Dai 2007). Drought severity increased in all monsoon regions and (in the case of Pinatubo) at the midlatitudes of both hemispheres.

4.5.3.3 Effects on the Biosphere

The cooler temperatures, reduced water availability, and changes in solar radiation affected plant life. Volcanic aerosols scatter solar shortwave radiation (see Sect. 3.3.1). Some of the scattered radiation reaches the earth's surface in the form of diffuse light. Therefore, after volcanic eruptions direct shortwave radiation decreases, but diffuse radiation increases. Diffuse radiation is important for photosynthesis and some forests responded to the eruption of Pinatubo by increasing their photosynthesis (Gu et al. 2003). Conversely, vegetation also responded to the volcanic cooling through browning. This was found in satellite data following the Pinatubo eruption (Jong et al. 2012). Hence, volcanic eruptions may affect the biosphere. In turn, this may link back to climate via water vapour or carbon cycle feedbacks (Farquhar and Roderick 2003). Thus, Pinatubo pointed scientists to study biological feedback processes.

In palaeoclimatology, the question arises whether tree ring width is able to capture the climatic effects of volcanic eruptions given the possible change in photosynthesis. However, volcanic eruptions do appear in tree ring chronologies (Anchukaitis et al. 2012), as is also shown in many instances in this section (e.g., Fig. 4.7).

Box 4.5 Climate Engineering

Stratospheric aerosols change the energy budget, affect the biosphere, and slow the water cycle. This important lesson was learned from Pinatubo (and other eruptions) and has relevance for another topic, climate engineering, which is briefly discussed in this box. Crutzen (2006) and Budyko (1974) among others have suggested that global warming could be counterbalanced by injecting sulphate aerosols into the stratosphere, mimicking the cooling effect of a volcanic eruption (also see Box 3.4, p. 162). Methods that aim to artificially counteract global warming are termed "climate engineering"; those methods that try to change the radiation balance (such as this one) are termed "solar radiation management". The effectiveness of sulphur injection depends highly on the particle size (see Sect. 3.3.1); larger particles are less efficient than smaller ones (Heckendorn et al. 2009). If the analogy with volcanic eruptions holds, climate engineering via stratospheric sulphur injection would slow the water cycle and weaken the monsoons, with unforeseeable consequences for many tropical and subtropical countries (Haywood et al. 2013).

Another aspect of this discussion is the change in stratospheric ozone. Provided sufficient chlorine, climate engineering with stratospheric sulphur injection would slow down the recovery of stratospheric ozone (Heckendorn et al. 2009).

4.5.3.4 Small Eruptions

When will the next volcanic eruption occur? Once it does, are we prepared to forecast the consequences? Eventually, an eruption the size of Pinatubo (or larger) will occur. In recent years, smaller volcanic eruptions have also caught the interest of researchers. These are the likely cause for the recent increase in sulphate aerosols in the stratosphere (Neely et al. 2013) and might have had an impact on global temperature (Solomon et al. 2011) as will be discussed in more detail in Sect. 4.5.9. The climatic consequences of volcanic eruptions, as exemplified here with Pinatubo, reveal the complexity of the machinery.

The Pinatubo eruption was followed by warming winters on the northern continents, which is a typical response of the climate system to a change in the stratospheric temperature gradient (see Sect. 3.3.1). Thus, it contributed to a sequence of warm European winters that are the topic of the next section.

4.5.4 *Warm European Winters, Increased Storminess and Positive NAO in the 1990s*

During the late 1980s and 1990s, a sequence of very mild winters surprised people in central Europe. Initiated by a steplike decrease from 1988 to 1989, snow suddenly became a rarity in Alpine ski resorts. Springs began earlier in the year, with typical vegetation stages such as bud burst or leaf unfolding occurring two weeks earlier in the 1990s than in the mid-1980s (Fig. 4.91). Similar changes were also found in other regions and for other variables. The lack of spring snow was not a local event but affected the entire European continent and is clearly visible on a hemispheric scale (Fig. 4.91; also see Brown and Robinson 2011).

There were further signs of a change in weather. A number of severe winter storms hit Europe and caused considerable damage. Winters were different than before and the topic of "global warming" caught the attention of Europeans for the first time.

In a popular science book first authored by Hartmut Grassl[25] (Grassl and Klingholz 1990), which I read as an undergraduate student, the authors argued that towards the end of the 20th century, when global warming will have been clearly attributed to human activity, one will be able, in hindsight, to date the time when global warming first became apparent. The authors argued that this turning point in climate would most likely be in the 1980s.

The first publication detecting an anthropogenic fingerprint in global temperature was published only a few years later (Hegerl et al. 1996). So, in this respect, Hartmut Grassl was right. Since then, many other studies have detected an anthropogenic influence in other variables and statistics. Reverting to the warm winters in Europe

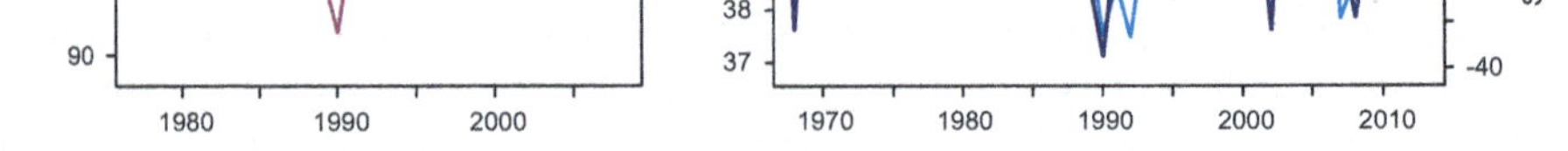

Fig. 4.91 (*left*) Phenological data from Grossaffoltern, Switzerland (Data from the BERNCLIM network operated by the Institute of Geography of the University of Bern). (*right*) Northern Hemisphere snow area and European snow mass anomalies in March (Data sources: ESA-GLOBSNOW, FMI, Rutgers University)

[25]Hartmut Grassl, *1940, is a German climate scientist who served as a long-term director of the Max-Planck Institute for Meteorology in Hamburg.

in the late 1980s and 1990s, however, another ingredient (in addition to greenhouse gases) needs to be considered. The NAO switched to a strongly positive mode. In fact, the trend in the NAO index from the 1960s to the 1990s was even exceptional from a long-term perspective (see Fig. 4.10). In the following, I will briefly look at the warm winters, storms, and NAO during these years.

4.5.4.1 Winter Storms

In 1990, the storms Vivian and Wiebke swept across Europe and caused immense damage. Other storms appeared during the subsequent winters. The strongest storm in Switzerland was Lothar in 1999 (Fig. 4.92). Lothar had developed from a small depression over the North Atlantic and was not well predicted (Wernli et al. 2002). Figure 4.93 (bottom right) shows the 10-m wind field for Switzerland at 12 UTC 26 December 1999, the peak of the storm, obtained from downscaling 20CR (see Sect. 2.9). Wind speeds were particularly high in the western and central parts of Switzerland and in the Alps.

Was the frequency of storms in the 1990s outstanding from a 120-yr perspective? Long time series of high percentiles of wintertime wind maxima in Zurich from 20CR and observations (Fig. 4.94) agree well with each other and with a damage-based storm catalogue (Stucki et al. 2014). In fact, the 1990s stand out as the

Fig. 4.92 Winterstorm "Lothar" caused large forest damages in Switzerland on 26 December 1999 (Photo: Hansrudolf Schneider, Berner Oberländer BOZ)

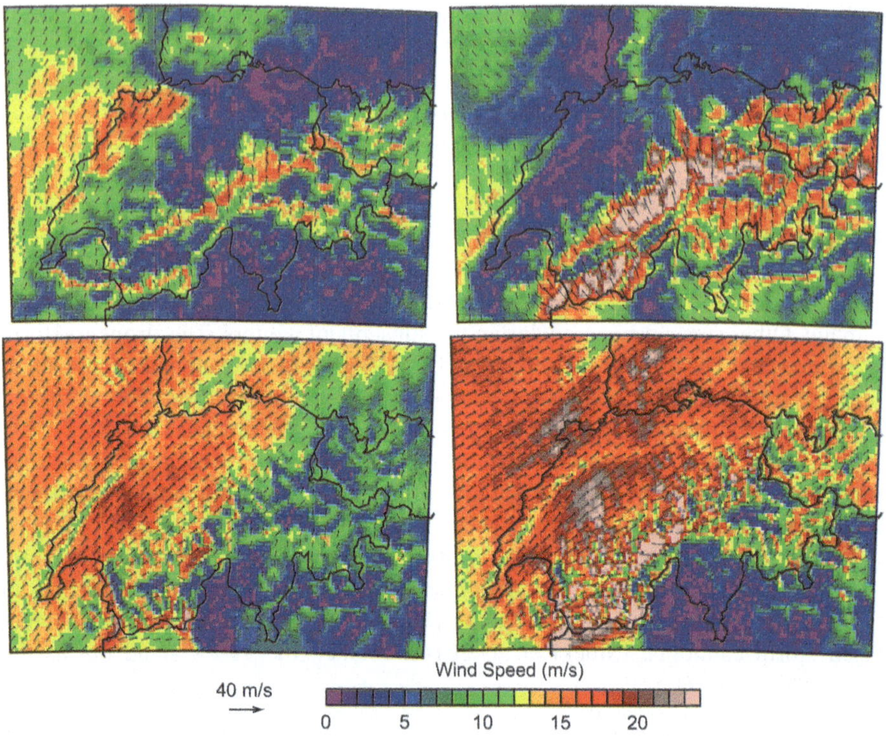

Fig. 4.93 Wind speed at 10 m for the peak phases of four storms in Switzerland obtained from downscaling the ensemble mean of 20CR with the WRF model down to a resolution of 3 km. (*top left*) 20 February 1879; (*top right*) 5 January 1919; (*bottom left*) 23 February 1935; (*bottom right*) 26 December 1999 (Downscaling by Silke Dierer)

most stormy decade of the entire record. Another stormy period was the early 20th century. This agrees with other studies on storminess in Europe, which find pronounced decadal variability with peaks in the late 19th century and in the 1990s, but no long-term trend (Cornes and Jones 2011; Dangendorf et al. 2014; Matulla et al. 2008; Wang et al. 2009b, 2011). Since the 1990s, storminess has returned to average values.

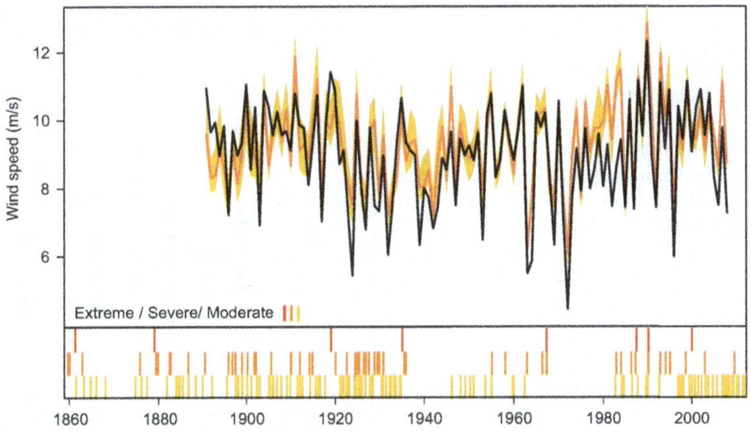

Fig. 4.94 (*top*) Time series of the 93rd percentile of daily maximum wind speed in Zurich per winter (December–February) since 1891 based on hourly observations (*black*) and four-times-daily data from 20CR (*red*: ensemble mean; *orange*: spread of ensemble members). Data are from Brönnimann et al. (2012c). (*bottom*) Extreme, severe, and moderate storms in Switzerland from a damage-based storm catalogue back to 1859 (Data from Stucki et al. 2014)

Box 4.6 A historical perspective to assess storm risks

How can we characterise storm risk in the presence of such strong decadal variability? A historical perspective helps. Documentary data are available for many past events and, with new reanalysis datasets such as 20CR, many storms can be studied quantitatively. Downscaling these storms from 20CR is a first step. Figure 4.93 shows downscaled wind speeds over Switzerland for the peak phases of four wind storms (1879, 1919, 1935, and 1999). Due to the different flow situations and the specific orography of Switzerland, the spatial patterns of wind maxima are different in each case. Downscaled wind speeds were highest over the Jura Mountains in 1879, north of the Alpine crest in 1919, over the Swiss Plateau in 1935, and over the Swiss Plateau and Alpine peaks in 1999. A large number of events are thus needed to characterise storm risks. Based on ca. 100 downscaled storms, a wind storm risk map has recently been produced for Switzerland (www.bafu.admin.ch/naturgefahren/01919/). The quantitative information can be used for damage modelling and comparison with historical damages. This example shows how modern quantitative methods in historical climatology can supplement traditional approaches.

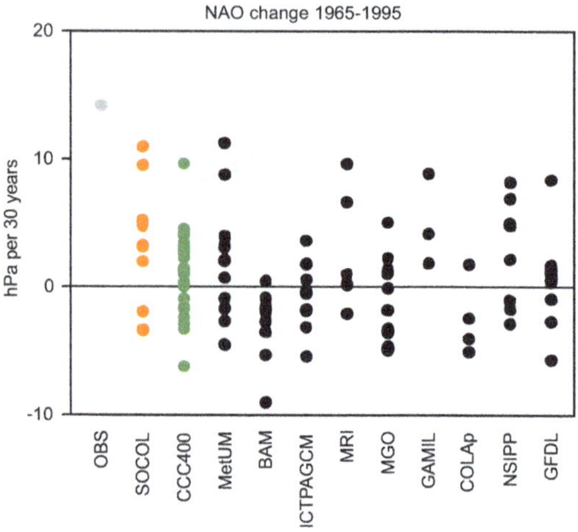

Fig. 4.95 Trend of the wintertime pressure difference (hPa) between the Azores and Iceland from 1965 to 1995 in different ensemble simulations (each dot represents one member) of GCMs (updated from Scaife et al. 2009)

4.5.4.2 Forced NAO?

What caused the sudden change to warm winters, more frequent wind storms, shift in snow cover, and responses of phenological variables? Can they be blamed on increasing greenhouse gases? There is strong evidence that many of these features (as well as others, such as the strengthening of the polar vortex addressed in Fig. 4.42) are related to a change in atmospheric circulation that can best be described with a trend in the NAO. In this sense, the example shown in Sect. 3.3.2 (Jan. 1990) was typical and represents the strengthening of the zonal circulation and weakening of the meridional stratospheric circulation.

So, how important were greenhouse gases and the NAO in explaining the warm and stormy winters? Perhaps these two questions cannot be separated. According to many climate models, global climate change is expected to lead to a more positive mode of the NAO (Collins et al. 2013). Thus, not only the global-mean temperatures, but also the circulation changes in the early 1990s fit with the model expectations. Some studies have even tried to attribute the pressure change over the North Atlantic to greenhouse gases (Gillett et al. 2005). However, this change in the NAO was unprecedented, and only a small fraction of the NAO change can be explained in that way. In fact, climate model simulations with prescribed sea-surface temperatures mostly fail at quantitatively reproducing past changes in the NAO index although most of these models reproduce a small trend (Fig. 4.95; updated from Scaife et al. 2009) and our CCC400 simulations are no exception. The CASTRO simulations, which include stratospheric processes and chemistry-climate interactions, reproduce the trend slightly better than some of the other models. However, the trend remains considerably weaker than in observations. Thus, internal variability contributed strongly to this trend. For summer climate in Europe, Sutton and Dong (2012) found that changes in the Atlantic Ocean (as captured by the AMO index) contributed to the European climate shift around 1990.

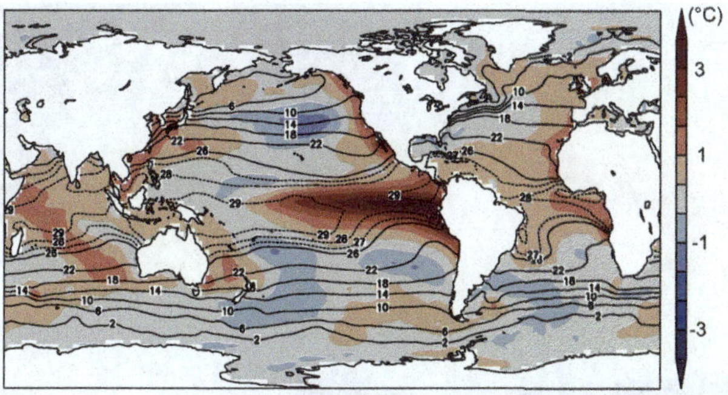

Fig. 4.96 Absolute values (contours) and anomalies (with respect to 1961–1990; colour shading) of sea-surface temperatures from December 1997 to February 1998 (HadISST data; adapted from Brönnimann 2007)

Was this the first time that global warming became evident, as forecasted by Hartmut Grassl? Probably Grassl was right. However, we have learned in the meantime—a point repeatedly made by Brückner—that decadal variability also makes a contribution.

4.5.5 The El Niño of 1997/1998, Forest Fires and the Atmospheric Brown Cloud

After the climate shift of 1976 (see Sect. 4.4.5), eastern Pacific El Niño events became more frequent and date line El Niños became rarer. Two particularly strong events are noteworthy: El Niño of 1982/1983 was the first to hit the headlines, paving El Niño's way to global public awareness. Following this event, progress was made to understand El Niño. The 10-yr Tropical Ocean and Global Atmosphere (TOGA) study of the World Climate Research Programme (WCRP) provided an observing system composed of moored buoys, satellites, and ships. Simple El Niño models were developed (Sect. 3.2.6) and the first successful forecasts of El Niño events were made (Barnett et al. 1988).

The other, equally famous El Niño event was the one of 1997/1998. Today, this is often discussed in the context of the peak in global-mean temperatures that was reached in 1998 following the event. In fact, the tropical Pacific lost large amounts of energy to the atmosphere. The event was a very strong "canonical" or eastern Pacific event. The seasonal anomalies of sea-surface temperature reached 5 °C near the coast of South America (Fig. 4.96).

El Niño of 1997/1998 and the subsequent La Niña events also had a large effect on sea level. Sea-surface heights from satellite altimetry data revealed east–west differences across the Pacific of several decimetres. During the subsequent La Niña event, the sea-surface height anomaly reversed.

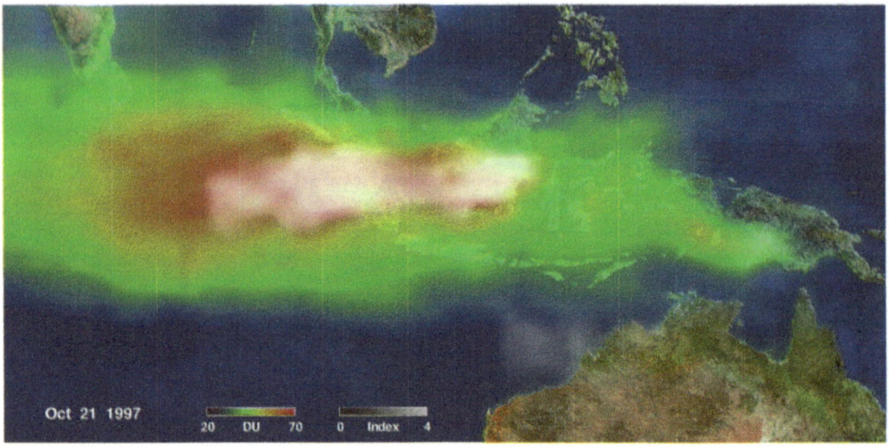

Fig. 4.97 TOMS aerosol index on 21 October 1997. *Green* and *red colours* indicate the presence of tropospheric aerosols (Data source: NASA)

4.5.5.1 Forest Fires

In Sect. 3.2.6, I described the most important El Niño teleconnections. Global climate anomalies in 1997/1998 showed many of the features expected for an eastern Pacific El Niño. Here, I highlight a further aspect related to El Niño-induced changes in precipitation. The decrease of precipitation in Indonesia led to drought and forest fires. The forest fires of 1997 (continuing into 1998) were among the worst in recent history. Their ignition was mostly related to the clearing of land for oil palm and timber cultivation and other farming techniques. However, El Niño-related drought meant that dry fuel was particularly abundant. Figure 4.97 shows the Total Ozone Mapping Spectrometer (TOMS) aerosol index on 21 October 1997. The aerosol cloud over Indonesia and stretching into the Indian Ocean stands out. The local population was affected by severe air pollution. Globally, the fires were a significant source of CO_2. Worldwide, around 25×10^6 hectares of land were affected by fire during the 1997 El Niño (Tacconi 2003).

This is another compelling example of the interaction among physical, chemical and biological mechanisms in the climate system. A climate event affects the biosphere, burning biomass affects atmospheric chemistry and composition, and changes in chemistry and composition (CO_2, aerosols) feed back to climate.

4.5.5.2 The Atmospheric Brown Cloud

Aerosols are not only a problem after El Niño events. Air pollution can reach critical values in Indian cities during the dry season, forming a cloud of haze and pollutants that covers large areas of South Asia and adversely affects the health of the large

Fig. 4.98 Atmospheric Brown Cloud (NASA image by Jeff Schmaltz, MODIS Rapid Response)

regional population (see Fig. 4.98; Ruchirawat et al. 2008). The cloud consists of aerosols, with a large contribution of black carbon, and other pollutants (Lelieveld et al. 2001). One of the main sources is biomass burning related to domestic cooking or agricultural burning. Another fraction comes from fossil fuel burning (e.g., from coal power plants; Gustafsson et al. 2009). The phenomenon is known as "Atmospheric Brown Cloud" or ABC (Ramanathan and Crutzen 2003).

Recent work has addressed the role of the ABC in climate. It affects chemistry, radiation, and the hydrological cycle over the region. Several aspects of the climate system are currently being studied with respect to the ABC. These include the Indian monsoon system and hurricanes over the Indian Ocean.

With respect to radiation, a particularly important component of the ABC is black carbon. It absorbs solar radiation and cools the ground, but heats the layers in which it resides. Large heating rates are found at 700 hPa, which can reach the magnitude of the greenhouse effect (Ramanathan et al. 2007) and change the vertical temperature structure. Effects of these aerosols on the Indian summer monsoon have been suspected by many authors, but the results are inconclusive. Likewise, the effect of aerosols on hurricanes is a matter of debate. It has been suggested that the Atmospheric Brown Cloud may amplify tropical cyclones over the Indian Ocean (Evan et al. 2011). However, Wang et al. (2014) found a weakening effect. Hence, some parts of the machinery still remain hidden.

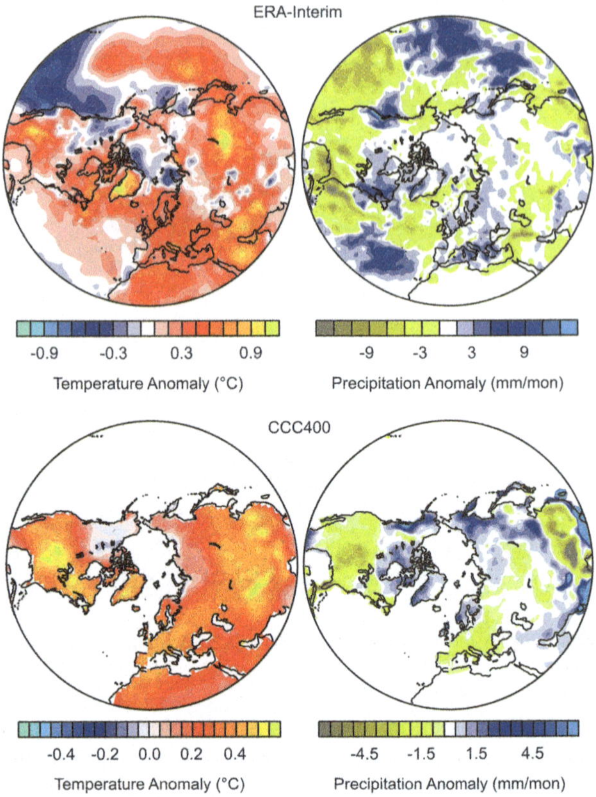

Fig. 4.99 Anomalies in (*left*) temperature and (*right*) precipitation in boreal summer (Jun.–Aug.) during 1998–2004 (relative to 1985–2005) (Data are from (*top*) ERA-Interim and (*bottom*) CCC400)

4.5.6 Droughts in Australia and the Northern Subtropics

4.5.6.1 Subtropical Drought 1998–2004

El Niño of 1997/1998 was followed by a pronounced, long-lasting La Niña. This can be described as another shift towards a climate state with more frequent La Niñas. At the same time, the North Atlantic remained warm because the AMO was in a positive phase. This constellation was conducive to drought in North America (also see Sect. 4.3.3). Figure 4.99 shows temperature and precipitation anomalies for the summers from 1998 to 2004. A band stretching across the northern subtropics, from the Pacific to the Middle East, experienced drought conditions: high temperatures and low precipitation.

Using model simulations, Hoerling and Kumar (2003) were able to relate this large-scale, multiannual drought to global sea-surface temperature anomalies. Similarly, CCC400 (Fig. 4.99, bottom) reproduced the precipitation deficit and temperature surplus over North America and large parts of Eurasia well. Although simulations with forced sea-surface temperatures do not allow the separation of ocean "forced" signals from internal atmospheric variability (which is reflected in sea-surface temperatures and thus "fed back" to the model atmosphere), the similar atmospheric circulation response to imposed sea-surface temperatures (and subsequently the precipitation distribution) by the model and observations strongly suggests an oceanic influence. This is particularly the case for large-scale sea-surface temperature anomalies and multiannual droughts, as shown in several other cases in this book. In fact, the sea-surface temperature pattern resembles that of the "Dust Bowl" and of the drought conditions in 2010–2013.

4.5.6.2 The Millenium Drought in Australia

Around the same time, Australia suffered from record drought conditions. The Big Dry or Millennium Drought, which lasted from 1995 to 2009 (i.e., it started before and ended after the Northern Hemisphere counterpart), was the worst drought since the beginning of measurements. Figure 4.100 shows precipitation and temperature anomalies for this event and for another large drought of the 20th century, the World War Two Drought of 1936–1945. Note that the Millennium Drought does not appear

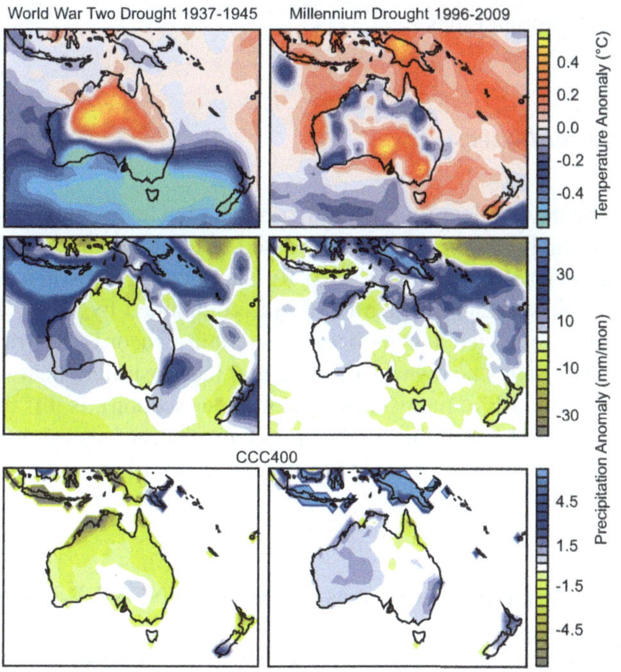

Fig. 4.100 Anomalies in (*top*) temperature and (*middle*, *bottom*) precipitation in December–February during the (*left*) World War Two Drought (1937–1945; relative to 1961–1990, 20CR) and (*right*) Millennium Drought (1996–2009; relative to 1985–2005, ERA-Interim). (*bottom*) Precipitation anomalies in CCC400 ensemble mean

very strong because most of the reference period chosen for this book (1985–2005) coincided with the drought. Figure 4.100 (bottom), as in the previous figure, shows the corresponding precipitation anomalies from CCC400 (land only). While a clear precipitation deficit is modelled for the World War Two Drought, the Millennium Drought is poorly reproduced.

The two droughts were quite different. The World War Two Drought affected northern Australia, whereas the Millennium Drought was strongest in southeastern Australia. Also, the temperature anomalies were different. Droughts in Australia can be related to several modes of variability. These include the Indian Ocean Dipole, ENSO, and the Southern Annular Mode (see Sects. 3.2.6 and 3.2.7 and Table 3.2). The Federation Drought, in the early 20th century, was related to a persistent El Niño. During the World War Two Drought, the Indian Ocean seemed to play an important role with possible contributions from ENSO and SAM (Verdon-Kidd and Kiem 2009). The Millennium Drought was partly related to an upward trend in the Southern Annular Mode and an associated poleward extension of the southern edge of the Hadley cell (Cai et al. 2014). Both features are qualitatively reproduced in climate models. Thus, some authors argue that the Millennium Drought had an anthropogenic contribution (Delworth and Zeng 2014). This quick analysis shows that despite a possible anthropogenic contribution, there is no blueprint for future Australian droughts. The three most prominent Australian droughts of the 20th century were of different types.

4.5.7 Megaheatwaves

At the start of the new millennium, a number of epochal heatwaves struck the globe. The 2003 European heatwave was characterised as a millennial event (Luterbacher et al. 2004; Schär et al. 2004). The 2010 Russian heatwave (which was linked to the Pakistan flood; see Lau and Kim 2012) accompanied by forest and bog fires (Fig. 4.101), again changed the map of temperature records (Barriopedro et al. 2011). The United States heatwaves of 2012 and the Australian heatwaves of 2009 and 2013 set new records at many sites (see previous section). Is this the expression of a new gear shift of the machinery? The previous section focused on droughts and underlying processes such as oceanic forcing. Here, I will focus on heatwaves and take on a statistical view.

4.5.7.1 Record-Breaking Heatwaves in the NCEP/NCAR Reanalysis

Although increasing trends in the occurrence of megaheatwaves are difficult to establish statistically, a simple illustration demonstrates the recent widespread occurrence of heatwaves. Figure 4.102 (based on daily NCEP/NCAR Reanalysis data) shows the year during which the maximum temperature was observed. Clearly, the dominating fraction of the globe is in orange and yellow, meaning that records were set during the last 15 years or so.

Heatwaves can be deadly. In addition to the suffocating heat, air pollutants often accumulate in the stagnant air masses. The elderly and people with poor health, including those suffering from cardiovascular and respiratory diseases, suffer. In extreme cases, heatwaves can lead to excess mortality (the premature death) of thousands of people. Heatwaves also affect the energy and water supply, transport, and other areas of life.

Strong heatwaves have occurred in the past (Fig. 4.11); the United States experienced frequent heatwaves during the "Dust Bowl" droughts (Sect. 4.3.3) and Europe during the late 1940s (Sect. 4.4.3). During the last 10 years, however, the world has seen heatwaves of a new dimension. Events of the same magnitude as the 2003 European heatwave or the 2010 Russian heatwave are expected to become more frequent, perhaps becoming the new norm by the end of the century (IPCC 2013). They are often depicted as a glimpse at the future climate.

4.5.7.2 Changes in Heatwaves

The frequency and severity of heatwaves has been extensively studied for Europe and North America. Exact results depend on the definition of heatwaves (threshold temperature or percentile, minimum duration, etc.). However, independent of the definition, Europe-based studies show a trend towards more-frequent and longer-lasting heatwaves over the past 130 years. According to Della-Marta et al. (2007), heatwave lengths have doubled in western Europe since the 19th century. For the Mediterranean region, Kuglitsch et al. (2010) found a positive trend in heatwave intensity, length, and number during the past 50 years. Changes in heatwave occurrence could be successfully attributed to anthropogenic influence (Fischer and Knutti 2015).

Fig. 4.101 (*top*) Bog fires in Germany, 1893, one of the driest springs on record in central Europe, rivalling that of 2007. During Brückner's time, the effect of the smoke of bog fires on atmosphere and climate was heavily discussed. (*bottom*) Bog fires in Russia during the 2010 heatwave (Getty images)

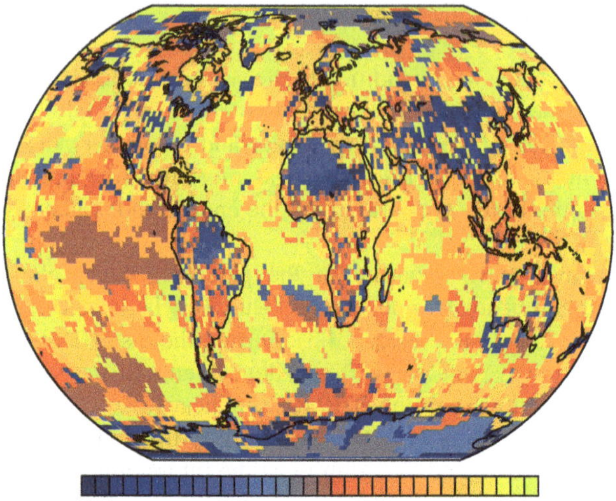

Fig. 4.102 The year during which the maximum daily mean temperature occurred from 1948 to 2013 (From NCEP/NCAR Reanalysis data)

Since 1951, Hartmann et al. (2013) report an increase in warm nights across the globe and an increase in warm days at most places. Clearly, the rising temperatures contribute to an increase in heatwave risk. An increase in temperature variability might contribute over certain midlatitudes regions. This is consistent with our CCC400 simulations, which show an increase in the distance between the maximum and mean temperatures with a rise in the mean over central Europe and North America (Fig. 4.12). Heatwaves are amplified by land–atmosphere interactions, which affect atmospheric circulation (Miralles et al. 2014) and can result in megaheatwaves.

The recent megaheatwaves show that the machinery, when forced, can produce new types of events. Or, as Wallace Broecker[26] put it:

> "The climate system is an angry beast and we are poking it with sticks."

4.5.8 *Recent Arctic Warming*

In early September 2013, the freight vessel "Nordic Orion" left Vancouver to transport coal to Finland via the Northwest Passage (Fig. 4.103). With the Arctic

[26]Wallace Broecker, *1931, is a chemical oceanographer and is most widely known for his work on the global ocean circulation (the "conveyor belt") and the carbon cycle. He is professor at the Lamont-Doherty Earth Observatory of Columbia University.

Fig. 4.103 The freight vessel Nordic Orion (Photo: Nordic Bulk Carriers)

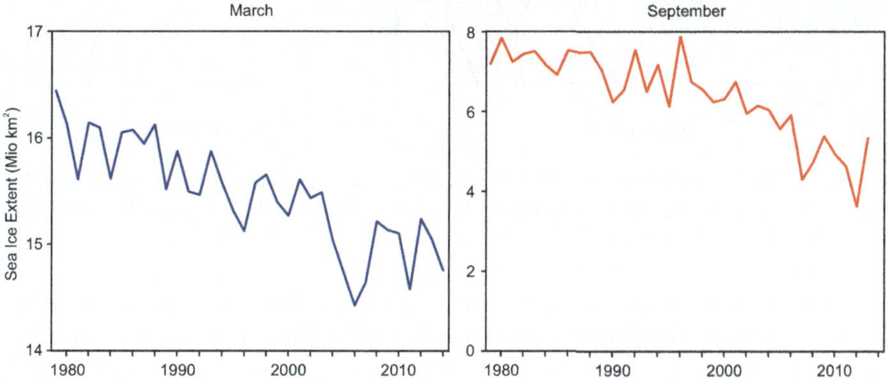

Fig. 4.104 Arctic sea ice extent in March and September, 1979–2014 (Data: NSIDC)

warming rapidly, commercial and economic interests in the region (e.g., as a sea route or a source of raw materials) increased. Today, the northern sea routes open almost regularly (although they are not normally used by ships as large as the Nordic Orion). The planting of a Russian flag in the seafloor at the North Pole underlines the strategic interests related to the opening of the Arctic. At the same time, environmental concerns have strengthened; the loss of a tanker at sea would introduce pollutants that the slow biological and chemical processes in the Arctic are ill-equipped to manage.

4.5.8.1 Record Sea Ice Loss

These new interests in the Arctic are the consequence of sea ice loss. In fact, sea ice has been disappearing rapidly—very rapidly. A record minimum sea ice area was reached in 2007 and again in 2012 (Fig. 4.104). Even though sea ice, at least in some areas, also decreased during the early 20th century warming (see Sect. 4.3.2; Fig. 4.45), the current degree is of a new dimension.

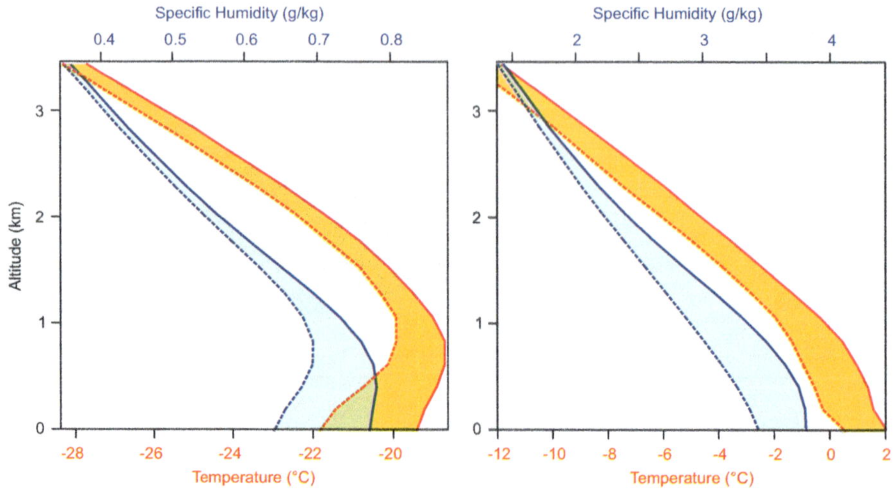

Fig. 4.105 Vertical profiles of specific humidity (*blue*) and temperature (*red*) averaged from 70° to 90°N in (*left*) January–March and (*right*) July–September. The *dashed* and *solid curves* give averages for 1979–1988 and 2004–2013, respectively (Data: ERA-Interim)

A strong reduction has also been found in multi-year ice (Comiso 2012). Is the loss of sea ice a potential tipping point (Lenton 2012)? Or, can Arctic sea ice recover rapidly (Ridley et al. 2012)? To answer these questions, good models are needed. Fortunately, sea ice has greatly improved in climate models. CMIP3 simulations could not reproduce the observed decrease of sea ice (Stroeve et al. 2007). CMIP5 simulations still underestimate the rate of the observed change, but come much closer (Stroeve et al. 2012). This also lends credibility to model-projected changes in sea ice. According to the IPCC Fifth Assessment Report, the loss of Arctic sea ice will continue. Depending on the emissions scenario, September sea ice will decrease by 43 % or will disappear by the end of the 21st century. Thus, it is not a question of "if", but "when" the first ice-free summer is observed in the Arctic.

4.5.8.2 Vertical Structure of Temperature and Moisture Changes

When ice is lost, the lower atmosphere warms. In fact, the temperature increase in the lower atmosphere over the Arctic is largest over regions of sea ice loss (Serreze et al. 2009), and in these regions it is largest in the lowermost layers of the atmosphere.

Figure 4.105 shows temperature and moisture profiles averaged over the Arctic (poleward of 70°N) during the late winter/early spring (Jan.–Mar.) and late summer/early autumn (Jul.–Sep.) of two decades, 1979–1988 and 2004–2013. This reveals that the temperature increase is largest near the ground, particularly in winter, but also in summer. The increase of summer temperature near the ground (2 °C) is equivalent to an altitude shift of 1 km. Also, water vapour increased in the lowest kilometre, which has relevance for feedback processes (see Sect. 3.4). Water vapour is a greenhouse gas that affects clouds, fuels storms, and transports energy. It is an element of Arctic amplification (Pithan and Mauritsen 2014; Screen and Simmonds 2010).

Thus, the vertical structure of the recent warming is different from the warming in the first half of the 20th century (see Sect. 4.2.2), which (according to very sparse measurements) was as large as (or larger than) at 700 hPa than near the ground.

4.5.8.3 Links with the Extratropics and Atmospheric Circulation

The warming of the Arctic is not just due to the local effect of greenhouse gases; it is intimately linked to midlatitude processes. Anomalous atmospheric circulation contributed to the sea ice loss in 2007 (Slingo and Sutton 2007). In addition, anomalous atmospheric circulation in the early summer of 2007–2012 (and expressed in the Arctic Dipole; Overland et al. 2012) contributed to the extreme temperature anomalies. Also, oceanic heat flux into the Arctic increased (Polyakov et al. 2013). Ding et al. (2014) related the warming of Greenland and eastern Canada to a change in the North Atlantic Oscillation, which they attributed to an atmospheric response to anomalous sea-surface temperatures in the tropical Pacific (similar as described in Chap. 3).

Conversely, it has also been suggested that changes in Arctic sea ice affect the midlatitudes and the NAO (and thus winters in central Europe and North America). In fact, some recent winters were cold and snow rich with very low NAO values.

A possible mechanism is sketched in Fig. 4.106. Open water during fall acts as a moisture source. This moisture can impact Siberian snow cover in fall (Wegmann et al. 2015), which has increased in recent years. More snow may lead to cooling and strengthen the trough over Siberia. Via changes in stratospheric circulation and subsequent downward propagation, the midlatitude circulation of the subsequent late winter may be affected (Cohen et al. 2014; see Sect. 3.1.4). The amplified Arctic warming and loss of sea ice also changes the temperature gradient to the midlatitude and might cause circulation over the Arctic to become more meridional. Even weather extremes at midlatitudes have been suggested to become more frequent, although this matter is up to debate (Francis and Vavrus 2012; Wallace et al. 2014).

In absolute terms, the Arctic Ocean may not be a relevant net moisture source for the Arctic atmosphere. However, the trend may have importance. Furthermore, more water vapour is transported to the Arctic from the midlatitudes (Zhang et al. 2013). With higher available latent energy, cyclones can develop more easily. Thus, the Arctic climate could become more dynamical (Jaiser et al. 2012). In August 2012, a massive cyclone (known as the Great Arctic cyclone; see Fig. 4.107) moved over the North Pole. It was later blamed by some scientists for the increased sea ice loss during that year. Will the machinery produce a new Arctic climate?

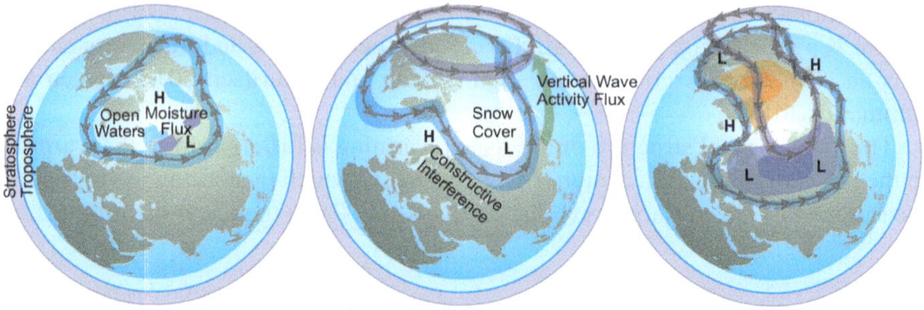

Fig. 4.106 Schematic of the processes coupling midlatitude weather in spring to Arctic sea ice loss for (*left*) early fall, (*middle*) late fall, and (*right*) midwinter (Adapted from Cohen et al. 2014)

Fig. 4.107 Satellite image of the Great Arctic cyclone of August 2012. Some scientists believe that this single storm played an important role in sea ice loss during that year (Mosaic of NASA/MODIS images, NASA Earth Observatory)

4.5.9 The Global Warming Hiatus

Global temperatures increased steeply during the 1990s, increasing the public interest and awareness of climate change. After the 1997/1998 El Niño, however, the temperature increase did not continue at the same pace. The slowdown of global warming, known as the "hiatus", caused discussion among the scientific community and the public. This is the most recent of the climatic changes since 1700 and will conclude this book. Here, I will provide a brief overview of this interesting event.

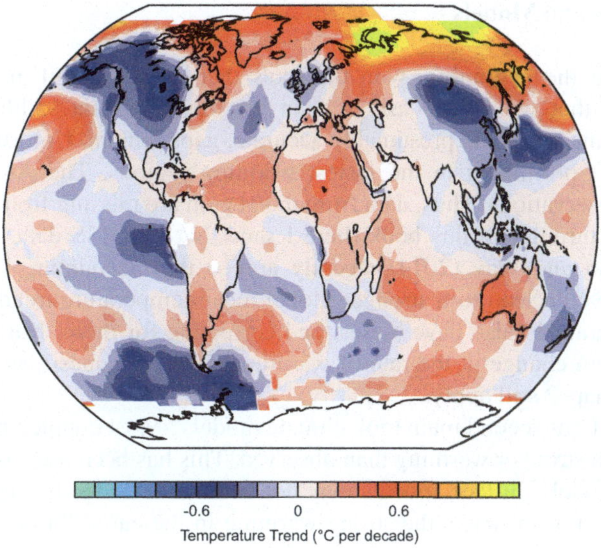

Fig. 4.108 Trend in annual-mean temperature from 1998 to 2013 (Data source: NASA-GISS, Hansen et al. (2010))

4.5.9.1 Trends

A look at the spatial pattern of temperature trends is telling (Fig. 4.108; note that I used NASA-GISS data rather than the HadCRUT4 data that was used in all previous plots). The strongest warming is found in the Siberian Arctic, Greenland, and Africa. Australia, where eight of the last 15 summers were extremely hot, also stands out. Conversely, cooling dominated over North America, East Asia, and some oceanic regions.

Clear regional patterns appear. For instance, the equatorial Atlantic reflects a change in the Atlantic Meridional Mode. In fact, precipitation in the Sahel region has increased over the period as would be expected from the heating gradient. However, a very clear pattern also emerges over the tropical Pacific: La Niña. Across the Pacific and North America, a wave pattern of positive and negative trends is seen. Consequently, Ding et al. (2014) relate the warming of the western Arctic to the tropical Pacific. Some of the other patterns are also consistent with La Niña teleconnections.

Is this behaviour of the machinery peculiar? Or, did the forcings follow an unexpected path? A plethora of studies has addressed these points and different hypotheses have been suggested. Thus, the apparent hiatus is not just a climatic event, but is suitable for concluding this book and recapitulating some of the features that were discussed in earlier sections.

4.5.9.2 Data and Models

Let us start at the beginning, with the tools that we described in Chap. 2. The first question that we should ask is: Can we trust the data? Although today's global observing system is presumably accurate, gaps remain. Unfortunately, these gaps are most prominent in the fastest warming regions. The Arctic is poorly covered by observations. Thus, data from this region are missing from HadCRUT4, which I used throughout this book (here I show NASA-GISS data, Hansen et al. 2010). Cowtan and Way (2014) recently argued that accounting for the Arctic contribution would lead to a stronger global-mean temperature trend. Recent sea-surface temperatures also have biases, and correcting for these, Karl et al. (2015) found no recent change in the rate of warming. Thus, the questions and concerns featured in Chap. 2 still apply.

What about our second main tool, climate models? Most coupled model simulations suggest a stronger warming than observed. This has been used as an argument to discount models. In fact, models may not be able to perfectly capture decadal-scale changes. For instance, the strong warming in the early 20th century and the subsequent stagnation are not reproduced (in terms of magnitude) by all model simulations. The "hiatus" is a good opportunity to critically assess models (Fyfe et al. 2013). However, models have also proven useful to study the hiatus. Stratifying models according to their ENSO phase yields a good agreement with observations (Risbey et al. 2014), and models have been used in various set-ups to address the underlying mechanisms (e.g., Kosaka and Xie 2013).

4.5.9.3 Statistical Perspective

Given the tools at hand, we can proceed in several directions. As in Chap. 1 of this book, we can take a statistical or a process-based perspective. Taking the statistical perspective leads us to the question: How unusual is it to find a strong trend in a short record? The historical record can help us (demonstrated in Fig. 1.3); it shows that short-term deviations of a trend from its expected path are common. According to Liebmann et al. (2010), the slowdown of the global temperature rise in the 21st century may have been unusual, but it was not unprecedented. Palaeoclimatological records help us to extend that perspective. Based on detrended temperature reconstructions, Crowley et al. (2014) estimated the rarity of deviations such as the recent one. They found that the last 10–15 years cannot be considered "unusual natural variability" in the context of the past 230 years. Analysing extreme ensemble members, Deser et al. (2012) showed that regional cooling is consistent with model projections. The examples presented in this book also demonstrate that decadal-scale variations have always occurred naturally. In fact, this was the core of Brückner's research.

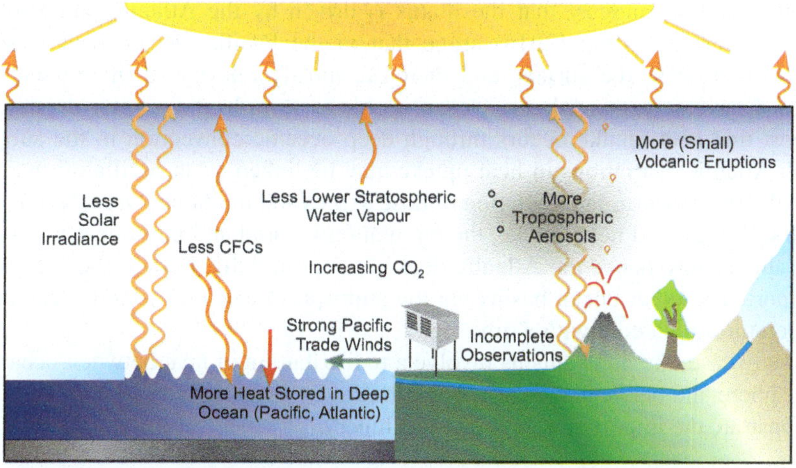

Fig. 4.109 Schematic of suggested mechanisms of the hiatus

4.5.9.4 Process-Based Perspective: Internal Variability

The statistical perspective implies that the hiatus is less surprising than initially thought. However, the question remains: What caused the hiatus? Was it a response of the machinery to some forcing? Or, was internal variability within the system responsible?

Arguments have been put forth for both theories (a schematic view is given in Fig. 4.109). Following the structure of Chap. 3, I will first address internal variability. Many authors ascribe the hiatus to an unusual behaviour of the Pacific Ocean and atmosphere. Since 1999, La Niña events have become more frequent. ENSO and PDO indices show a clear change, and the Pacific Walker circulation has been strengthening (Fig. 4.42). The changes in the Pacific Ocean affect global atmospheric circulation and global temperatures through altering the pathways of heat transport.

The belief that internal processes are responsible posits that incoming energy follows the expected trajectory, but this energy does not remain in the atmosphere. Processes within the system shift the energy away from the atmosphere and into the ocean. In fact, this is the case for La Niña-like situations (Mayer et al. 2014). Meehl et al. (2011) and England et al. (2014) and others provide evidence that heat uptake into the subsurface Pacific Ocean has been increased in the past 15 years. Although of a limited spatial extent, this increased heat sink can affect the global mean temperature via changing atmospheric circulation, as was found by forcing a coupled model with tropical Pacific sea-surface temperatures (Kosaka and Xie 2013).

Other authors suggest that the hiatus is driven by the Atlantic and Southern Ocean. Chen and Tung (2014) argue that in the Pacific Ocean, heat is stored relatively close to the surface and therefore mostly relevant at interannual time scales. Conversely, the Atlantic Ocean is able to store heat a greater depth. They suggest that heat uptake occurs through deep oceanic convection in the subpolar North Atlantic. Variations in heat uptake may be linked to an oscillatory mode of the Atlantic overturning circulation (whose expression might be the AMO). During phases of increased heat uptake, the circulation is rapid and more heat is supplied from the tropics, hence the Atlantic does not cool initially. Rather, the heat deficit is exported to other ocean basins via the atmosphere and the Atlantic sea-surface temperatures only cool with a time lag.

In any case, the mechanisms involve a redistribution of heat between ocean and atmosphere and between different ocean basins by means of the atmosphere. They demonstrate the important role of the machinery.

4.5.9.5 Changes in the Forcing

In addition to internal mechanisms, we should consider the possibility of external forcings. Following the structure of Chap. 3, I will explore volcanic, solar, greenhouse gas, and aerosol forcings.

Again, consider the fit of the temperature curve to the (extrapolated) forcings. During 1983–1998, temperatures in HadCRUT4 increased by 0.3 °C decade^{-1}. Then, during 1998–2013, the trend decreased to 0.08 °C decade^{-1}. According to our fit, greenhouse gases contributed a warming of 0.34 °C decade^{-1}, which was only partly offset by other forcings (total warming 0.31 °C decade^{-1}). Was internal variability responsible for the slowdown? Or, are our (extrapolated) forcings inaccurate?

Perhaps both. No major volcanic eruption has occurred since Pinatubo in 1991. In my extrapolation of stratospheric aerosols, stable background aerosol concentration is assumed. However, satellite data suggest a recent increase in background aerosols. Initially, this increase was thought to originate from tropospheric pollution. Specifically, it was ascribed to East Asian aerosol emissions, which have rapidly increased. However, recent studies find this contribution to be small (Neely et al. 2013). More probable is that the increase in background aerosols originated from a sequence of small volcanic eruptions (Sect. 4.5.3). These eruptions might have produced a negative volcanic forcing and contributed to the hiatus, consistent with the suggestions of Solomon et al. (2011). Accounting for the volcanic aerosols changes since Pinatubo also reduces the discrepancy between models and observations (Santer et al. 2014).

The second main natural forcing, the sun, might have also contributed. The last sunspot cycle was peculiar and ended with a pronounced and prolonged minimum, whereas the ongoing cycle has only reached a small maximum. Could the sun have been a contributor? Perhaps, but arguably minor. From the maximum to minimum of a solar cycle (i.e., from 2001 to 2009), there can be a contribution of 0.15 °C (e.g., Foster and Rahmstorf 2011; Lean and Rind 2008). Correspondingly, a reduced contribution should occur for the 1998–2013 period. In my simple fit for 1998–2013, the contribution is 0.04 °C—sizeable, but not the main player.

Other papers relate the hiatus to changes in non-CO_2 greenhouse gases. One candidate is water vapour in the lowermost stratosphere, which increased in the 1990s and then suddenly began to decrease around 2001. Around the tropopause, water vapour has the strongest greenhouse effect (also see Sects. 3.3 and 3.4). According to Solomon et al. (2010), this may have contributed to the slow down of warming. In addition to water vapour, well-mixed greenhouse gases may have also contributed. Estrada et al. (2013) speculate that the decline in CFCs and in the methane growth rate might have contributed to a slow down of the forced signal.

Finally, tropospheric aerosols need to be considered. The "global brightening" may already be over. In addition to effects on air quality, the particularly strong increase in East Asian aerosols may have had an effect on global climate. There is evidence for a "renewed dimming" in East Asia, which may help to explain the regional cooling and imprints on regional sea-surface temperatures.

In this book, we have analysed many ways the machinery can operate, how it reacts to forcings, and how it can generate internal variability on decadal time scales. Perhaps, the hiatus is too recent to be assessed. However, from the discussion in this book, is it really surprising? When expressed relative to the expected greenhouse gas-forced warming trajectory, it is a cold period that might already have ended when this book appears or continue for another few years. Is it just one of Brückner's cool periods?

This book shows that the climate system has also deviated from its forced trajectory in the past. This is because the machinery produces decadal variability. Likely, the future will show the climate system return back to its trajectory in a phase of accelerated warming. After all, hiatus literally means "a pause or break in continuity in a sequence or activity". It is just a matter of time until the climate system reverts to its global warming trend.

Chapter 5
Conclusions

Eduard Brückner, an active climate scientist 125 years ago, studied climatic changes back to 1700. He found that the climate system exhibits variations on multidecadal time scales—a bold statement to make at that time. However, he could not see inside the machinery; the mechanisms remained hidden. Considerable progress has been made since Brückner's time, particularly in recent years. Thanks to new analytical and sampling techniques, new proxies have emerged that provide information on poorly understood aspects of the climate state. Modelling capabilities have improved and the use of simulations to study past climate has become commonplace. Furthermore, the increasing use of powerful numerical techniques such as data assimilation has triggered new data recovery efforts (see Chap. 2). If this momentum continues, we may soon be able to study continental- to hemispheric-scale weather back to the 1780s.

Because of these efforts, we have a new perception of climatic changes back to 1700. Temperature variability is increasingly well reconstructed and understood. Changes in the dynamics of the system are more difficult to study (Shepherd 2014), but new techniques and datasets provide some insight. Changes in the strength of the midlatitude westerlies, the Hadley circulation, and the width of the tropical belt can be explored at least to some extent and with the help of climate models, for the last 50–300 years. Furthermore, progress made by the scientific community over the last decade has improved our understanding of the response of the machinery to external forcings such as volcanic eruptions (see Chap. 3), with palaeodata playing an important role (e.g., Sigl et al. 2015).

Sizable uncertainties remain because we have only begun to see the machinery behind climatic changes. In this sense, "machinery" can also be used to describe our perception of climate—an enormously complex sequence of physical mechanisms, many of which we have not yet understood.

Progress has also been made in our understanding of the multidecadal variability of temperature. This was one of Brückner's main contributions to climate science. During the last 140 years, multidecadal variability overlaid the general, forced

S. Brönnimann, *Climatic Changes Since 1700*, Advances in Global Change Research 55, DOI 10.1007/978-3-319-19042-6_5

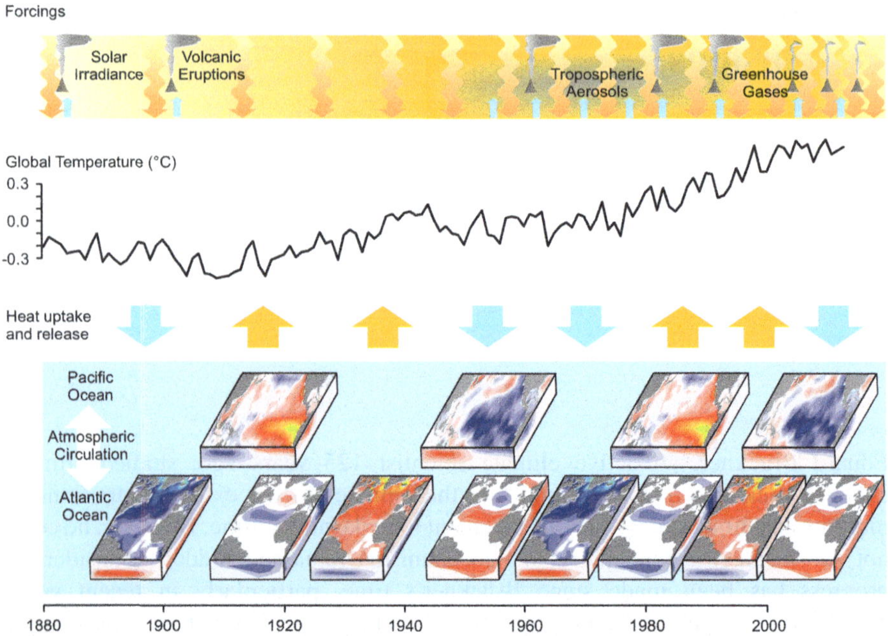

Fig. 5.1 (*top*) Radiative forcings since 1880, annual-mean global surface air temperatures (NASA/GISS Land–Ocean Temperature Index; Hansen et al. 2010). (*bottom*) Schematic of ocean surface and subsurface temperature anomalies (Adapted from Brönnimann 2015)

warming trend. This took the form of several phases of accelerated or slowed warming. The most prominent of these is arguably the ongoing "hiatus" in global warming. A conceptual view of our understanding of these phases is given in Fig. 5.1 (Brönnimann 2015). Their timing is synchronous with changes in external forcings of the climate system, that is, greenhouse gases, solar irradiance, volcanic activity, and tropospheric aerosols.

However, the forcings do not fully explain the global temperature variations. Studies of the recent hiatus reveal that both earth's energy balance and energy redistribution within the climate system play a role. Currently, the deep oceans—rather than the atmosphere—are warming at an accelerated rate. This has brought the distribution of heat within the system to the forefront of research (see Sect. 4.5.9). As part of El Niño–Southern Oscillation, the tropical Pacific Ocean may store large amounts of heat. This heat is then exchanged with tropical and extratropical basins. Via atmospheric circulation, temperatures around the globe adjust to the rhythm of the tropical Pacific. In this way, the low-frequency expression of El Niño–Southern Oscillation, the Pacific Decadal Oscillation, may be a pacemaker of accelerated or slowed warming. This includes the early 20th century warming (Thompson et al. 2015), the mid-century stagnation, the late 20th century accelerated warming, and the current hiatus.

Alternatively, perhaps the Atlantic and Southern Oceans are the real drivers of change in deep-ocean heat uptake on decadal time scales, while the tropical Pacific affects heat storage mainly from interannual to multiannual scales (Chen and Tung 2014; see Sect. 4.5.9). An oscillatory mode such as the Atlantic meridional overturning circulation (whose expression in sea-surface temperature is the Atlantic multidecadal oscillation) could have also presided over past global temperature variations. This is explored in Sect. 4.3.5 in the context of the early 20th century warming. A strong circulation means that more heat is subducted into the subpolar Atlantic. Concurrently, warm waters from the tropics are transported northward. Thus, despite the increased heat sink, North Atlantic sea-surface temperatures do not cool. Hence, heat uptake and Atlantic sea-surface temperatures are phase shifted (see Fig. 5.1, bottom). This may not only hold for the hiatus (Chen and Tung 2014), but also for earlier phases of accelerated or slowed warming. Is the Pacific and/or the Atlantic Ocean in the driver's seat? Either way, the atmosphere acts as the machinery to transfer energy into the region of uptake.

Is Brückner's 35-year cycle the expression of changes in the tropical Pacific and the Atlantic Oceans? His global temperature reconstruction, which ends just prior to the data in Fig. 5.1, shares similarities with indices of the Atlantic Multidecadal Oscillation (in the 18th century) and the Pacific Decadal Oscillation (in the 19th century) as represented in MC data (Fig. 5.2). Were other past warm phases (e.g., from 1790 to 1804; Sect. 4.2.3), which could only partly be ascribed to forcings, driven by Pacific sea-surface temperatures? Further work on climatic changes since 1700 will hopefully provide greater insight into these phases of accelerated or slowed warming.

This book has shown how we can learn from past climate events and how we can learn from the same events when viewed in a different context. In addition to revisiting past climate, we revisited earlier scientific works, for which Brückner's book (Fig. 5.3) is just one example. Similarly, we anticipate that future scientists will revisit our work with a different perspective. Although large efforts in data rescue are currently underway, we will continuously revisit historical data, applying

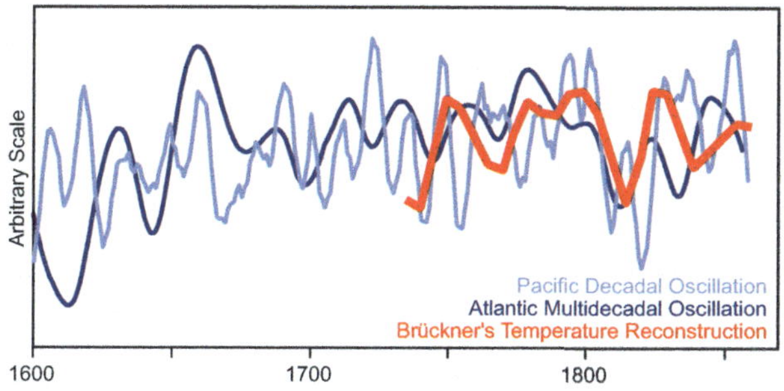

Fig. 5.2 Indices of the Atlantic Multidecadal Oscillation and the Pacific Decadal Oscillation (from Fig. 4.7) and Brückner's global temperature reconstruction (From Fig. 4.3)

Fig. 5.3 Brückner's book "Klima-Schwankungen seit 1700"

new techniques, and exploiting information in yet unknown ways. Perhaps we will find answers to questions about past climate in photographs of stellar spectra, observations of cloud motion, or UV radiation records in fossil pigments. In any case, the foundation for future work has been laid by those who devoted their time and energy to observe their environment in the 18th and 19th centuries—and today. My deepest respect goes to those who made the observations on which our science has grown.

References

Abram NJ, Gagan MK, Cole JE, Hantoro WS, Mudelsee M (2008) Recent intensification of tropical climate variability in the Indian Ocean. Nat Geosci 1:849–853
Abram NJ, Mulvaney R, Wolff EW et al (2013) Acceleration of snow melt in an Antarctic Peninsula ice core during the twentieth century. Nat Geosci 6:404–411
Aceituno P, Prieto MR, Solari M et al (2009) The 1877–1878 El Niño episode: associated impacts in South America. Clim Change 92:389–416
Adams JB, Mann ME, Ammann CM (2003) Proxy evidence for an El Niño-like response to volcanic forcing. Nature 426:274–278
Adler RF, Huffman GJ, Chang A et al (2003) The version-2 global precipitation climatology project (GPCP) monthly precipitation analysis (1979–Present). J Hydrometeorol 4:1147–1167
Ahmed M, Anchukaitis K, Buckley BM et al (2013) Continental-scale temperature variability during the past two millennia. Nat Geosci 6:339–346
Allan RJ, Compo GP, Stone R, Luterbacher J, Brönnimann S (2011) The international atmospheric circulation reconstructions over the earth (ACRE) initiative. Bull Am Meteorol Soc 92:1421–1425
Allen RJ, Sherwood SC, Norris JR, Zender CS (2012) Recent Northern Hemisphere tropical expansion primarily driven by black carbon and tropospheric ozone. Nature 485:350–354
Allen RJ, Norris JR, Kovilakam M (2014) Influence of anthropogenic aerosols and the Pacific Decadal Oscillation on tropical belt width. Nat Geosci 7:270–274
Alvarez LW, Alvarez W, Asaro F, Michel HV (1980) Extraterrestrial cause for the Cretaceous-Tertiary extinction. Science 208:1095–1108
Ammann CM, Joos F, Schimel DS, Otto-Bliesner BL, Tomas RA (2007) Solar influence on climate during the past millennium: Results from transient simulations with the NCAR Climate System Model. Proc Natl Acad Sci USA 104:3713–3718
Anchukaitis KJ, Buckley BM, Cook ER et al (2010) Influence of volcanic eruptions on the climate of the Asian monsoon region. Geophys Res Lett 37:L22703
Anchukaitis KJ, Breitenmoser P, Briffa KR et al (2012) Tree rings and volcanic cooling. Nat Geosci 5:836–837
Anderson DM, Overpeck JT, Gupta AK (2002) Increase in the Asian southwest monsoon during the past four centuries. Science 297:596–599
Anderson DM, Mauk EM, Wahl ER et al (2013) Global warming in an independent record of the past 130 years. Geophys Res Lett 40:189–193
Andersson ME, Verronen PT, Rodger CJ, Clilverd MA, Seppälä A (2014) Missing driver in the sun-earth connection from energetic electron precipitation impacts mesospheric ozone. Nat Commun 5:5197

S. Brönnimann, *Climatic Changes Since 1700*, Advances in Global Change Research 55, DOI 10.1007/978-3-319-19042-6

Andrews DG, Holton JR, Leovy CB (1987) Middle atmosphere dynamics. Academic, San Diego

Anet JG, Muthers S, Rozanov EV et al (2014) Impact of solar vs. volcanic activity variations on tropospheric temperatures and precipitation during the Dalton Minimum. Clim Past 10:921–938

Ångström A (1916) Über die Gegenstrahlung der Atmosphäre (On the counter-radiation of the atmosphere). Meteorol Z 33:529–538. (Translated and edited by Volken E, Brönnimann S, Philipona R, Meteorol Z 22:761–769, 2013)

Ångström A (1935) Teleconnections of climatic changes in present time. Geogr Ann 17:242–258

Annan JD, Hargreaves JC (2012) Identification of climatic state with limited proxy data. Clim Past 8:1141–1151

Arfeuille F, Luo BP, Heckendorn P et al (2013) Modeling the stratospheric warming following the Mt. Pinatubo eruption: uncertainties in aerosol extinctions. Atmos Chem Phys 13:11221–11234

Arfeuille F, Weisenstein D, Mack H et al (2014) Volcanic forcing for climate modeling: a new microphysics-based data set covering years 1600–present. Clim Past 10:359–375

Aronova E, Baker KS, Oreskes N (2010) Big science and big data in biology: from the International Geophysical Year through the International Biological Program to the Long Term Ecological Research (LTER) network, 1957–present. Hist Stud Nat Sci 40:183–224

Arrhenius S (1896) On the influence of carbonic acid in the air upon the temperature of the ground. Philos Mag 41:237–276

Ashok K, Behera SK, Rao SA, Weng H, Yamagata T (2007) El Niño Modoki and its possible teleconnection. J Geophys Res 112:C11007

Assmann R (1891) Ein Apparat zur Ventilation des feuchten Thermometers (A device for the ventilation of the wet-bulb thermometer). Meteorol Z 8:15–24. (Translated and edited by Volken E, Brönnimann S, Meteorol Z 21:423–430, 2012)

Assmann R (1907) Über die Existenz eines wärmeren Luftstromes in der Höhe von 10 bis 15 km. Sitzber K Preuss Aka 24:495–504

Atkinson P, Lloyd C (eds) (2011) geoENV VII – Geostatistics for Environmental Applications. Quantitative Geology and Geostatistics. Springer, Dordrecht/Heidelberg/London/New York

Auchmann R, Brönnimann S (2012) A physics-based correction model for homogenizing sub-daily temperature series. J Geophys Res 117:D17119

Auchmann R, Brönnimann S, Breda L et al (2012) Extreme climate, not extreme weather: the summer of 1816 in Geneva, Switzerland. Clim Past 8:325–335

Auer I, Böhm R, Jurkovic A et al (2007) HISTALP–historical instrumental climatological surface time series of the Greater Alpine Region. Int J Climatol 27:17–46

Austin J, Butchart N, Shine KP (1992) Possibility of an Arctic ozone hole in a doubled-CO_2 climate. Nature 360:221–225

Baldwin MP, Dunkerton TJ (2001) Stratospheric harbingers of anomalous weather regimes. Science 294:581–584

Ball WT, Unruh YC, Krivova NA et al (2012) Reconstruction of total solar irradiance 1974–2009. Astron Astrophys 541:A27

Bândă N, Krol M, van Weele M, van Noije T, Röckmann T (2013) Analysis of global methane changes after the 1991 Pinatubo volcanic eruption. Atmos Chem Phys 13:2267–2281

Barnett TP, Graham NE, Cane MA et al (1988) On the prediction of the El Niño of 1986–1987. Science 241:192–196

Barnston AG, Livezey RE (1987) Classification, seasonality and persistence of low-frequency atmospheric circulation patterns. Mon Wather Rev 115:1083–1126

Barriopedro D, Fischer EM, Luterbacher J, Trigo RM, García-Herrera R (2011) The hot summer of 2010: Redrawing the temperature record map of Europe. Science 332:220–224

Barrow J (1819) Physikalisch-geografische Nachrichten aus dem nördlichen Polarmeer. Ann Phys 62:157–166

Baumgartner A (1950) Niederschlagsschwankungen und Dürregefährdung mit Bezug auf den Waldbau. Forstwissenschaftliches Centralblatt 69:636–662

Behringer W (2007) Kulturgeschichte des Klimas: von der Eiszeit bis zur globalen Erwärmung. Beck'sche Reihe, Beck

Beisbart C (2012) How can computer simulations produce new knowledge? Eur J Philos Sci 2:395–434

Bengtsson L, Kanamitsu M, Kållberg P, Uppala S (1982) FGGE 4-dimensional data assimilation at ECMWF. Bull Am Meteorol Soc 63:29–43

Bhartia PK (2009) Role of satellite measurements in the discovery of stratospheric ozone depletion. In: Twenty years of ozone decline. Springer, pp 183–189

Bhend J, Franke J, Folini D, Wild M, Brönnimann S (2012) An ensemble-based approach to climate reconstructions. Clim Past 8:963–976

Bichet A, Folini D, Wild M, Schär C (2014) Enhanced Central European summer precipitation in the late 19th century: A link to the Tropics. Q J R Meteorol Soc 140:111–123

Bindoff NL, Stott PA, AchutaRao KM et al (2013) Detection and attribution of climate change: From global to regional. In: Stocker T, Qin D, Plattner GK et al (eds) Climate change 2013: The physical science basis. Contribution of Working Group I to the Fifth Assessment Report of the Intergovernmental Panel on Climate Change. Cambridge University Press, Cambridge/New York

Biondi F, Gershunov A, Cayan DR (2001) North Pacific decadal climate variability since 1661. J Clim 14:5–10

Birkeland BJ (1930) Temperaturvariationen auf Spitzbergen. Meteorol Z 47:2

Birner T (2010) Recent widening of the tropical belt from global tropopause statistics: sensitivities. J Geophys Res 115:D23109

Bjerknes V (1904) Das Problem der Wettervorhersage, betrachtet vom Standpunkte der Mechanik und der Physik (The problem of weather prediction, considered from the viewpoints of mechanics and physics). Meteorol Z 21:1–7. (Translated and edited by Volken E, Brönnimann S, Meteorol Z 18:663–667, 2009)

Bjerknes J (1964) Atlantic air-sea interaction. Adv Geophys 10:1–82

Bjerknes J (1966) A possible response of the atmospheric Hadley circulation to equatorial anomalies of ocean temperature. Tellus 18:820–829

Black DE, Abahazi MA, Thunell RC et al (2007) An 8-century tropical Atlantic SST record from the Cariaco basin: baseline variability, twentieth-century warming, and Atlantic hurricane frequency. Paleoceanography 22:PA4204

Black E, Sutton R (2007) The influence of oceanic conditions on the hot European summer of 2003. Clim Dyn 28:53–66

Blanford HF (1884) On the connexion of the Himalaya snowfall with dry winds and seasons of drought in India. Proc R Soc London 37:3–22

Blunden J, Arndt DS (2012) State of the climate in 2011. Bull Am Meteorol Soc 93:S1–S282

Bodeker GE, Hassler B, Young PJ, Portmann RW (2013) A vertically resolved, global, gap-free ozone database for assessing or constraining global climate model simulations. Earth Syst Sci Data 5:31–43

Bodenmann T, Brönnimann S, Hadorn GH, Krüger T, Weissert H (2011) Perceiving, explaining, and observing climatic changes: an historical case study of the "year without a summer" 1816. Meteorol Z 20:577–587

Böhm R, Jones PD, Hiebl J et al (2010) The early instrumental warm-bias: a solution for long central European temperature series 1760–2007. Clim Change 101:41–67

Bojkov RD (2012) International Ozone Commission: history and activities. IAMAS Publication Series p 100

Bölsche W (1919) Eiszeit und Klimawechsel. Kosmos, Gesellschaft der Naturfreunde, Franckh'sche Verlagsbuchhandlung, Stuttgart

Bony S, Stevens B, Frierson DMW, Jakob C, Kageyama M, Pincus R, Shepherd TG, Sherwood SC, Siebesma AP, Sobel AH, Watanabe M, Webb MJ (2015) Clouds, circulation and climate sensitivity. Nat Geosci 8:261–268

Boos WR, Kuang Z (2010) Dominant control of the South Asian monsoon by orographic insulation versus plateau heating. Nature 463:218–222

Booth BBB, Dunstone NJ, Halloran PR, Andrews T, Bellouin N (2012) Aerosols implicated as a prime driver of twentieth-century North Atlantic climate variability. Nature 484:228–232

Bordoni S, Schneider T (2008) Monsoons as eddy-mediated regime transitions of the tropical overturning circulation. Nat Geosci 1:515–519
Teisserenc de Bort LP (1881) Etude sur l'hiver de 1879–80 et recherches sur la position des centres d'action de l'atmosphere dans les hivers anormaux. Annales du Bureau central météorologique de France 4:17–62
Teisserenc de Bort LP (1902) Variations de la temperature de l'air libre dans la zone comprise 8 km et 13 km d'altitude. C R Hebd Seances Acad Sci 24:987–989
Boucher O, Randall D, Artaxo P et al (2013) Clouds and aerosols. In: Stocker T, Qin D, Plattner GK et al (eds) Climate change 2013: The physical science basis. Contribution of Working Group I to the Fifth Assessment Report of the Intergovernmental Panel on Climate Change. Cambridge University Press, Cambridge/New York
Braconnot P, Harrison SP, Kageyama M et al (2012) Evaluation of climate models using palaeoclimatic data. Nat Clim Change 2:417–424
Bradley RS, Hughes MK, Diaz HF (2003) Climate in medieval time. Science 302:404–405
Bradley RS, Vuille M, Diaz HF, Vergara W (2006) Threats to water supplies in the tropical Andes. Science 312:1755–1756
Braganza K, Gergis JL, Power SB, Risbey JS, Fowler AM (2009) A multiproxy index of the El Niño–Southern Oscillation, AD 1525–1982. J Geophys Res 114:D05106
Brandes HW (1819) Einige Resultate aus der Witterungs-Geschichte des Jahres 1783, und Bitte um Nachrichten aus jener Zeit. Ann Phys-Berlin 61:421–426
Brasseur GP (2008) Creating knowledge from the confrontation of observations and models: the case of stratospheric ozone. Adv Glob Change Res 33:303–316
Brázdil R, Pfister C, Wanner H, von Storch H, Luterbacher J (2005) Historical climatology in Europe – the state of the art. Clim Change 70:363–430
Brázdil R, Dobrovolný P, Trnka M et al (2013) Droughts in the Czech Lands, 1090–2012 AD. Clim Past 9:1985–2002
Breitenmoser P, Beer J, Brönnimann S et al (2012) Solar and volcanic fingerprints in tree-ring chronologies over the past 2000 years. Palaeogeogr Palaeoclimatol Palaeoecol 313:127–139
Breitenmoser P, Brönnimann S, Frank D (2014) Forward modelling of tree-ring width and comparison with a global network of tree-ring chronologies. Clim Past 10:437–449
Briffa KR, Jones PD, Schweingruber FH, Osborn TJ (1998) Influence of volcanic eruptions on Northern Hemisphere summer temperature over the past 600 years. Nature 393:450–455
Brohan P, Kennedy JJ, Harris I, Tett SFB, Jones PD (2006) Uncertainty estimates in regional and global observed temperature changes: a new data set from 1850. J Geophys Res 111:D12106
Brohan P, Allan R, Freeman JE et al (2009) Marine observations of old weather. Bull Am Meteorol Soc 90:219–230
Bromwich DH, Nicolas JP, Monaghan AJ et al (2013) Central West Antarctica among the most rapidly warming regions on Earth. Nat Geosci 6:139–145
Brönnimann S (2002) Ozon in der Atmosphäre. Haupt
Brönnimann S (2007) Impact of El Niño–Southern Oscillation on European climate. Rev Geophys 45:RG3003
Brönnimann S (2009) Early twentieth-century warming. Nat Geosci 2:735–736
Brönnimann S (2012) Climatic Data. In: Philander SG, Golson JG (eds) Encyclopedia of global warming & climate change, 2nd edn. SAGE, Thousand Oaks, pp 307–309
Brönnimann S (2015) Pacemakers of warming. Nat Geosci 8:87–98
Brönnimann S, Compo GP (2012) Ozone highs and associated flow features in the first half of the twentieth century in different data sets. Meteorol Z 21:49–59
Brönnimann S, Stickler A (2013) Aerological observations in the tropics in the early twentieth century. Meteorol Z 22:349–358
Brönnimann S, Luterbacher J, Staehelin J et al (2004) Extreme climate of the global troposphere and stratosphere in 1940–42 related to El Niño. Nature 431:971–974
Brönnimann S, Annis JL, Vogler C, Jones PD (2007a) Reconstructing the Quasi-Biennial Oscillation back to the early 1900s. Geophys Res Lett 34:L22805

Brönnimann S, Xoplaki E, Casty C, Pauling A, Luterbacher J (2007b) ENSO influence on Europe during the last centuries. Clim Dyn 28:181–197
Brönnimann S, Stickler A, Griesser T et al (2009a) Exceptional atmospheric circulation during the "Dust Bowl". Geophys Res Lett 36:L08802
Brönnimann S, Stickler A, Griesser T et al (2009b) Variability of large-scale atmospheric circulation indices for the Northern Hemisphere during the past 100 years. Meteorol Z 18:379–396
Brönnimann S, Grant AN, Compo GP et al (2012a) A multi-data set comparison of the vertical structure of temperature variability and change over the Arctic during the past 100 years. Clim Dyn 39:1577–1598
Brönnimann S, Griesser T, Stickler A (2012b) A gridded monthly upper-air data set from 1918 to 1957. Clim Dyn 38:475–493
Brönnimann S, Martius O, von Waldow H et al (2012c) Extreme winds at northern mid-latitudes since 1871. Meteorol Z 21:13–27
Brönnimann S, Bhend J, Franke J et al (2013a) A global historical ozone data set and prominent features of stratospheric variability prior to 1979. Atmos Chem Phys 13:9623–9639
Brönnimann S, Franke J, Breitenmoser P et al (2013b) Transient state estimation in paleoclimatology using data assimilation. PAGES News 21:74–75
Brönnimann S, Mariani I, Schwikowski M, Auchmann R, Eichler A (2013c) Simulating the temperature and precipitation signal in an Alpine ice core. Clim Past 9:2013–2022
Brooks CEP (1922) The evolution of climate [Preface by Simpson, G.C.]. Benn Brothers, London
Brown RD, Robinson DA (2011) Northern hemisphere spring snow cover variability and change over 1922–2010 including an assessment of uncertainty. The Cryosphere 5:219–229
Brückner E (1890) Klimaschwankungen seit 1700 nebst Bemerkungen über die Klimaschwankungen der Diluvialzeit. E. D. Hölzel, Wien and Olmütz
Brugnara Y, Brönnimann S, Luterbacher J, Rozanov E (2013) Influence of the sunspot cycle on the Northern Hemisphere wintertime circulation from long upper-air data sets. Atmos Chem Phys 13:6275–6288
Brugnara Y, Auchmann R, Brönnimann S, Allan RJ, Auer I, Barriendos M, Bergström H, Bhend J, Brázdil R, Compo GP, Cornes RC, Dominguez-Castro F, van Engelen AFV, Filipiak J, Holopainen J, Jourdain S, Kunz M, Luterbacher J, Maugeri M, Mercalli L, Moberg A, Mock CJ, Pichard G, Rezníckovà L, van der Schrier G, Slonosky V, Ustrnul Z, Valente MA, Wypych A, Yin X (2015) A collection of sub-daily pressure and temperature observations for the early instrumental period with a focus on the "year without a summer" 1816, Clim Past (in press)
Bryson RA, Kutzbach JE (1968) Air Pollution. Resource Paper No. 2. ERIC
Budyko MI (1974) Change of climate (Izmenenie klimata). Hydrometeoizdat, Leningrad
Büntgen U, Frank DC, Nievergelt D, Esper J (2006) Summer temperature variations in the European Alps, AD 755-2004. J Clim 19:5606–5623
Büntgen U, Tegel W, Nicolussi K et al (2011) 2500 years of European climate variability and human susceptibility. Science 331:578–582
Büntgen U, Trnka M, Krusic PJ, Kyncl T, Kyncl J, Nicolussi K, Luterbacher J, Zorita E, Charpentier Ljungqvist F, Auer I, Konter O, Schneider L, Tegel W, Stepanek P, Brönnimann S, Hellmann L, Nievergelt D, Esper J (2015) Tree-ring amplification of the early-19th century summer cooling in Central Europe. J Clim 28:5272–5288
Bünti JL (1973) Chronik des Johann Laurentz Bünti, Landamman, 1661–1736. Beiträge zur Geschichte Nidwaldens 34:1–418
Burgers G, Jin FF, van Oldenborgh GJ (2005) The simplest ENSO recharge oscillator. Geophys Res Lett 32:L13706
Burt S (2012) The Weather Observer's Handbook. Cambridge University Press, Cambridge
Cai W, Cowan T (2007) Impacts of increasing anthropogenic aerosols on the atmospheric circulation trends of the Southern Hemisphere: An air-sea positive feedback. Geophys Res Lett 34:L23709
Cai W, Zheng XT, Weller E et al (2013) Projected response of the Indian Ocean Dipole to greenhouse warming. Nat Geosci 6:999–1007

Cai W, Purich A, Cowan T, van Rensch P, Weller E (2014) Did climate change–induced rainfall trends contribute to the Australian Millennium Drought? J Clim 27:3145–3168

Calanca P (2007) Climate change and drought occurrence in the Alpine region: How severe are becoming the extremes? Glob Plan Change 57:151–160

Callendar GS (1938) The artificial production of carbon dioxide and its influence on temperature. Q J R Meteorol Soc 64:223–240

Calogovic J, Albert C, Arnold F et al (2010) Sudden cosmic ray decreases: No change of global cloud cover. Geophys Res Lett 37:L03802

Carril AF, Gualdi S, Cherchi A, Navarra A (2008) Heatwaves in Europe: Areas of homogeneous variability and links with the regional to large-scale atmospheric and SSTs anomalies. Clim Dyn 30:77–98

Cassidy D (1985) Meteorology in Mannheim: The Palatine Meteorological Society, 1780–1795. Sudhoffs Archiv 69:8–25

Cassou C, Terray L, Phillips AS (2005) Tropical Atlantic influence on European heat waves. J Clim 18:2805–2811

Casty C, Wanner H, Luterbacher J, Esper J, Böhm R (2005) Temperature and precipitation variability in the European Alps since 1500. Int J Climatol 25:1855–1880

Casty C, Raible CC, Stocker TF, Wanner H, Luterbacher J (2007) A European pattern climatology 1766–2000. Clim Dyn 29:791–805

Caussinus H, Mestre O (2004) Detection and correction of artificial shifts in climate series. J R Stat Soc C 53:405–425

CCMVal SPARC, Eyring V, Shepherd T, Waugh D (2010) SPARC report on the evaluation of chemistry–climate models, SPARC Rep. 5, WCRP-132. Tech. rep., SPARC

Chang FC, Wallace JM (1987) Meteorological conditions during heat waves and droughts in the United States Great Plains. Mon Weather Rev 115:1253–1269

Chang P, Ji L, Li H (1997) A decadal climate variation in the tropical Atlantic Ocean from thermodynamic air-sea interactions. Nature 385:516–518

Charlson RJ, Lovelock JE, Andreae MO, Warren SG (1987) Oceanic phytoplankton, atmospheric sulphur, cloud albedo and climate. Nature 326:655–661

Charney JG, Fjörtoft R, von Neumann J (1950) Numerical integration of the barotropic vorticity equation. Tellus 2:237–254

Chen GS, Liu Z, Kutzbach JE (2014) Reexamining the barrier effect of Tibetan Plateau on the South Asian summer monsoon. Clim Past 10:1269–1275

Chen X, Tung KK (2014) Varying planetary heat sink led to global-warming slowdown and acceleration. Science 345:897–903

Chenoweth M (2003) The 18th century climate of Jamaica: derived from the journals of Thomas Thistlewood, 1750–1786. American Philosophical Society, Philadelphia

Chenoweth M (2009) Daily synoptic weather map analysis of the New England cold wave and snowstorms of 5 to 11 June 1816. In: Historical climate variability and impacts in North America. Springer, Dordrecht, pp 107–121

Chenoweth M, Divine D (2008) A document-based 318-year record of tropical cyclones in the Lesser Antilles, 1690–2007. Geochem Geophys Geosyst 9:Q08013

Chenoweth M, Divine D (2012) Tropical cyclones in the lesser antilles: descriptive statistics and historical variability in cyclone energy, 1638–2009. Clim Change 113:583–598

Christensen JH, Krishna Kumar K, Aldrian E et al (2013) Climate phenomena and their relevance for future regional climate change. In: Stocker T, Qin D, Plattner GK et al (eds) Climate change 2013: The physical science basis. Contribution of Working Group I to the Fifth Assessment Report of the Intergovernmental Panel on Climate Change, Cambridge University Press, Cambridge/New York

Christiansen B, Ljungqvist FC (2012) The extra-tropical Northern Hemisphere temperature in the last two millennia: reconstructions of low-frequency variability. Clim Past 8:765–786

Ciais P, Reichstein M, Viovy N et al (2005) Europe-wide reduction in primary productivity caused by the heat and drought in 2003. Nature 437:529–533

Cicerone RJ, Stolarski RS, Walters S (1974) Stratospheric ozone destruction by man-made chlorofluoromethanes. Science 185:1165–1167
Cobb KM, Westphal N, Sayani HR et al (2013) Highly variable El Niño–Southern Oscillation throughout the Holocene. Science 339:67–70
Cohen JL, Furtado JC, Barlow MA, Alexeev VA, Cherry JE (2012) Arctic warming, increasing snow cover and widespread boreal winter cooling. Environ Res Lett 7:014007
Cohen J, Screen JA, Furtado JC et al (2014) Recent Arctic amplification and extreme mid-latitude weather. Nat Geosci 7:627–637
Colbert AJ, Soden BJ (2012) Climatological variations in North Atlantic tropical cyclone tracks. J Clim 25:657–673
Collins M, Knutti R, Arblaster J et al (2013) Long-term climate change: projections, commitments and irreversibility. In: Stocker T, Qin D, Plattner GK et al (eds) Climate change 2013: The physical science basis. Contribution of Working Group I to the Fifth Assessment Report of the Intergovernmental Panel on Climate Change. Cambridge University Press, Cambridge/New York
Comiso JC (2012) Large decadal decline of the Arctic multiyear ice cover. J Clim 25:1176–1193
Compo GP, Whitaker JS, Sardeshmukh PD et al (2011) The Twentieth Century Reanalysis project. Q J R Meteorol Soc 137:1–28
Cook BI, Miller RL, Seager R (2008) Dust and sea surface temperature forcing of the 1930s "Dust Bowl" drought. Geophys Res Lett 35:L08710
Cook BI, Miller RL, Seager R (2009) Amplification of the North American "Dust Bowl" drought through human-induced land degradation. Proc Natl Acad Sci USA 106:4997–5001
Cook ER, Krusic PJ (2003) The North American Drought Atlas. American Geophysical Union, Fall Meeting 2003, abstract number GC52A-01
Cook ER, D'Arrigo RD, Cole JE, Stahle DW, Villalba R (2000) Tree-ring records of past ENSO variability and forcing. In: Diaz HF, Markgraf V (eds) El Niño and the Southern Oscillation: Multiscale variability and global and regional impacts. Cambridge University Press, Cambridge/New York, pp 297–323
Cook ER, D'Arrigo RD, Mann ME (2002) A well-verified, multiproxy reconstruction of the winter North Atlantic Oscillation Index since AD 1400. J Clim 15:1754–1764
Cook ER, Krusic PJ, Jones PD (2003) Dendroclimatic signals in long tree-ring chronologies from the Himalayas of Nepal. Int J Climatol 23:707–732
Cook ER, Woodhouse CA, Eakin CM, Meko DM, Stahle DW (2004) Long-term aridity changes in the western United States. Science 306:1015–1018
Cook ER, Buckley BM, Palmer JG et al (2006) Millennia-long tree-ring records from Tasmania and New Zealand: A basis for modelling climate variability and forcing, past, present and future. J Quat Sci 21:689–699
Cook ER, Anchukaitis KJ, Buckley BM et al (2010) Asian Monsoon Failure and Megadrought During the Last Millennium. Science 328:486–489
Cornes RC, Jones PD (2011) An examination of storm activity in the northeast Atlantic region over the 1851–2003 period using the EMULATE gridded MSLP data series. J Geophys Res 116:D16110
Cowtan K, Way RG (2014) Coverage bias in the HadCRUT4 temperature series and its impact on recent temperature trends. Q J R Meteorol Soc 140:1935–1944
Croll J (1870) XII. On ocean-currents, Part I: Ocean-currents in relation to the distribution of heat over the globe. Philos Mag 39:81–106
Croll J (1875) Climate and time in their geological relations. A theory of secular changes of the Earth's climate. Daldy, Isbister, & Co, London
Crowley TJ (2000) Causes of climate change over the past 1000 years. Science 289:270–277
Crowley TJ, Zielinski G, Vinther B et al (2008) Volcanism and the little ice age. PAGES News 16:22–23
Crowley TJ, Obrochta SP, Liu J (2014) Recent global temperature "plateau" in the context of a new proxy reconstruction. Earth's Future 2:281–294

Crutzen PJ (1970) The influence of nitrogen oxides on the atmospheric ozone content. Q J R Meteorol Soc 96:320–325

Crutzen PJ (2006) Albedo enhancement by stratospheric sulfur injections: a contribution to resolve a policy dilemma? Clim Change 77:211–220

Crutzen PJ, Birks JW (1982) Atmosphere after a nuclear war: Twilight at noon. Ambio 11:114–125

Czaja A, Blunt N (2011) A new mechanism for ocean–atmosphere coupling in midlatitudes. Q J R Meteorol Soc 137:1095–1101

Daley R (1993) Atmospheric data analysis. Cambridge University Press, Cambridge/New York/Melbourne

von Danckelman A (1884) Die Bewölkungsverhältnisse des südwestlichen Afrikas (Cloud-conditions in Southwest Africa). Meteorol Z 1:301–311. (Translated and edited by Volken E, Lehmann K, Brönnimann S, Meteorol Z 18:341–348, 2009)

Dangendorf S, Müller-Navarra S, Jensen J et al (2014) North Sea storminess from a novel storm surge record since ad 1843. J Clim 27:3582–3595

D'Arrigo R, Wilson R, Tudhope A (2008) The impact of volcanic forcing on tropical temperatures during the past four centuries. Nat Geosci 2:51–56

D'Arrigo R, Palmer J, Ummenhofer C, Kyaw NN, Krusic P (2013) Myanmar monsoon drought variability inferred by tree rings over the past 300 years: Oinkages to ENSO. PAGES Newsletter 21:50–51

Davis M (2001) Late Victorian holocausts: El Niño famines and the making of the Third World. Verso, London

de Jong R, Björck S, Björkman L, Clemmensen LB (2006) Storminess variation during the last 6500 years as reconstructed from an ombrotrophic peat bog in Halland, southwest Sweden. J Quat Sci 21:905–919

De Marchi L (1895a) Le cause dell'era glaciale. Premiato dal R. Istituto Lombardo, Pavia

De Marchi L (1895b) Le cause dell'era glaciale: ricerca teorica delle condizioni che determinano l'attuale distribuzione delle temperature e delle pioggie sulla superficie terrestre e che possono averla modificata nei precedenti periodi geologici. Fusi

Dee DP, Uppala SM, Simmons AJ et al (2011) The ERA-Interim reanalysis: configuration and performance of the data assimilation system. Q J R Meteorol Soc 137:553–597

Delaygue G, Bard E (2011) An Antarctic view of Beryllium-10 and solar activity for the past millennium. Clim Dyn 36:2201–2218

Della-Marta PM, Wanner H (2006) A method of homogenizing the extremes and mean of daily temperature measurements. J Clim 19:4179–4197

Della-Marta PM, Haylock MR, Luterbacher J, Wanner H (2007) Doubled length of western European summer heat waves since 1880. J Geophys Res 112:D15103

Delworth TL, Greatbatch RJ (2000) Multidecadal thermohaline circulation variability driven by atmospheric surface flux forcing. J Clim 13:1481–1495

Delworth TL, Knutson TR (2000) Simulation of early 20th century global warming. Science 287:2246–2250

Delworth TL, Mann ME (2000) Observed and simulated multidecadal variability in the Northern Hemisphere. Clim Dyn 16:661–676

Delworth DL, Zeng F (2014) Regional rainfall decline in Australia attributed to anthropogenic greenhouse gases and ozone levels. Nat Geosci 7:583–587

Deser C, Knutti R, Solomon S, Phillips AS (2012) Communication of the role of natural variability in future North American climate. Nat Clim Change 2:775–779

Devereux S (2000) Famine in the twentieth century. IDS Working Paper, vol 105. Institute of Development Studies, Brighton

DeWeaver ET, Bitz CM, Tremblay LB (2008) Arctic sea ice decline: observations, projections, mechanisms, and implications. Geophysical Monograph, vol 180. AGU, Washington, DC

Diatta S, Fink AH (2014) Statistical relationship between remote climate indices and West African monsoon variability. Int J Climatol 34:3348–3367

Diaz HF, McCabe GJ (1999) A possible connection between the 1878 yellow fever epidemic in the southern United States and the 1877–78 El Niño episode. Bull Am Meteorol Soc 80:21–27

Diaz HF, Trigo R, Hughes MK et al (2011) Spatial and temporal characteristics of climate in medieval times revisited. Bull Am Meteorol Soc 92:1487–1500
Digby W (1878) The famine campaign in southern India. Longmans, Green & Co, London
Ding Q, Wallace JM, Battisti DS et al (2014) Tropical forcing of the recent rapid Arctic warming in northeastern Canada and Greenland. Nature 509:209–212
Dirren S, Hakim GJ (2005) Toward the assimilation of time-averaged observations. Geophys Res Lett 32:L04804
Dobrovolný P, Moberg A, Brázdil R et al (2010) Monthly, seasonal and annual temperature reconstructions for Central Europe derived from documentary evidence and instrumental records since AD 1500. Clim Change 101:69–107
Dobrovolný P, Brázdil R, Trnka M, Kotyza O, Valášek H (2015) Precipitation reconstruction for the Czech Lands, AD 1501–2010. Int J Climatol 35:1–14
Dobson G (1956) Origin and distribution of the polyatomic molecules in the atmosphere. Proc R Soc A 230:187–193
Dobson GMB, Harrison DN (1926) Measurements of the amount of ozone in the Earth's atmosphere and its relation to other geophysical conditions. Proc R Soc A 110:660–693
Dörries M (2003) Global science: the eruption of Krakatau. Endeavour 27:113–116
Dove WH (1852) Die Verbreitung der Wärme auf der Oberfläche der Erde: erläutert durch Isothermen, thermische Isanomalen und Temperaturcurven. Reimer, Berlin
Driscoll S, Bozzo A, Gray LJ, Robock A, Stenchikov G (2012) Coupled Model Intercomparison Project 5 (CMIP5) simulations of climate following volcanic eruptions. J Geophys Res 117:D17105
Dufour L (1870) Notes sur le problème de la variation du climat. Bulletin de la Société Vaudoise des Sciences naturelles 10:360–436
Dunstone NJ, Smith DM, Booth BBB, Hermanson L, Eade R (2013) Anthropogenic aerosol forcing of Atlantic tropical storms. Nat Geosci 6:534–539
Eddy JA (1976) The maunder minimum. Science 192:1189–1202
Edwards PN (2010) A vast machine: Computer models, climate data, and the politics of global warming. MIT, Cambridge
Efthymiadis D, Jones PD, Briffa KR et al (2006) Construction of a 10-min-gridded precipitation data set for the Greater Alpine Region for 1800–2003. J Geophys Res 111:1–22
Ehrenheim F (1824) Om climaternes rörlighet. Norstedt
Eichler A, Schwikowski M, Gäggeler HW et al (2000) Glaciochemical dating of an ice core from upper Grenzgletscher (4200 m a.s.l.). J Glaciol 46:507–515
Eichler A, Brütsch S, Olivier S, Papina T, Schwikowski M (2009a) A 750 year ice core record of past biogenic emissions from Siberian boreal forests. Geophys Res Lett 36:L18813
Eichler A, Olivier S, Henderson K et al (2009b) Temperature response in the Altai region lags solar forcing. Geophys Res Lett 36:L01808
Eichler A, Tinner W, Brütsch S et al (2011) An ice-core based history of Siberian forest fires since AD 1250. Quat Sci Rev 30:1027–1034
Emden R (1913) Über Strahlungsgleichgewicht und atmosphärische Strahlung. Sitzungsberichte der Bayerischen Akademie der Wissenschaften, Math. Phys Klasse, pp 55–142
Emeis S (2012) Assmann's development of aspiration psychrometers. Meteorol Z 21:431–435
Emile-Geay J, Cobb KM, Mann ME, Wittenberg AT (2013) Estimating Central Equatorial Pacific SST variability over the past millennium. Part II: reconstructions and implications. J Clim 26:2329–2352
Enfield DB, Mestas-Nuñez AM, Trimble PJ (2001) The Atlantic Multidecadal Oscillation and its relation to rainfall and river flows in the continental U.S. Geophys Res Lett 28:2077–2080
England MH, McGregor S, Spence P et al (2014) Recent intensification of wind-driven circulation in the Pacific and the ongoing warming hiatus. Nat Clim Change 4:222–227
Ermolli I, Matthes K, Dudok de Wit T et al (2013) Recent variability of the solar spectral irradiance and its impact on climate modelling. Atmos Chem Phys 13:3945–3977
Esper J, Cook ER, Schweingruber FH (2002) Low-frequency signals in long tree-ring chronologies for reconstructing past temperature variability. Science 295:2250–2253

Estrada F, Perron P, Martínez-López B (2013) Statistically derived contributions of diverse human influences to twentieth-century temperature changes. Nat Geosci 6:1050–1055

Evan AT, Kossin JP, Ramanathan V et al (2011) Arabian Sea tropical cyclones intensified by emissions of black carbon and other aerosols. Nature 479:94–97

Ewen T, Brönnimann S, Annis J (2008) An extended Pacific–North American index from upper-air historical data back to 1922. J Clim 21:1295–1308

Exner FM (1913) Über monatliche Witterungsanomalien auf der nördlichen Erdhälfte im Winter. Sitzber Kais Akad Wiss 122:1165–1241

Fabry C, Buisson H (1913) L'absorption de l'ultraviolet par l'ozone et la limite du spectre solaire. J Phys Rad, Série 5 3:196–206

Fagan BM (2000) The Little Ice Age: How climate made history, 1300–1850. Basic Books, New York

Farman JC, Gardiner BG, Shanklin JD (1985) Large losses of total ozone in Antarctica reveal seasonal ClOx/NOx interaction. Nature 315:207–210

Farquhar GD, Roderick ML (2003) Pinatubo, diffuse light, and the carbon cycle. Science 299:1997–1998

Fasullo J (2004) A stratified diagnosis of the Indian monsoon–Eurasian snow cover relationship. J Clim 17:1110–1122

Fasullo JT, Trenberth KE (2008a) The annual cycle of the energy budget. Part I: Global mean and land–ocean exchanges. J Clim 21:2297–2312

Fasullo JT, Trenberth KE (2008b) The annual cycle of the energy budget. Part II: Meridional structures and poleward transports. J Clim 21:2313–2325

Ferrel W (1856) An essay on the winds and the currents of the oceans. Nashv J Med Surg 11:287–301, 375–389

Feulner G (2011) Are the most recent estimates for Maunder Minimum solar irradiance in agreement with temperature reconstructions? Geophys Res Lett 38:L16706

Fischer EM, Knutti R (2014) Detection of spatially aggregated changes in temperature and precipitation extremes. Geophys Res Lett 41:547–554

Fischer EM, Knutti R (2015) Anthropogenic contribution to global occurrence of heavy-precipitation and high-temperature extremes. Nat Clim Change 5:560–565

Fischer EM, Luterbacher J, Zorita E et al (2007a) European climate response to tropical volcanic eruptions over the last half millennium. Geophys Res Lett 34:L05707

Fischer EM, Seneviratne SI, Lüthi D, Schär C (2007b) Contribution of land-atmosphere coupling to recent European summer heat waves. Geophys Res Lett 34:L06707

Fischer AM, Schraner M, Rozanov E et al (2008) Interannual-to-decadal variability of the stratosphere during the 20th century: ensemble simulations with a chemistry-climate model. Atmos Chem Phys 8:7755–7777

Flato G, Marotzke J, Abiodun B et al (2013) Evaluation of climate models. In: Stocker T, Qin D, Plattner GK et al (eds) Climate change 2013: The physical science basis. Contribution of Working Group I to the Fifth Assessment Report of the Intergovernmental Panel on Climate Change. Cambridge/New York

Fleming J (2007) The Callendar effect: the life and work of Guy Stewart Callendar (1898–1964). Amer Meteor Soc, Boston

Fogt RL, Perlwitz J, Monaghan AJ et al (2009) Historical SAM variability. Part II: twentieth-century variability and trends from reconstructions, observations, and the IPCC AR4 Models. J Clim 22:5346–5365

Folland C, Parker DE, Colman A (1999) Large scale modes of ocean surface temperature since the late nineteenth century. In: Navarra A (ed) Beyond El Niño: decadal and interdecadal climate variability. Springer, New York, pp 73–102

Folland CK, Knight J, Linderholm HW et al (2009) The summer North Atlantic Oscillation: Past, present, and future. J Clim 22:1082–1103

Forster PM, Thompson DWJ, Baldwin MP et al (2011) Stratospheric changes and climate, chapter 4. In: Scientific Assessment of Ozone Depletion: 2010, Global ozone research and monitoring project-report no. 52, World Meteorological Organization, Geneva

Foster G, Rahmstorf S (2011) Global temperature evolution 1979–2010. Env Res Lett 6:044022

Francis JA, Vavrus SJ (2012) Evidence linking Arctic amplification to extreme weather in mid-latitudes. Geophys Res Lett 39:L06801

Frank D, Büntgen U, Böhm R, Maugeri M, Esper J (2007) Warmer early instrumental measurements versus colder reconstructed temperatures: shooting at a moving target. Quat Sci Rev 26:3298–3310

Frank DC, Esper J, Raible CC, Büntgen U, Trouet V, Stocker B, Joos F (2010) Ensemble reconstruction constraints on the global carbon cycle sensitivity to climate. Nature 463:527–532

Franke J, Gonzàlez-Rouco J, Frank D, Graham EN (2011) 200 years of European temperature variability: insights from and tests of the proxy surrogate reconstruction analog method. Clim Dyn 37:133–150

Franke J, Frank D, Raible CC, Esper J, Brönnimann S (2013) Spectral biases in tree-ring climate proxies. Nat Clim Change 3:360–364

Franklin B (1784) Meteorological imaginations and conjectures. Manch Lit Philos Soc Mem Proc 2:373–377

Frenger I, Gruber N, Knutti R, Munnich M (2013) Imprint of Southern Ocean eddies on winds, clouds and rainfall. Nat Geosci 6:608–612

Frierson DMW, Hwang YT, Fučkar NS et al (2013) Contribution of ocean overturning circulation to tropical rainfall peak in the Northern Hemisphere. Nat Geosci 6:940–944

Fritts HC (1976) Tree rings and climate. Academic, London

Fu C, Diaz HF, Dong D, Fletcher JO (1999) Changes in atmospheric circulation over Northern Hemisphere oceans associated with the rapid warming of the 1920s. Int J Climatol 19:581–606

Fueglistaler S, Dessler AE, Dunkerton TJ et al (2009) Tropical tropopause layer. Rev Geophys 47:RG1004

Funke B, Baumgaertner A, Calisto M et al (2011) Composition changes after the "Halloween" solar proton event: The High-Energy Particle Precipitation in the Atmosphere (HEPPA) model versus MIPAS data intercomparison study. Atmos Chem Phys 11:9089–9139

Fyfe JC, Gillett NP, Zwiers FW (2013) Overestimated global warming over the past 20 years. Nat Clim Change 3:767–769

Gaetani M, Pohl B, Douville H, Fontaine B (2011) West African Monsoon influence on the summer Euro-Atlantic circulation. Geophys Res Lett 38:L09705

Galton F (1863) Meteorographica, or methods of mapping the weather. Macmillan, London

Galton F (1888) Co-relations and their measurement, chiefly from anthropometric data. Proc R Soc Lond 45:135–145

Gao C, Robock A, Ammann C (2008) Volcanic forcing of climate over the past 1500 years: An improved ice core-based index for climate models. J Geophys Res 113:D23111

Garcia-Herrera R, Diaz H, Garcia R et al (2008) A chronology of El Niño events from primary documentary sources in northern Peru. J Clim 21:1948–1962

Garden D (2008) El Niño, irrigation dams and stopbanks: Examining the repercussions of the 1876–78 El Niño in Australia and New Zealand. Hist Meteorol 4:1

Garfinkel CI, Hartmann DL (2008) Different ENSO teleconnections and their effects on the stratospheric polar vortex. J Geophys Res 113:D18114

Garrett TJ, Zhao C (2006) Increased Arctic cloud longwave emissivity associated with pollution from mid-latitudes. Nature 440:787–789

Gates WL (1992) AMIP: The atmospheric model intercomparison project. Bull Am Meteorol Soc 73:1962–1970

Gergis J, Gallant AJE, Braganza K et al (2012) On the long-term context of the 1997–2009 'Big Dry' in South-Eastern Australia: insights from a 206-year multi-proxy rainfall reconstruction. Clim Change 111:923–944

Giannini A, Saravanan R, Chang P (2003) Oceanic forcing of Sahel rainfall on interannual to interdecadal time scales. Science 302:1027–1030

Giannini A, Biasutti M, Verstraete MM (2008) A climate model-based review of drought in the Sahel: desertification, the re-greening and climate change. Glob Plan Change 64:119–128

Gibson JK, Kållberg P, Uppala S et al (1997) ERA Description, ECMWF Re-analysis Final Report Series, 1. Tech. rep., ECMWF, Reading, UK

Giese BS, Slowey NC, Ray S et al (2010) The 1918/19 El Niño. Bull Am Meteorol Soc 91:177–183

Gill AE (1980) Some simple solutions for heat-induced tropical circulation. Q J R Meteorol Soc 106:447–462

Gillett NP, Thompson DWJ (2003) Simulation of recent Southern Hemisphere climate change. Science 302:273–275

Gillett NP, Zwiers FW, Weaver AJ, Stott PA (2003) Detection of human influence on sea-level pressure. Nature 422:292–294

Gillett NP, Allan RJ, Ansell TJ (2005) Detection of external influence on sea level pressure with a multi-model ensemble. Geophys Res Lett 32:L19714

Gimmi U, Luterbacher J, Pfister C, Wanner H (2007) A method to reconstruct long precipitation series using systematic descriptive observations in weather diaries: The example of the precipitation series for Bern, Switzerland (1760–2003). Theor Appl Climatol 87:185–199

Giorgetta MA, Bengtsson L, Arpe K (1999) An investigation of QBO signals in the east Asian and Indian monsoon in GCM experiments. Clim Dyn 15:435–450

Glacken CJ (1967) Traces on the Rhodian shore: Nature and culture in Western thought from ancient times to the end of the eighteenth century. University of California Press, Berkeley

Glaser R (2008) Klimageschichte Mitteleuropas – 1200 Jahre Wetter, Klima, Katastrophen. WBG, Darmstadt

Glaser R, Riemann D (2009) A thousand-year record of temperature variations for Germany and Central Europe based on documentary data. J Quat Sci 24:437–449

Gleckler PJ, Wigley TML, Santer BD et al (2006) Volcanoes and climate: Krakatoa's signature persists in the ocean. Nature 439:675–675

Glur L, Wirth SB, Büntgen U et al (2013) Frequent floods in the European Alps coincide with cooler periods of the past 2500 years. Sci Rep 3:2770

Goldsmith P, Tuck AF, Foot JS, Simmons EL, Newson RL (1973) Nitrogen oxides, nuclear weapon testing, concorde and stratospheric ozone. Nature 244:545–551

Gong DY, Luterbacher J (2008) Variability of the low-level cross-equatorial jet of the western Indian Ocean since 1660 as derived from coral proxies. Geophys Res Lett 35:L01705

Goosse H, Arzel O, Luterbacher J et al (2006) The origin of the European" Medieval Warm Period". Clim Past 2:99–113

Goosse H, Crespin E, de Montety A et al (2010) Reconstructing surface temperature changes over the past 600 years using climate model simulations with data assimilation. J Geophys Res 115:D09108

Graf HF, Zanchettin D (2012) Central Pacific El Niño, the "subtropical bridge," and Eurasian climate. J Geophys Res 117:D01102

Graham NE, Hughes MK, Ammann CM et al (2007) Tropical Pacific–mid-latitude teleconnections in medieval times. Clim Change 83:241–285

Graham NE, Ammann CM, Fleitmann D, Cobb KM, Luterbacher J (2011) Support for global climate reorganization during the "Medieval Climate Anomaly". Clim Dyn 37:1217–1245

Grant AN, Brönnimann S, Ewen T, Griesser T, Stickler A (2009) The early twentieth century warm period in the European Arctic. Meteorol Z 18:425–432

Grassl H, Klingholz R (1990) Wir Klimamacher. Büchergilde Gutenberg, Frankfurt a. M.

Gray WM (1984) Atlantic seasonal hurricane frequency. Part I: El Niño and 30 mb quasi-biennial oscillation influences. Mon Weather Rev 112:1649–1668

Gray ST, Graumlich LJ, Betancourt JL, Pederson GT (2004) A tree-ring based reconstruction of the Atlantic Multidecadal Oscillation since 1567 AD. Geophys Res Lett 31:L12205

Gray LJ, Beer J, Geller M et al (2010) Solar influences on climate. Rev Geophys 48:RG4001

Gray LJ, Scaife AA, Mitchell DM, Osprey S, Ineson S, Hardiman S, Butchart N, Knight J, Sutton R, Kodera K (2013) A lagged response to the 11 year solar cycle in observed winter Atlantic/European weather patterns. J Geophys Res 118:13405–13420

Griesser T, Brönnimann S, Grant A et al (2010) Reconstruction of global monthly upper-level temperature and geopotential height fields back to 1880. J Clim 23:5590–5609

Groissmayr FB (1949) Die grosse säkulare Klimawende um 1940 und das Katastrophenjahr 1947 in Zentraleuropa. Ber. Dt. Wetterdienst. US-Zone, Nr. 10, 39 pp
Grove RH (1995) Green imperialism: Colonial expansion, tropical island Edens and the origins of environmentalism, 1600–1860. Cambridge University Press, Cambridge
Grütter J (2014) Droughts in Europe 1945–1959: relation to atmospheric circulation and the role of sea surface temperature. MSc thesis, Institute of Geography University of Bern
Grütter J, Lehmann S, Auchmann R, Martius O, Brönnimann S (2013) The heatwaves in Switzerland in summer 1947. In: Brönnimann S, Martius O (eds) Weather extremes during the past 140 years, Geographica Bernensia G89, Bern, pp 69–80
Gu L, Baldocchi DD, Wofsy SC et al (2003) Response of a deciduous forest to the Mount Pinatubo eruption: Enhanced photosynthesis. Science 299:2035–2038
Guevara-Murua A, Williams CA, Hendy EJ, Rust AC, Cashman KV (2014) Observations of a stratospheric aerosol veil from a tropical volcanic eruption in December 1808: is this the Unknown ~1809 eruption? Clim Past 10:1707–1722
Gulev SK, Latif M, Keenlyside N, Park W, Koltermann KP (2013) North Atlantic Ocean control on surface heat flux on multidecadal timescales. Nature 499:464–467
Gustafsson Ö, Kruså M, Zencak Z et al (2009) Brown clouds over South Asia: Biomass or fossil fuel combustion? Science 323:495–498
Hadley G (1735) Concerning the cause of the general trade-winds. Philos Trans 39:58–62
Hadley OL, Kirchstetter TW (2012) Black-carbon reduction of snow albedo. Nat Clim Change 2:437–440
Haigh JD (2003) The effects of solar variability on the Earth's climate. Philos Trans R Soc A 361:95–111
Haigh JD, Winning AR, Toumi R, Harder JW (2010) An influence of solar spectral variations on radiative forcing of climate. Nature 467:696–699
Hakim GJ, Annan J, Brönnimann S et al (2013) Overview of data assimilation methods. PAGES News 21:72–73
Häkkinen S (1999) A simulation of thermohaline effects of a great salinity anomaly. J Clim 12:1781–1795
Häkkinen S, Rhines PB, Worthen DL (2011) Atmospheric blocking and Atlantic Multidecadal Ocean variability. Science 334:655–659
Halley E (1686) An historical account of the trade winds, and monsoons, observable in the seas between and near the tropicks, with an attempt to assign the phisical cause of the said winds. Philos Trans 16:153–168
Halley E (1694) An account of the watery circulation of the sea, and of the cause of springs. Philos Trans 142:468–472
Ham YG, Kug JS, Park JY, Jin FF (2013) Sea surface temperature in the north tropical Atlantic as a trigger for El Niño/Southern Oscillation events. Nat Geosci 6:112–116
Hamilton K (2012) Sereno Bishop, Rollo Russell, Bishop's Ring and the discovery of the "Krakatoa Easterlies". Atmos Ocean 50:169–175
Hammer T (2000) Desertifikation im Sahel Lösungskonzepte der dritten Generation. Geographische Rundschau 52:4–11
Hann J (1883) Handbuch der Klimatologie. Verlag J Engelhorn
Hann J (1887) Atlas der Meteorologie. J. Perthes, Gotha
Hann J (1890) Zur Witterungsgeschichte von Nord-Grönland, Westküste. Meteorol Z 7:109–115
Hann J (1897) Handbuch der Klimatologie. Allgemeine Klimatologie, vol. 1, 2nd edn.. Verlag J. Engelhorn, Stuttgart
Hann J (1908–1911) Handbuch der Klimatologie, 3 Vols. Engelhorn, Stuttgart
Hansen J, Lacis A, Ruedy R, Sato M (1992) Potential climate impact of Mount Pinatubo eruption. Geophys Res Lett 19:215–218
Hansen J, Ruedy R, Sato M, Lo K (2010) Global surface temperature change. Rev Geophys 48:RG4004
Hao ZX, Zheng JY, Wu GF, Zhang XZ, Ge QS (2010) 1876–1878 severe drought in North China: Facts, impacts and climatic background. Chinese Sci Bull 55:3001–3007

Harington CR (1992) The year without a summer? World climate in 1816. Canadian Museum of Nature, Ottawa

Harris I, Jones PD, Osborn TJ, Lister DH (2014) Updated high-resolution grids of monthly climatic observations–the CRU TS3. 10 Dataset. Int J Climatol 34:623–642

Hartmann B, Wendler G (2005) The significance of the 1976 Pacific climate shift in the climatology of Alaska. J Clim 18:4824–4839

Hartmann DL (1994) Global physical climatology. International Geophysics Series. Academic, San Diego

Hartmann DL, Wallace JM, Limpasuvan V, Thompson DW, Holton JR (2000) Can ozone depletion and global warming interact to produce rapid climate change? Proc Natl Acad Sci USA 97:1412–1417

Hartmann DL, Klein Tank AMG, Rusticucci M et al (2013) Observations: atmosphere and surface. In: Stocker T, Qin D, Plattner GK et al (eds) Climate change 2013: The physical science basis. Contribution of Working Group I to the Fifth Assessment Report of the Intergovernmental Panel on Climate Change. Cambridge University Press, Cambridge/New York

Hasselmann K (1993) Optimal fingerprints for the detection of time-dependent climate change. J Clim 6:1957–1971

Haylock MR, Hofstra N, Klein Tank AMG et al (2008) A European daily high-resolution gridded data set of surface temperature and precipitation for 1950–2006. J Geophys Res 113:1–12

Haywood JM, Jones A, Bellouin N, Stephenson D (2013) Asymmetric forcing from stratospheric aerosols impacts Sahelian rainfall. Nat Clim Change 3:660–665

Heckendorn P, Weisenstein D, Fueglistaler S et al (2009) The impact of geoengineering aerosols on stratospheric temperature and ozone. Env Res Lett 4:045108

Hegerl G, Zwiers F (2011) Use of models in detection and attribution of climate change. WIREs Clim Change 2:570–591

Hegerl GC, von Storch H, Hasselmann K et al (1996) Detecting greenhouse-gas-induced climate change with an optimal fingerprint method. J Clim 9:2281–2306

Hegerl GC, Crowley TJ, Allen M et al (2007) Detection of human influence on a new, validated 1500-year temperature reconstruction. J Clim 20:650–666

Hegerl G, Luterbacher J, Gonzalez-Rouco F et al (2011) Influence of human and natural forcing on European seasonal temperatures. Nat Geosci 4:99–103

Held IM, Delworth TL, Lu J, Findell KL, Knutson TR (2005) Simulation of Sahel drought in the 20th and 21st centuries. Proc Natl Acad Sci USA 102:17891–17896

Helfand HM, Schubert SD (1995) Climatology of the simulated Great Plains low-level jet and its contribution to the continental moisture budget of the United States. J Clim 8:784–806

Hengl T (2009) A practical guide to geostatistical mapping. University of Amsterdam, Amsterdam

Herceg D, Sobel AH, Sun L (2007) Regional modeling of decadal rainfall variability over the Sahel. Clim Dyn 29:89–99

Herschel W (1801) Observations tending to investigate the nature of the sun, in order to find the causes or symptoms of its variable emission of light and heat; with remarks on the use that may possibly be drawn from solar observations. Philos Trans 91:265–318

Herzog M, Graf HF (2010) Applying the three-dimensional model ATHAM to volcanic plumes: dynamic of large co-ignimbrite eruptions and associated injection heights for volcanic gases. Geophys Res Lett 37:L19807

Hiebl J (2006) The early instrumental climate period (1760–1860) in Europe. Evidence from the Alpine region and Southern Scandinavia Diploma thesis, Geogr Inst, University of Vienna

Higgins RW, Yao Y, Yarosh ES, Janowiak JE, Mo KC (1997) Influence of the Great Plains low-level jet on summertime precipitation and moisture transport over the central United States. J Clim 10:481–507

Hildebrandsson HH (1897) Quelques recherches sur les centres d'action de l'atmosphére. Kongl svenska Vetensk Akad Handl 29, 36 pp

Hirschi M, Seneviratne SI, Alexandrov V et al (2011) Observational evidence for soil-moisture impact on hot extremes in southeastern Europe. Nat Geosci 4:17–21

Hirschi E, Auchmann R, Martius O, Brönnimann S (2013) The 1945–1949 droughts in Switzerland. In: Brönnimann S, Martius O (eds) Weather extremes during the past 140 years, Geographica Bernensia G89, Bern, pp 81–90
Hoerling M, Kumar A (2003) The perfect ocean for drought. Science 299:691–694
Hoinka K (1997) The tropopause: discovery, definition, and demarcation. Meteorol Z 6:281–303
Holton JR (2004) An introduction to dynamic meteorology, 4th edn. Elsevier/Academic, Burlington
Holton JR, Tan HC (1980) The influence of the equatorial Quasi-Biennial Oscillation on the global circulation at 50 mb. J Atmos Sci 37:2200–2208
Holton JR, Haynes PH, McIntyre ME et al (1995) Stratosphere-troposphere exchange. Rev Geophys 33:403–439
Holzhauser H (2010) Zur Geschichte des Gornergletschers. Ein Puzzle aus historischen Dokumenten und fossilen Hölzern aus dem Gletchervorland. Geographica Bernensia, G 84, Geographisches Institut der Universität Bern
Honda M, Nakamura H, Ukita J, Kousaka I, Takeuchi K (2001) Interannual seesaw between the Aleutian and Icelandic lows. Part I: Seasonal dependence and life cycle. J Clim 14:1029–1042
Horel JD, Wallace JM (1981) Planetary-scale atmospheric phenomena associated with the Southern Oscillation. Mon Weather Rev 109:813–829
Hoyt DV (1979) The Smithsonian astrophysical observatory solar constant program. Rev Geophys 17:427–458
Hu Y, Zhou C, Liu J (2011) Observational evidence for poleward expansion of the Hadley circulation. Advances in Atmospheric Sciences 28:33–44
Hu Y, Tao L, Liu J (2013) Poleward expansion of the Hadley circulation in CMIP5 simulations. Adv Atmos Sci 30:790–795
Huang B, Hu Z-Z, Kinter JL, Wu Z, Kumar A (2012) Connection of stratospheric QBO with global atmospheric general circulation and tropical SST. Part I: methodology and composite life cycle. Clim Dyn 38:1–23
Hudson RD, Andrade MF, Follette MB, Frolov AD (2006) The total ozone field separated into meteorological regimes–Part II: Northern Hemisphere mid-latitude total ozone trends. Atmos Chem Phys 6:5183–5191
von Humboldt A (1845) Kosmos. Entwurf einer physischen Weltbeschreibung, vol 1. Cotta'scher Verlag, Stuttgart/Tübingen
Humphreys WJ (1913) Volcanic dust and other factors in the production of climatic changes, and their possible relation to ice ages. Bull Mount Weather Obs 6:1–34
Hunt JCR (1998) Lewis fry Richardson and his contributions to mathematics, meteorology, and models of conflict. Annu Rev Fluid Mech 30:xiii–xxxv
Hurrell JW (1995) Decadal trends in the North Atlantic Oscillation. Science 269:676–679
Hurrell JW, Kushnir Y, Ottersen G, Visbeck M (2003) An overview of the North atlantic oscillation. In: JW H, Kushnir Y, Ottersen G, Visbeck M (eds) The North Atlantic Oscillation: climatic significance and environmental impact. Geophysical Monograph Series 134, AGU, Washington, DC
Hurt RD (1981) The Dust Bowl: An agricultural and social history. Nelson-Hall, Chicago
Ideler JL (1832) Über die angeblichen Veränderungen des Klima. Annalen der Erd-, Völker- und Staatenkunde (Berghaus' Annalen) 5:417–471
Ineson S, Scaife AA (2009) The role of the stratosphere in the European climate response to El Niño. Nat Geosci 2:32–36
Ineson S, Scaife AA, Knight JR et al (2011) Solar forcing of winter climate variability in the Northern Hemisphere. Nat Geosci 4:753–757
IPCC (2007) Climate change 2007: Synthesis report. Contribution of Working Groups I, II and III to the Fourth Assessment Report of the Intergovernmental Panel on Climate Change [Core Writing Team, Pachauri RK, Reisinger A (eds)]. IPCC, Geneva

IPCC (2012) Managing the Risks of Extreme Events and Disasters to Advance Climate Change Adaptation. A special report of Working Groups I and II of the Intergovernmental Panel on Climate Change [Field CB, Barros V, Stocker TF, Qin D, Dokken DJ, Ebi KL, Mastrandrea MD, Mach KJ, Plattner G-K, Allen SK, Tignor M, Midgley PM (eds)]. Cambridge University Press, Cambridge/New York

IPCC (2013) Climate change 2013: The physical science basis. Contribution of Working Group I to the Fifth Assessment Report of the Intergovernmental Panel on Climate Change. [Stocker TF, Qin D, Plattner G-K, Tignor M, Allen SK, Boschung J, Nauels A, Xia Y, Bex V, Midgley PM (eds)]. Cambridge University Press, Cambridge/New York

Isaksson E, Pohjola V, Jauhiainen T et al (2001) A new ice-core record from Lomonosovfonna, Svalbard: Viewing the 1920–97 data in relation to present climate and environmental conditions. J Glaciol 47:335–345

Isotta FA, Frei C, Weilguni V et al (2013) The climate of daily precipitation in the Alps: Development and analysis of a high-resolution grid dataset from pan-Alpine rain-gauge data. Int J Climatol 34:1657–1675

Iwi AM, Hermanson L, Haines K, Sutton RT (2012) Mechanisms linking volcanic aerosols to the Atlantic Meridional Overturning Circulation. J Clim 25:3039–3051

Izumo T, Vialard J, Lengaigne M et al (2010) Influence of the state of the Indian Ocean Dipole on the following year's El Niño. Nat Geosci 3:168–172

Jacobeit J, Glaser R, Nonnenmacher M, Stangl H (2004) Mitteleuropäische Hochwasserentwicklung im Kontext atmosphärischer Zirkulationsschwankungen. In: Gönnert G et al (eds) Proc. Klimaänderung und Küstenschutz, pp 31–41

Jacques-Coper M, Garreaud RD (2015) Characterization of the 1970s climate shift in South America. Int J Climatol 35:2164–2179

Jaiser R, Dethloff K, Handorf D, Rinke A, Cohen J (2012) Impact of sea ice cover changes on the Northern Hemisphere atmospheric winter circulation. Tellus A 64:11595

Janicot S, Harzallah A, Fontaine B, Moron V (1998) West African monsoon dynamics and eastern equatorial Atlantic and Pacific SST anomalies (1970–88). J Clim 11:1874–1882

Jansen E, Overpeck J, Briffa KR et al (2007) Palaeoclimate. In: Solomon S, Qin D, Manning M et al (eds) Climate change 2007: The physical science basis. Contribution of Working Group I to the Fifth Assessment Report of the Intergovernmental Panel on Climate Change. Cambridge University Press, Cambridge/New York

Jefferson T (1787) Notes on the State of Virginia. J Stockdale, London, 382 pp

Johannessen OM, Bengtsson L, Miles MW et al (2004) Arctic climate change: observed and modelled temperature and sea-ice variability. Tellus A 56:328–341

Johnston H, Whitten G, Birks J (1973) Effect of nuclear explosions on stratospheric nitric oxide and ozone. J Geophys Res 78:6107–6135

Jones JM, Fogt RL, Widmann M et al (2009a) Historical SAM variability. Part I: Century-length seasonal reconstructions. J Clim 22:5319–5345

Jones PD, Jonsson T, Wheeler D (1997) Extension to the North Atlantic Oscillation using early instrumental pressure observations from Gibraltar and south-west Iceland. Int J Climatol 17:1433–1450

Jones PD, Briffa KR, Osborn TJ et al (2009b) High-resolution palaeoclimatology of the last millennium: a review of current status and future prospects. Holocene 19:3–49

Jones PD, Lister DH, Osborn TJ et al (2012) Hemispheric and large-scale land-surface air temperature variations: An extensive revision and an update to 2010. J Geophys Res 117:D05127

Jong R, Verbesselt J, Schaepman ME, Bruin S (2012) Trend changes in global greening and browning: contribution of short-term trends to longer-term change. Global Change Biol 18:642–655

Joseph R, Zeng N (2011) Seasonally modulated tropical drought induced by volcanic aerosol. J Clim 24:2045–2060

Jouzel J (2013) A brief history of ice core science over the last 50 yr. Clim Past 9:2525–2547

Juckes MN, Allen MR, Briffa KR et al (2007) Millennial temperature reconstruction intercomparison and evaluation. Clim Past 3:591–609

Jungclaus JH, Koenigk T (2010) Low-frequency variability of the arctic climate: The role of oceanic and atmospheric heat transport variations. Clim Dyn 34:265–279

Kalnay E, Kanamitsu M, Kistler R et al (1996) The NCEP/NCAR 40-year reanalysis project. Bull Am Meteorol Soc 77:437–471

Kanamitsu M, Ebisuzaki W, Woollen J et al (2002) NCEP–DOE AMIP-II reanalysis (R-2). Bull Am Meteorol Soc 83:1631–1643

Kang SM, Polvani LM, Fyfe JC, Sigmond M (2011) Impact of polar ozone depletion on subtropical precipitation. Science 332:951–954

Kang S, Yang B, Qin C et al (2013) Extreme drought events in the years 1877–1878, and 1928, in the southeast Qilian Mountains and the air–sea coupling system. Quat Int 283:85–92

Karl TR, Arguez A, Huang B, Lawrimore JH, McMahon JR, Menne MJ, Peterson TC, Vose RS, Zhang H-M (2015) Possible artifacts of data biases in the recent global surface warming hiatus. Science 348:1469–1472

Karoly DJ (1989) Southern Hemisphere circulation features associated with El Niño-Southern Oscillation events. J Clim 2:1239–1252

Katz RW, Brown BG (1992) Extreme events in a changing climate: variability is more important than averages. Clim Change 21:289–302

Kawatani Y, Hamilton K (2013) Weakened stratospheric Quasibiennial Oscillation driven by increased tropical mean upwelling. Nature 497:478–481

Keenlyside NS, Latif M (2007) Understanding equatorial Atlantic interannual variability. J Clim 20:131–142

Kellogg WW (1987) Mankind's impact on climate: The evolution of an awareness. Clim Change 10:113–136

Kennedy JJ, Rayner NA, Smith RO, Parker DE, Saunby M (2011) Reassessing biases and other uncertainties in sea surface temperature observations measured in situ since 1850: 1. Measurement and sampling uncertainties. J Geophys Res 116:D14103

Kidston J, Scaife AA, Hardiman SC, Mitchell DM, Burchart N, Baldwin MP, Gray LJ (2015) Stratospheric influence on tropospheric jet streams, storm tracks and surface weather. Nat Geosci 8:433–440

Kirkby J, Curtius J, Almeida J et al (2011) Role of sulphuric acid, ammonia and galactic cosmic rays in atmospheric aerosol nucleation. Nature 476:429–433

Kistler R, Collins W, Saha S et al (2001) The NCEP–NCAR 50–year reanalysis: Monthly means CD–ROM and documentation. Bull Am Meteorol Soc 82:247–267

Klingaman WK, Klingaman NP (2013) The year without summer: 1816 and the Volcano that Darkened the World and Changed History. St. Martin's Press, New York, 338 pp

Knight JR, Allan RJ, Folland CK, Vellinga M, Mann ME (2005) A signature of persistent natural thermohaline circulation cycles in observed climate. Geophys Res Lett 32:L20708

Knight JR, Folland CK, Scaife AA (2006) Climate impacts of the Atlantic Multidecadal Oscillation. Geophys Res Lett 33:L17706

Kodera K, Kuroda Y (2002) Dynamical response to the solar cycle. J Geophys Res 107:4749

Komhyr WD, Barrett EW, Slocum G, Weickmann HK (1971) Physical sciences: atmospheric total ozone increase during the 1960s. Nature 232:390–391

Konare A, Zakey AS, Solmon F et al (2008) A regional climate modeling study of the effect of desert dust on the West African monsoon. J Geophys Res 113,:D12206

Kopp G, Lean JL (2011) A new, lower value of total solar irradiance: evidence and climate significance. Geophys Res Lett 38:L01706

Köppen W (1881) Über mehrjährige Perioden der Witterung–III. Mehrjährige Änderungen der Temperatur 1841 bis 1875 in den Tropen der nördlichen und südlichen gemässigten Zone, an den Jahresmitteln untersucht. Zeitschrift der Österreichischen Gesellschaft für Meteorologie Bd XVI:141–150

Kosaka Y, Xie SP (2013) Recent global-warming hiatus tied to equatorial Pacific surface cooling. Nature 501:403–407

Koster RD, Dirmeyer PA, Guo Z et al (2004) Regions of strong coupling between soil moisture and precipitation. Science 305:1138–1140

Krämer D (2015) Menschen grasten nun mit dem Vieh. Die letzte grosse Hungerkrise der Schweiz 1816/17. Schwabe, Basel, 527 pp

Kripalani RH, Kulkarni A (1997) Climatic impact of El Niño/La Niña on the Indian monsoon: a new perspective. Weather 52:39–46

Krivova NA, Vieira LEA, Solanki SK (2010) Reconstruction of solar spectral irradiance since the Maunder minimum. J Geophys Res 115:A12112

Krüger T (2013) Discovering the ice ages: international reception and consequences for a historical understanding of climate. Brill, Leiden

Krüger K, Toohey M (2015) Did the Tambora eruption influence the Southern Annular Mode? In: Volcanoes, climate and society, 7–10 April, Bern, Switzerland

Kuglitsch FG, Toreti A, Xoplaki E et al (2010) Heat wave changes in the eastern Mediterranean since 1960. Geophys Res Lett 37:L04802

Kuglitsch FG, Auchmann R, Bleisch R et al (2012) Break detection of annual Swiss temperature series. J Geophys Res 117:D13105

Kumar KK, Rajagopalan B, Cane MA (1999) On the weakening relationship between the Indian Monsoon and ENSO. Science 284:2156–2159

Kumar KK, Rajagopalan B, Hoerling M, Bates G, Cane M (2006) Unraveling the mystery of Indian Monsoon failure during El Niño. Science 314:115–119

Labitzke K, van Loon H (1999) The stratosphere: phenomena, history, and relevance. Springer, Berlin/New York

Labitzke K, Kunze M, Brönnimann S (2006) Sunspots, the QBO and the stratosphere in the North Polar Region–20 years later. Meteorol Z 15:355–363

Laepple T, Werner M, Lohmann G (2011) Synchronicity of antarctic temperatures and local solar insolation on orbital timescales. Nature 471:91–94

Lamb H (1965) The early medieval warm epoch and its sequel. Palaeogeogr Palaeoclimatol Palaeoecol 1:13–37

Lamb HH (1969) The new look of climatology. Nature 223:1209–1215

Lamb HH (1972) Climate: present, past and future. Fundamentals and climate now, vol 1. Methuen, London

Lamb HH (1977) Climate: present, past and future. Climatic history and the future, vol 2. Methuen, London

Langematz U, Claussnitzer A, Matthes K, Kunze M (2005) The climate during the Maunder Minimum: a simulation with the Freie Universität Berlin climate middle atmosphere model (FUB-CMAM). J Atmos Sol-Terr Phys 67:55–69

Laprise R (1992) The resolution of global spectral models. Bull Am Meteorol Soc 73:1453–1454

Larkin NK, Harrison D (2005) Global seasonal temperature and precipitation anomalies during El Niño autumn and winter. Geophys Res Lett 32:L16705

Lau WKM, Kim KM (2012) The 2010 Pakistan flood and Russian heat wave: teleconnection of hydrometeorological extremes. J Hydrometeorol 13:392–403

Lauwaet D, Lipzig NPM, Ridder K (2009) The effect of vegetation changes on precipitation and mesoscale convective systems in the Sahel. Clim Dyn 33:521–534

Lavigne F, Degeai JP, Komorowski JC et al (2013) Source of the great AD 1257 mystery eruption unveiled, Samalas volcano, Rinjani Volcanic Complex, Indonesia. Proc Natl Acad Sci USA 110:16742–16747

Le Roy Ladurie E (1971) Times of feast, times of famine: a history of climate since the year 1000. Doubleday, New York

Le Treut H, Somerville R, Cubasch U et al (2007) Historical overview of climate change. In: Solomon S, Qin D, Manning M et al (eds) Climate change 2007: The physical science basis. Contribution of Working Group I to the Fifth Assessment Report of the Intergovernmental Panel on Climate Change. Cambridge University Press, Cambridge/New York

Lean J (2000) Evolution of the Sun's spectral irradiance since the Maunder minimum. Geophys Res Lett 27:2425–2428

Lean JL, Rind DH (2008) How natural and anthropogenic influences alter global and regional surface temperatures: 1889 to 2006. Geophys Res Lett 35:L18701

Lean JL, Rind DH (2009) How will Earth's surface temperature change in future decades? Geophys Res Lett 36:L15708

Leclercq PW, Oerlemans J (2012) Global and hemispheric temperature reconstruction from glacier length fluctuations. Clim Dyn 38:1065–1079

Lehmann PN (2013) The threat of the desert: debates on climate change in the late 19th century. Presented at the conference climate and beyond, 24 Jan 2013, Bern

Lehner F, Raible CC, Stocker TF (2012) Testing the robustness of a precipitation proxy-based North Atlantic Oscillation reconstruction. Quat Sci Rev 45:85–94

Lehner F, Born A, Raible CC, Stocker TF (2013) Amplified inception of European Little Ice Age by sea ice-ocean-atmosphere feedbacks. J Clim 26:7586–7602

Lejenäs H (1989) The severe winter in Europe 1941–42: the large-scale circulation, cut-off lows, and blocking. Bull Am Meteorol Soc 70:271–281

Lelieveld J, Crutzen P, Ramanathan V et al (2001) The Indian Ocean experiment: widespread air pollution from South and Southeast Asia. Science 291:1031–1036

Lenton A, Codron F, Bopp L et al (2009) Stratospheric ozone depletion reduces ocean carbon uptake and enhances ocean acidification. Geophys Res Lett 36:L12606

Lenton TM (2012) Arctic climate tipping points. Ambio 41:10–22

Li J, Xie SP, Cook ER et al (2011) Interdecadal modulation of El Niño amplitude during the past millennium. Nat Clim Change 1:114–118

Li J, Xie SP, Cook ER et al (2013) El Niño modulations over the past seven centuries. Nat Clim Change 3:822–826

Lieberman V (2003) Strange parallels: Southeast Asia in global context, c. 800–1830. Integration on the Mainland. Cambridge University Press, New York/Cambridge

Liebmann B, Dole RM, Jones C, Bladé I, Allured D (2010) Influence of choice of time period on global surface temperature trend estimates. Bull Am Meteorol Soc 91:1485–1491

Linkin ME, Nigam S (2008) The North Pacific Oscillation-West Pacific teleconnection pattern: mature-phase structure and winter impacts. J Clim 21:1979–1997

Ljungqvist FC (2010) A new reconstruction of temperature variability in the extra-tropical Northern Hemisphere during the last two millennia. Geogr Ann A 92:339–351

Loader NJ, Jalkanen R, Mccarroll D, Moberg A (2011) Spring temperature variability in Northern Fennoscandia AD 1693–2011. J Quat Sci 26:566–570

Lockwood M (2012) Solar influence on global and regional climates. Surv Geophys 33:503–534

van Loon H, Meehl GA (2008) The response in the Pacific to the sun's decadal peaks and contrasts to cold events in the Southern Oscillation. J Atmos Sol-Terr Phys 70:1046–1055

Lorenz EN (1963) Deterministic nonperiodic flow. J Atmos Sci 20:130–141

Lorrey A, Fauchereau N, Stanton C et al (2013) The Little Ice Age climate of New Zealand reconstructed from Southern Alps cirque glaciers: a synoptic type approach. Clim Dyn 42:3039–3060

Lovelock JE, Margulis L (1974) Atmospheric homeostasis by and for the biosphere: the Gaia hypothesis. Tellus 26:2–10

Luterbacher J, Pfister C (2015) The year without a summer. Nat Geosci 8:246–248

Luterbacher J, Rickli R, Xoplaki E et al (2001) The late Maunder minimum (1675–1715)–a key period forstudying decadal scale climatic change in Europe. Clim Change 49:441–462

Luterbacher J, Xoplaki E, Dietrich D et al (2002) Reconstruction of sea level pressure fields over the Eastern North Atlantic and Europe back to 1500. Clim Dyn 18:545–561

Luterbacher J, Dietrich D, Xoplaki E, Grosjean M, Wanner H (2004) European seasonal and annual temperature variability, trends, and extremes since 1500. Science 303:1499–1503

Lutz HJ (1956) Ecological effects of forest fires in the interior of Alaska. US Department of Agriculture

MacDonald GM, Case RA (2005) Variations in the Pacific Decadal Oscillation over the past millennium. Geophys Res Lett 32:L08703

Maher N, McGregor S, England MH, Sen Gupta A (2015) Effects of volcanism on tropical variability. Geophys Res Lett 42. doi:10.1002/2015GL064751

McCarthy GD, Haigh ID, Hirschi JJ-M, Grist JP, Smeed DA (2015) Ocean impact on decadal Atlantic climate variability revealed by sea-level observations. Nature 521:508–510

McGregor S, Timmermann A (2011) The effect of explosive tropical volcanism on ENSO. J Clim 34:2178–2190

Manabe S, Stouffer RJ (1980) Sensitivity of a global climate model to an increase of CO_2 concentration in the atmosphere. J Geophys Res 85:5529–5554

Manabe S, Wetherald RT (1967) Thermal equilibrium of the atmosphere with a given distribution of relative humidity. J Atmos Sci 24:241–259

Manatsa D, Morioka Y, Behera SK, Yamagata T, Matarira CH (2013) Link between Antarctic ozone depletion and summer warming over Southern Africa. Nat Geosci 6:934–939

Manley G (1974) Central England temperatures: monthly means 1659 to 1973. Q J R Meteorol Soc 100:389–405

Mann A (1790) Ueber die allmählichen Veränderungen der Temperatur und des Bodens in verschiedenen Climaten, nebst Untersuchungen über die Ursachen dieser Veränderungen. Historia et Commentationes Academiae Theodoro-Palatinat 6 Physicum Mannheimii:82–111 (reprinted in abbreviated form in J Phys (Grens's J) 2: 231–244, 1790)

Mann ME (2002) Little Ice Age. In: Munn RE (ed) Encyclopedia of global environmental change. The earth system: physical and chemical dimensions of global environmental change, vol 1. Wiley, Chichester/New York, pp 504–509

Mann ME, Bradley RS, Hughes MK (1999) Northern hemisphere temperatures during the past millennium: inferences, uncertainties, and limitations. Geophys Res Lett 26:759–762

Mann ME, Cane MA, Zebiak SE, Clement A (2005) Volcanic and solar forcing of the tropical Pacific over the past 1000 years. J Clim 18:447–456

Mann ME, Zhang Z, Hughes MK et al (2008) Proxy-based reconstructions of hemispheric and global surface temperature variations over the past two millennia. Proc Natl Acad Sci USA 105:13252–13257

Mann ME, Woodruff JD, Donnelly JP, Zhang Z (2009a) Atlantic hurricanes and climate over the past 1,500 years. Nature 460:880–883

Mann ME, Zhang Z, Rutherford S et al (2009b) Global signatures and dynamical origins of the Little Ice Age and Medieval Climate Anomaly. Science 326:1256–1260

Manney GL, Santee ML, Rex M et al (2011) Unprecedented Arctic ozone loss in 2011. Nature 478:469–475

Mantua NJ, Hare SR (2002) The Pacific Decadal Oscillation. J Oceanogr 58:35–44

Mantua NJ, Hare SR, Zhang Y, Wallace JM, Francis RC (1997) A Pacific interdecadal climate oscillation with impacts on salmon production. Bull Am Meteorol Soc 78:1069–1079

Manzini E, Giorgetta M, Esch M, Kornblueh L, Roeckner E (2006) The influence of sea surface temperatures on the Northern winter stratosphere: ensemble simulations with the MAECHAM5 model. J Clim 19:3863–3881

Marlon JR, Bartlein PJ, Carcaillet C et al (2008) Climate and human influences on global biomass burning over the past two millennia. Nat Geosci 1:697–702

Marshall GJ (2003) Trends in the Southern annular mode from observations and reanalyses. J Clim 16:4134–4143

Martin-Puertas C, Matthes K, Brauer A et al (2012) Regional atmospheric circulation shifts induced by a grand solar minimum. Nat Geosci 5:397–401

Masson-Delmotte V, Schulz M, Abe-Ouchi A et al (2013) Information from paleoclimate archives. In: Stocker T, Qin D, Plattner GK et al (eds) Climate change 2013: The physical science basis. Contribution of Working Group I to the Fifth Assessment Report of the Intergovernmental Panel on Climate Change. Cambridge University Press, Cambridge/New York

Matulla C, Schöner W, Alexandersson H, von Storch H, Wang XL (2008) European storminess: late nineteenth century to present. Clim Dyn 31:125–130
Mauelshagen F (2010) Klimageschichte der Neuzeit. WBG (Wissenschaftliche Buchgesellschaft)
Mayer M, Haimberger L, Balmaseda MA (2014) On the energy exchange between tropical ocean basins related to ENSO. J Clim 27:6393–6403
McCabe GJ, Palecki MA, Betancourt JL (2004) Pacific and Atlantic Ocean influences on multidecadal drought frequency in the United States. Proc Natl Acad Sci USA 101:4136–4141
McGregor S, Timmermann A, Timm O (2010) A unified proxy for ENSO and PDO variability since 1650. Clim Past 6:1–17
McKee TB, Doesken NJ, Kleist J (1993) The relationship of drought frequency and duration to time scales. In: Proceedings of the 8th conference on applied climatology, vol 17. American Meteorological Society, Boston, pp 179–183
McVicar TR, Roderick ML, Donohue RJ et al (2012) Global review and synthesis of trends in observed terrestrial near-surface wind speeds: implications for evaporation. J Hydrol 416:182–205
Meadows DH, Meadows DL, Randers J, Behrens III WW (1972) Limits to Growth. Universe Books, New York, 205 pp
Meehl GA, Arblaster JM, Matthes K, Sassi F, van Loon H (2009) Amplifying the Pacific climate system response to a small 11-year solar cycle forcing. Science 325:1114–1118
Meehl GA, Arblaster JM, Fasullo JT, Hu A, Trenberth KE (2011) Model-based evidence of deep-ocean heat uptake during surface-temperature hiatus periods. Nat Clim Change 1:360–364
Meier N, Rutishauser T, Pfister C, Wanner H, Luterbacher J (2007) Grape harvest dates as a proxy for Swiss April to August temperature reconstructions back to AD 1480. Geophys Res Lett 34:L20705
Meinardus W (1898) Der Zusammenhang des Winterklimas in Mittel- und Nordwest-Europa mit dem Golfstrom. Z d Ges f Erdkunde in Berlin 23:183–200
Meldrum C (1872) On a periodicity in the frequency of cyclones in the Indian Ocean south of the equator. Nature 6:357–358
Miller AJ, Cayan DR, Barnett TP, Graham NE, Oberhuber JM (1994) The 1976–77 climate shift of the Pacific Ocean. Oceanography 7:21–26
Miller GH, Geirsdottir A, Zhong Y et al (2012) Abrupt onset of the Little Ice Age triggered by volcanism and sustained by sea-ice/ocean feedbacks. Geophys Res Lett 39:L02708
Miralles DG, Teuling AJ, van Heerwaarden CC, de Arellano JVG (2014) Mega-heatwave temperatures due to combined soil desiccation and atmospheric heat accumulation. Nat Geosci 7:345–349
Mo KC, Paegle JN (2001) The Pacific–South American modes and their downstream effects. Int J Clim 21:1211–1229
Moberg A, Alexandersson H, Bergström H, Jones PD (2003) Were Southern Swedish summer temperatures before 1860 as warm as measured? Int J Clim 23:1495–1521
Moberg A, Sonechkin DM, Holmgren K, Datsenko NM, Karlen W (2005) Highly variable Northern Hemisphere temperatures reconstructed from low- and high-resolution proxy data. Nature 433:613–617
Moffa-Sánchez P, Born A, Hall IR, Thornalley DJ, Barker S (2014) Solar forcing of North Atlantic surface temperature and salinity over the past millennium. Nat Geosci 7:275–278
Montzka SA, Krol M, Dlugokencky E et al (2011) Small interannual variability of global atmospheric hydroxyl. Science 331:67–69
Moore GWK, Holdsworth G, Alverson K (2002) Climate change in the North Pacific region over the past three centuries. Nature 420:401–403
Morice CP, Kennedy JJ, Rayner NA, Jones PD (2012) Quantifying uncertainties in global and regional temperature change using an ensemble of observational estimates: the HadCRUT4 data set. J Geophys Res 117:D08101
Müller R (2009) A brief history of stratospheric ozone research. Meteorol Z 18:3–24
Müller P (2010) Constructing climate knowledge with computer models. WIREs Clim Change 1:565–580

Muthers S, Anet JG, Raible CC et al (2014a) Northern hemispheric winter warming pattern after tropical volcanic eruptions: sensitivity to the ozone climatology. J Geophys Res 119:1340–1355

Muthers S, Anet JG, Stenke A et al (2014b) The coupled atmosphere-chemistry-ocean model SOCOL-MPIOM. Geosci Model Dev 7:2157–2179

Myhre G, Shindell D, Bréon F et al (2013) Anthropogenic and natural radiative forcing. In: Stocker T, Qin D, Plattner GK et al (eds) Climate change 2013: The physical science basis. Contribution of Working Group I to the Fifth Assessment Report of the Intergovernmental Panel on Climate Change. Cambridge University Press, Cambridge/New York

Namias J (1950) The index cycle and its role in the general circulation. J Meteorol 7:130–139

Namias J (1982) Anatomy of great plains protracted heat waves (especially the 1980 US summer drought). Mon Weather Rev 110:824–838

Neely RR, Toon OB, Solomon S et al (2013) Recent anthropogenic increases in SO_2 from Asia have minimal impact on stratospheric aerosol. Geophys Res Lett 40:999–1004

Neu U (2008) Is recent major hurricane activity normal? Nature 451:E5–E5

Neukom R, Luterbacher J, Villalba R et al (2010) Multi-centennial summer and winter precipitation variability in Southern South America. Geophys Res Lett 37:L14708

Neukom R, Luterbacher J, Villalba R et al (2011) Multiproxy summer and winter surface air temperature field reconstructions for Southern South America covering the past centuries. Clim Dyn 37:35–51

Neukom R, Gergis J, Karoly DJ et al (2014a) Inter-hemispheric temperature variability over the past millennium. Nat Clim Change 4:362–367

Neukom R, Nash DJ, Endfield GH et al (2014b) Multi-proxy summer and winter precipitation reconstruction for Southern Africa over the last 200 years. Clim Dyn 42:2713–2726

Newman PA, Nash ER, Rosenfield JE (2001) What controls the temperature of the Arctic stratosphere during the spring? J Geophys Res 106:19999–20010

Nicholls N (2008) Recent trends in the seasonal and temporal behaviour of the El Niño–Southern oscillation. Geophys Res Lett 35:L19703

Nicholson S (2000) Land surface processes and Sahel climate. Rev Geophys 38:117–139

Nicholson SE (2001) Climatic and environmental change in Africa during the last two centuries. Clim Res 17:123–144

Nicolussi K, Patzelt G (2001) Untersuchungen zur holozänen Gletscherentwicklung von Pasterze und Gepatschferner (Ostalpen). Z Gletscherk Glazialgeol 36:1–87

Nobre P, Shukla J (1996) Variations of sea surface temperature, wind stress, and rainfall over the tropical Atlantic and South America. J Clim 9:2464–2479

Nordli Ø, Przybylak R, Ogilvie AE, Isaksen K (2014) Long-term temperature trends and variability on Spitsbergen: The extended Svalbard Airport temperature series, 1898–2012. Polar Res 33:21349

North GR, Wu Q, Stevens MJ (2004) Detecting the 11-year solar cycle in the surface temperature field. Geophys Monogr Ser 141:251–259

Norton-Griffiths M, Rydén P (eds) (1989) The IUCN Sahel studies, vol 1. IUCN Regional Office for Eastern Africa, Nairobi

Nussbaumer SU, Deline P, Vincent C, Zumbühl HJ (2012) Mer de Glace. Art & science. Atelier ésope, Chamonix

Nyberg J, Malmgren BA, Winter A et al (2007) Low Atlantic hurricane activity in the 1970s and 1980s compared to the past 270 years. Nature 447:698–701

Oerlemans J (2005) Extracting a climate signal from 169 glacier records. Science 308:675–677

Oglesby RJ (1991) Springtime soil moisture, natural climatic variability, and North American drought as simulated by the NCAR community climate model 1. J Clim 4:890–897

Ohmura A, Lang H (1989) Secular variation of global radiation in Europe. In: Lenoble J, Geleyn JF (eds) IRS'88: Current problems in atmospheric radiation, A. Deepak Publishing, Hampton, pp 298–301

Ohvril H, Teral H, Neiman L et al (2009) Global dimming and brightening versus atmospheric column transparency, Europe, 1906–2007. J Geophys Res 114:D00D12 van Oldenborgh GJ, de Laat ATJ, Luterbacher J, Ingram WJ, Osborn TJ (2013) Claim of solar influence is on thin ice: are 11-year cycle solar minima associated with severe winters in Europe? Environ Res Lett 8:024014

Olivier S, Schwikowski M, Brütsch S et al (2003) Glaciochemical investigation of an ice core from Belukha glacier, Siberian Altai. Geophys Res Lett 30:2019

Onogi K, Tsutsui J, Koide H et al (2007) The JRA-25 reanalysis. J Meteorol Soc Jpn 85:369–432

Oort AH, Yienger JJ (1996) Observed interannual variability in the Hadley circulation and its connection to ENSO. J Clim 9:2751–2767

Oppenheimer C (2003) Climatic, environmental and human consequences of the largest known historic eruption: Tambora volcano (Indonesia) 1815. Prog Phys Geogr 27:230–259

Oppenheimer C (2015) Eruption politics. Nat Geosci 8:244–245

Ori A, Bischoff T, Schneider T (2015) Seasonal and interannual variations of the energy flux equator and ITCZ. Part I: zonally averaged ITCZ position. J Clim (submitted)

Ortega P, Lehner F, Swingedouw D, Masson-Delmotte V, Raible CC, Casado M, Yiou P (2015) A model-tested North Atlantic oscillation reconstruction for the past millennium. Nature 523:71–74

Osborn TJ, Briffa KR (2004) The real color of climate change. Science 306:621–622

Overland JE, Turet P (1994) Variability of the atmospheric energy flux across 70 N computed from the GFDL data set. In: Johannessen OM, Muench RD, Overland JE (eds) The polar oceans and their role in shaping the global environment. Geophysical Monograph Series, vol 85. AGU, Washington, DC, pp 313–325

Overland JE, Francis JA, Hanna E, Wang M (2012) The recent shift in early summer Arctic atmospheric circulation. Geophys Res Lett 39:L19804

Overpeck J, Hughen K, Hardy D et al (1997) Arctic environmental change of the last four centuries. Science 278:1251–1256

Paeth H (2007) Human activity and climate change in Africa. Adv Glob Change Res 33:209–220

Päivärinta SM, Seppälä A, Andersson ME et al (2013) Observed effects of solar proton events and sudden stratospheric warmings on odd nitrogen and ozone in the polar middle atmosphere. J Geophys Res 118:6837–6848

Pall P, Aina T, Stone DA et al (2011) Anthropogenic greenhouse gas contribution to flood risk in England and Wales in autumn 2000. Nature 470:382–385

Paret O (1947) Droht Europa eine Trockenheit? Stuttgarter Rundschau

Parker DE, Legg T, Folland CK (1992) A new daily central England temperature series, 1772–1991. Int J Clim 12:317–342

Pauling A, Luterbacher J, Casty C, Wanner H (2006) Five hundred years of gridded high-resolution precipitation reconstructions over Europe and the connection to large-scale circulation. Clim Dyn 26:387–405

Penck A, Brückner E (1909) Die Alpen im Eiszeitalter, vol 3. Tauchnitz Leipzig

Pepin N, Bradley RS, Diaz HF, Baraer M, Caceres EB, Forsythe N, Fowler H, Greenwood G, Hashmi MZ, Liu XD, Miller JR, Ning L, Ohmura A, Palazzi E ,Rangwala I, Schöner W, Severskiy I, Shahgedanova M, Wang MB, Williamson SN, Yang DQ (2015) Elevation-dependent warming in mountain regions of the world. Nat Clim Change 5:424–430

Persson A, Langen PL, Ditlevsen P, Vinther BM (2011) The influence of precipitation weighting on interannual variability of stable water isotopes in Greenland. J Geophys Res 116:D20120

Petoukhov V, Semenov VA (2010) A link between reduced Barents-Kara sea ice and cold winter extremes over northern continents. J Geophys Res 115:D21111

Pettersson O (1914) Climatic variations in historic and prehistoric time. Ur Svenska Hydrografisk-Biologiska Kommisionens Skrifter 5, 26 pp

Pfahl S, Sirocko F, Seelos K et al (2009) A new windstorm proxy from lake sediments: a comparison of geological and meteorological data from Western Germany for the period 1965–2001. J Geophys Res 114:D18106

Pfister C (1975) Agrarkonjunktur und Witterungsverlauf im westlichen Schweizer Mittelland 1755–1797. Ein Beitrag zur Umwelt- und Wirtschaftsgeschichte des 18. Jahrhunderts. Lang, Bern

Pfister C (1982) An analysis of the Little Ice Age climate in Switzerland and its consequences for agricultural production. In: Wigley TML, Ingram MJ, Farmer G (eds.): Climate and history. Studies in past climates and their impact on man. Cambridge University Press, New York, pp 297–323
Pfister C (1995) Das 1950er Syndrom: Der Weg in die Konsumgesellschaft. Haupt, Bern
Pfister C (1999) Wetternachhersage: 500 Jahre Klimavariationen und Naturkatastrophen (1496–1995). Haupt, Bern
Pfister C (2009) Die "Katastrophenlüke" des 20. Jahrhunderts und der Verlust traditionalen Risikobewusstseins. Gaia 18:239–246
Pfister C, Brändli D (1999) Rodungen im Gerbirge - Überschwemmungen im Vorland: Ein Deutungsmuster macht Karriere. In: Sieferle RP, Breuninger H (eds) Natur-Bilder. Wahrnehmungen von Natur und Umwelt in der Geschichte, Campus Verlag, Frankfurt/Main and New York, pp 297–323
Pictet MA (1818) Résumé des observations météorologiques faites ă l'Observatoire de Paris en 1817. Bibliothèque Universelle, Sciences et Arts 3:83–95
Pithan F, Mauritsen T (2014) Arctic amplification dominated by temperature feedbacks in contemporary climate models. Nat Geosci 7:181–184
Plass GN (1956) The carbon dioxide theory of climatic change. Tellus 8:140–154
Poli P, Hersbach H, Tan D, Dee D, Thépaut J-N, Simmons A, Peubey C, Laloyaux P, Komori T, Berrisford P, Dragani R, Trémolet Y, Hólm E, Bonavita M, Isaksen L, Fisher M (2013) The data assimilation system and initial performance evaluation of the ECMWF pilot reanalysis of the 20th-century assimilating surface observations only (ERA-20C). ERA report series, 14, ECMWF, UK
Polvani L (2010) The stratosphere: dynamics, transport, and chemistry. Geophysical Monograph series, vol 190. AGU, Washington, DC
Polyakov IV, Alekseev GV, Bekryaev RV et al (2003) Long-term ice variability in Arctic marginal seas. J Clim 16:2078–2085
Polyakov IV, Alekseev GV, Timokhov LA et al (2004) Variability of the intermediate Atlantic water of the Arctic Ocean over the last 100 years. J Clim 17:4485–4497
Polyakov IV, Bhatt US, Walsh JE et al (2013) Recent oceanic changes in the Arctic in the context of long-term observations. Ecol Appl 23:1745–1764
Pongratz J, Reick C, Raddatz T, Claussen M (2008) A reconstruction of global agricultural areas and land cover for the last millennium. Glob Biogeochem Cycles 22:GB3018
Power SB, Kociuba G (2011) What caused the observed twentieth-century weakening of the Walker circulation? J Clim 24:6501–6514
Rahmstorf S, Box JE, Feulner G, Mann ME, Robinson A, Rutherford S, Schaffernicht EJ (2015) Exceptional twentieth-century slowdown in Atlantic ocean overturning circulation. Nat Clim Change 5:475–480
Rajagopalan B, Molnar P (2013) Signatures of Tibetan Plateau heating on Indian summer monsoon rainfall variability. J Geophys Res 118:1170–1178
Ramanathan V, Crutzen P (2003) New directions: atmospheric brown "clouds". Atmos Environ 37:4033–4035
Ramanathan V, Ramana MV, Roberts G et al (2007) Warming trends in Asia amplified by brown cloud solar absorption. Nature 448:575–578
Ramella Pralungo L, Haimberger L (2014) A "Global Radiosonde and tracked-balloon Archive on Sixteen Pressure levels" (GRASP) going back to 1905 – Part 2: homogeneity adjustments for pilot balloon and radiosonde wind data. Earth Syst Sci Data 6:297–316
Ramirez E, Hoffmann G, Taupin J et al (2003) A new Andean deep ice core from Nevado Illimani (6350 m), Bolivia. Earth Planet Sci Lett 212:337–350
Rasmusson EM (1967) Atmospheric water vapor transport and the water balance of North America: Part I. Characteristics of the water vapor flux field. Mon Weather Rev 95:403–426
Rasmusson EM, Wallace JM (1983) Meteorological aspects of the El Niño/southern oscillation. Science 222:1195–1202

Rasool SI, Schneider SH (1971) Atmospheric carbon dioxide and aerosols: effects of large increases on global climate. Science 173:138–141

Ravishankara AR, Daniel JS, Portmann RW (2009) Nitrous oxide (N_2O): the dominant ozone-depleting substance emitted in the 21st century. Science 326:123–125

Rayner NA, Brohan P, Parker DE et al (2006) Improved analyses of changes and uncertainties in sea surface temperature measured in situ since the mid-nineteenth century: the HadSST2 dataset. J Clim 19:446–469

Reed RJ, Campbell WJ, Rasmussen LA, Rogers DG (1961) Evidence of a downward-propagating, annual wind reversal in the equatorial stratosphere. J Geophys Res 66:813–818

Revelle R, Suess HE (1957) Carbon dioxide exchange between atmosphere and ocean and the question of an increase of atmospheric CO_2 during the past decades. Tellus 9:18–27

Richardson LF (1922) Weather prediction by numerical processes. Cambridge University Press, London, 262 pp

Richter I, Behera SK, Masumoto Y et al (2013) Multiple causes of interannual sea surface temperature variability in the equatorial Atlantic Ocean. Nat Geosci 6:43–47

Ridley JK, Lowe JA, Hewitt HT (2012) How reversible is sea ice loss? The Cryosphere 6:193–198

Ridley HE, Asmerom Y, Baldini JUL, Breitenbach SFM, Aquino VV, Prufer KM, Culleton BJ, Polyak V, Lechleitner FA, Kennett DJ, Zhang M, Marwan N, Macpherson CG, Baldini LM, Xiao T, Peterkin JL, Awe J, Gerald H, Haug GH (2015) Aerosol forcing of the position of the intertropical convergence zone since ad 1550. Nat Geosci 8:195–200

Rienecker MM, Suarez MJ, Gelaro R et al (2011) MERRA: NASA's modern-era retrospective analysis for research and applications. J Clim 24:3624–3648

Risbey JS, Lewandowsky S, Langlais C et al (2014) Well-estimated global surface warming in climate projections selected for ENSO phase. Nat Clim Change 4:835–840

Robock A (2000) Volcanic eruptions and climate. Rev Geophys 38:191–219

Rodwell MJ, Hoskins BJ (1996) Monsoons and the dynamics of deserts. Q J R Meteorol Soc 122:1385–1404

Rodwell MJ, Hoskins BJ (2001) Subtropical anticyclones and summer monsoons. J Clim 14:3192–3211

Rodysill JR, Russell JM, Crausbay SD et al (2013) A severe drought during the last millennium in East Java, Indonesia. Quat Sci Rev 80:102–111

Rogers JC (1981) The North Pacific oscillation. J Climatol 1:39–57

Rogers JC (1985) Atmospheric circulation changes associated with the warming over the Northern North Atlantic in the 1920s. J Clim Appl Meteorol 24:1303–1310

Rogers JC, van Loon H (1979) The seasaw in winter temperature between Greenland and Northern Europe. Part II: some oceanic and atmospheric effects in middle and high latitudes. Mon Weather Rev 107:509–519

Rohde R, Muller R, Jacobson R et al (2013) Berkeley earth temperature averaging process. Geoinform Geostat: Overv 1:2

Rotstayn LD, Lohmann U (2002) Tropical rainfall trends and the indirect aerosol effect. J Clim 15:2103–2116

Ruchirawat M, Autrup H, Barregard L et al (2008) Impacts of atmospheric brown clouds on human health, Part III. In: Atmospheric brown clouds: regional assessment report with focus on Asia, United Nations Environment Programme, Nairobi

Ruddiman WF (2003) The anthropogenic greenhouse era began thousands of years ago. Clim Change 61:261–293

Russell A, McGregor GR, Marshall GJ (2006) 340 years of atmospheric circulation characteristics reconstructed from an Eastern Antarctic Peninsula ice core. Geophys Res Lett 33:L08702

Rutishauser T, Luterbacher J, Defila C, Frank D, Wanner H (2008) Swiss spring plant phenology 2007: Extremes, a multi-century perspective, and changes in temperature sensitivity. Geophys Res Lett 35:L05703

Sabeerali CT, Rao SA, Ajayamohan RS, Murtugudde R (2012) On the relationship between Indian summer monsoon withdrawal and Indo-Pacific SST anomalies before and after 1976/1977 climate shift. Clim Dyn 39:841–859

Sachs JP, Sachse D, Smittenberg RH et al (2009) Southward movement of the Pacific intertropical convergence zone AD 1400–1850. Nat Geosci 2:519–525
Saha S, Moorthi S, Pan HL et al (2010) The NCEP climate forecast system reanalysis. Bull Am Meteorol Soc 91:1015–1057
Santer BD, Bonfils C, Painter JF et al (2014) Volcanic contribution to decadal changes in tropospheric temperature. Nat Geosci 7:185–189
Sarmiento JL, Gruber N (2006) Ocean biogeochemical dynamics. Princeton University Press, Princeton
Sassi F, Kinnison D, Boville B, Garcia R, Roble R (2004) Effect of El Niño–Southern oscillation on the dynamical, thermal, and chemical structure of the middle atmosphere. J Geophys Res 109:D17108
Sato M, Hansen JE, McCormick MP, Pollack JB (1993) Stratospheric aerosol optical depths, 1850–1990. J Geophys Res 98:22987–22994
Saunders K, Hodgson DA, McMurtrie S, Grosjean M (2015) A diatom-conductivity transfer function for reconstructing past changes in the Southern Hemisphere westerly winds over the Southern ocean. J Q Sci 30:464–477
Scaife AA, Kucharski F, Folland CK et al (2009) The CLIVAR C20C project: selected twentieth century climate events. Clim Dyn 33:603–614
Schär C, Vidale PL, Lüthi D et al (2004) The role of increasing temperature variability in European summer heatwaves. Nature 427:332–336
Scherhag R (1939a) Die Erwärmung des Polargebiets. Annalen der Hydrographie und maritimen Meteorologie 67:57–67
Scherhag R (1939b) Die gegenwärtige Milderung der Winter und ihre Ursachen. Annalen der Hydrographie und maritimen Meteorologie 67:292–302
Scherhag R (1948) Neue Methoden der Wetteranalyse und Wetterprognose. Springer, Berlin
Scherhag R (1958) Das "Berliner Phänomen" und das Geophysikalische Jahr. Beilage zur Berliner Wetterkarte 137
Schlesinger ME, Ramankutty N (1994) An oscillation in the global climate system of period 65–70 years. Nature 367:723–726
Schmauss A (1948) Droht Europa eine Trockenheit. Allg Forstz 3:33–36
Schmocker-Fackel P, Naef F (2010a) Changes in flood frequencies in Switzerland since 1500. Hydrol Earth Syst Sci 14:1581–1594
Schmocker-Fackel P, Naef F (2010b) More frequent flooding? Changes in flood frequency in Switzerland since 1850. J Hydrol 381:1–8
Schneider T (2006) The general circulation of the atmosphere. Ann Rev Earth Planet Sci 34:655–688
Schneider T, O'Gorman PA, Levine XJ (2010) Water vapor and the dynamics of climate changes. Rev Geophys 48:RG3001
Schneider T, Bischoff T, Haug GH (2014a) Migrations and dynamics of the intertropical convergence zone. Nature 513:45–53
Schneider U, Becker A, Finger P, Meyer-Christoffer A, Ziese M, Bruno Rudolf B (2014b) GPCC's new land surface precipitation climatology based on quality-controlled in situ data and its role in quantifying the global water cycle. Theor Appl Climatol 115:15–40
Schraner M, Rozanov E, Schnadt Poberaj C et al (2008) Technical note: Chemistry-climate model SOCOL: version 2.0 with improved transport and chemistry/microphysics schemes. Atmos Chem Phys 8:5957–5974
van der Schrier G, Barkmeijer J (2005) Bjerknes' hypothesis on the coldness during ad 1790–1820 revisited. Clim Dyn 25:537–553
van der Schrier G, Barkmeijer J (2007) North American 1818–1824 drought and 1825–1840 pluvial and their possible relation to the atmospheric circulation. J Geophys Res 112:D13102
Schubert SD, Rood RB, Pfaendtner J (1993) An assimilated dataset for earth science applications. Bull Am Meteorol Soc 74:2331–2342
Schubert SD, Suarez MJ, Pegion PJ, Koster RD, Bacmeister JT (2004a) Causes of long-term drought in the US great plains. J Clim 17:485–503

Schubert SD, Suarez MJ, Pegion PJ, Koster RD, Bacmeister JT (2004b) On the cause of the 1930s Dust Bowl. Science 303:1855–1859
Schurer AP, Hegerl GC, Mann ME, Tett SF, Phipps SJ (2013) Separating forced from chaotic climate variability over the past millennium. J Clim 26:6954–6973
Schurer AP, Tett SF, Hegerl GC (2014) Small influence of solar variability on climate over the past millennium. Nat Geosci 7:104–108
Screen JA, Simmonds I (2010) The central role of diminishing sea ice in recent Arctic temperature amplification. Nature 464:1334–1337
Seager R (2007) The turn of the century North American drought: global context, dynamics, and past analogs. J Clim 20:5527–5552
Seager R, Burgman R (2011) Medieval hydroclimate revisited. PAGES Newsl 19:10–11
Seager R, Battisti DS, Yin J et al (2002) Is the Gulf stream responsible for Europe's mild winters? Q J R Meteorol Soc 128:2563–2586
Seager R, Kushnir Y, Ting M et al (2008) Would advance knowledge of 1930s SSTs have allowed prediction of the Dust Bowl drought? J Clim 21:3261–3281
Seaquist JW, Hickler T, Eklundh L, Ardö J, Heumann BW (2009) Disentangling the effects of climate and people on Sahel vegetation dynamics. Biogeosciences 6:469–477
Seidel DJ, Fu Q, Randel WJ, Reichler TJ (2008) Widening of the tropical belt in a changing climate. Nat Geosci 1:21–24
Semenov VA, Bengtsson L (2003) Modes of the wintertime Arctic temperature variability. Geophys Res Lett 30:1781
Seneviratne SI, Stöckli R (2008) The role of land-atmosphere interactions for climate variability in Europe. Adv Glob Change Res 33:179–193
Seneviratne SI, Lüthi D, Litschi M, Schar C (2006) Land-atmosphere coupling and climate change in Europe. Nature 443:205–209
Seneviratne SI, Corti T, Davin EL et al (2010) Investigating soil moisture–climate interactions in a changing climate: a review. Earth-Sci Rev 99:125–161
Serreze MC, Barry RG (2011) Processes and impacts of Arctic amplification: a research synthesis. Glob Plan Change 77:85–96
Serreze MC, Barrett AP, Stroeve JC, Kindig DN, Holland MM (2009) The emergence of surface-based Arctic amplification. The Cryosphere 3:11–19
Shanahan TM, Overpeck JT, Anchukaitis KJ et al (2009) Atlantic forcing of persistent drought in West Africa. Science 324:377–380
Shapiro AI, Schmutz W, Rozanov E et al (2011) A new approach to the long-term reconstruction of the solar irradiance leads to large historical solar forcing. Astron Astrophys 529:A67
Shaw WN (1905) The pulse of the atmospheric circulation. Nature 73:175–177
Sheets H (1974) Disaster in the desert: Failures of international relief in the West African drought. Carnegie endowment for international peace. Humanitarian policy studies. Special report, Carnegie Endowment for International Peace
Sheffield J, Andreadis KM, Wood EF, Lettenmaier DP (2009) Global and continental drought in the second half of the twentieth century: severity-area-duration analysis and temporal variability of large-scale events. J Clim 22:1962–1981
Shepherd TG (2014) Atmospheric circulation as a source of uncertainty in climate change projections. Nat Geosci 7:703–708
Shepherd TG, Plummer DA, Scinocca JF et al (2014) Reconciliation of halogen-induced ozone loss with the total-column ozone record. Nat Geosci 7:443–449
Shindell DT, Faluvegi G (2002) An exploration of ozone changes and their radiative forcing prior to the chlorofluorocarbon era. Atmos Chem Phys 2:363–374
Shindell DT, Schmidt GA, Mann ME, Rind D, Waple A (2001) Solar forcing of regional climate change during the Maunder minimum. Science 294:2149–2152
Shindell DT, Schmidt GA, Mann ME, Faluvegi G (2004) Dynamic winter climate response to large tropical volcanic eruptions since 1600. J Geophys Res 109:D05104
Siedler G, Griffies SM, Gould J, Church JA (2013) Ocean circulation and climate: a 21st century perspective. International Geophysics. Academic Press, Amsterdam, 868 pp

Siegenthaler U, Oeschger H (1978) Predicting future atmospheric carbon dioxide levels. Science 199:388–395

Sigl M, McConnell JR, Layman L, Maselli O, McGwire K, Pasteris D, Dahl-Jensen D, Steffensen JP, Edwards R, Mulvaney R (2013) A new bipolar ice core record of volcanism from WAIS divide and NEEM and implications for climate forcing of the last 2000? years. J Geophys Res 118:1151–1169

Sigl M, Winstrup M, McConnell JR, Welten KC, Plunkett G, Ludlow F, Büntgen U, Caffee M, Chellman N, Dahl-Jensen D, Fischer H, Kipfstuhl S, Kostick C, Maselli OJ, Mekhaldi F, Mulvaney R, Muscheler R, Pasteris DR, Pilcher JR, Salzer M, Schüpbach S, Steffensen JP, Vinther B, Woodruff TE (2015) Timing and climate forcing of volcanic eruptions for the past 2,500 years. Nature 523:543–549

Sigmond M, Scinocca JF, Kharin VV, Shepherd TG (2013) Enhanced seasonal forecast skill following stratospheric sudden warmings. Nat Geosci 6:98–102

Sime LC, Lang N, Thomas ER, Benton AK, Mulvaney R (2011) On high-resolution sampling of short ice cores: dating and temperature information recovery from Antarctic Peninsula virtual cores. J Geophys Res 116:D20117

Sirocko F, Brunck H, Pfahl S (2012) Solar influence on winter severity in central Europe. Geophys Res Lett 39:L16704

Sitnov SA (2009) Influence of the 11-year solar cycle on the effects of the equatorial quasi-biennial oscillation, manifesting in the extratropical Northern atmosphere. Clim Dyn 32:1–17

Skamarock WC, Klemp JB, Dudhia J et al (2005) A description of the advanced research WRF version 2. Technical report, DTIC Document

Slingo J, Sutton R (2007) Sea-ice decline due to more than warming alone. Nature 450:27–27

Smerdon JE (2012) Climate models as a test bed for climate reconstruction methods: pseudoproxy experiments. WIREs Clim Change 3:63–77

Smith SR, Bourassa MA, Long M (2011) Pirate attacks affect Indian Ocean climate research. EOS Trans Am Geophys Union 92:225–226

Soden BJ, Wetherald RT, Stenchikov GL, Robock A (2002) Global cooling after the eruption of Mount Pinatubo: A test of climate feedback by water vapor. Science 296:727–730

Solomon S (1999) Stratospheric ozone depletion: A review of concepts and history. Rev Geophys 37:275–316

Solomon S, Garcia RR, Rowland FS, Wuebbles DJ (1986) On the depletion of Antarctic ozone. Nature 321:755–758

Solomon S, Rosenlof KH, Portmann RW et al (2010) Contributions of stratospheric water vapor to decadal changes in the rate of global warming. Science 327:1219–1223

Solomon S, Daniel J, Neely R et al (2011) The persistently variable "background" stratospheric aerosol layer and global climate change. Science 333:866–870

Sonklar K (1858) Über den Zusammenhang der Gletscherschwankungen mit den meteorologischen Verhältnissen. Sitzber Akad Wiss Wien 32, 169–206

Staehelin J, Thudium J, Buehler R, Volz-Thomas A, Graber W (1994) Trend in surface ozone concentrations at Arosa (Switzerland). Atmos Environ 28:75–87

Stehr N, von Storch H (eds) (2000) Eduard Brückner – the sources and consequences of climate change and climate variability in historical times. Kluwer Academic, Dordrecht

Steig EJ, Schneider DP, Rutherford SD et al (2009) Warming of the Antarctic ice-sheet surface since the 1957 International Geophysical Year. Nature 457:459–462

Steig EJ, Ding Q, White JW et al (2013) Recent climate and ice-sheet changes in West Antarctica compared with the past 2,000 years. Nat Geosci 6:372–375

Steiner D, Pauling A, Nussbaumer SU et al (2008) Sensitivity of European glaciers to precipitation and temperature–two case studies. Clim Change 90:413–441

Steinhilber F, Beer J, Fröhlich C (2009) Total solar irradiance during the Holocene. Geophys Res Lett 36:L19704

Steinhilber F, Abreu JA, Beer J et al (2012) 9,400 years of cosmic radiation and solar activity from ice cores and tree rings. Proc Natl Acad Sci USA 109:5967–5971

Stenchikov G, Delworth TL, Ramaswamy V et al (2009) Volcanic signals in oceans. J Geophys Res 114:D16104

Stephens GL, Li J, Wild M et al (2012) An update on earth's energy balance in light of the latest global observations. Nat Geosci 5:691–696

Stephenson D, Wanner H, Brönnimann S, Luterbacher J (2003) The history of scientific research on the North Atlantic oscillation. In: Hurrell J, Kushnir Y, Ottersen G, Visbeck M (eds) The North Atlantic oscillation: Climatic significance and environmental impact, Geophysical Monograph Series 134. AGU, Washington, DC, pp 37–50

Stephenson DB, Hannachi A, O'Neill A (2004) On the existence of multiple climate regimes. Q J R Meteorol Soc 130:583–605

Stevens B, Feingold G (2009) Untangling aerosol effects on clouds and precipitation in a buffered system. Nature 461:607–613

Stewart MM, Grosjean M, Kuglitsch FG, Nussbaumer SU, von Gunten L (2011) Reconstructions of late Holocene paleofloods and glacier length changes in the Upper Engadine, Switzerland (ca. 1450 BC–AD 420). Palaeogeogr Palaeoclimatol 311:215–223

Stickler A, Brönnimann S (2011) Significant bias of the NCEP/NCAR and twentieth-century reanalyses relative to pilot balloon observations over the West African Monsoon region (1940–1957). Q J R Meteorol Soc 137:1400–1416

Stickler A, Grant AN, Ewen T et al (2010) The comprehensive historical upper-air network. Bull Am Meteorol Soc 91:741–751

Stier P, Feichter J, Kinne S et al (2005) The aerosol-climate model ECHAM5-HAM. Atmos Chem Phys 5:1125–1156

Stocker TF, Qin D, Plattner GK et al (2013) Technical summary. In: Stocker TF, Qin D, Plattner GK et al (eds) Climate change 2013: The physical science basis. Contribution of Working Group I to the Fifth Assessment Report of the Intergovernmental Panel on Climate Change. Cambridge University Press, Cambridge/New York

Stommel H, Stommel E (1979) The year without a summer. Sci Am 240:176–186

von Storch H, Stehr N (2000) Climate change in perspective. Nature 405:615

von Storch H, Zorita E, Jones JM et al (2004) Reconstructing past climate from noisy data. Science 306:679–682

Stothers RB (1984) The great Tambora eruption in 1815 and its aftermath. Science 224:1191–1198

Stothers RB (2000) Climatic and demographic consequences of the massive volcanic eruption of 1258. Clim Change 45:361–374

Stothers RB (2001) Major optical depth perturbations to the stratosphere from volcanic eruptions: Stellar extinction period, 1961–1978. J Geophys Res 106:2993–3003

Stroeve J, Holland MM, Meier W, Scambos T, Serreze M (2007) Arctic sea ice decline: faster than forecast. Geophys Res Lett 34:L09501

Stroeve JC, Kattsov V, Barrett A et al (2012) Trends in Arctic sea ice extent from CMIP5, CMIP3 and observations. Geophys Res Lett 39:L16502

Stucki P, Rickli R, Brönnimann S et al (2012) Weather patterns and hydro-climatological precursors of extreme floods in Switzerland since 1868. Meteorol Z 21:531–550

Stucki P, Brönnimann S, Martius O et al (2014) A catalog of high-impact windstorms in Switzerland since 1859. Nat Hazards Earth Syst Sci 14:2867–2882

Sturm C, Zhang Q, Noone D (2010) An introduction to stable water isotopes in climate models: benefits of forward proxy modelling for paleoclimatology. Clim Past 6:115–129

Suarez MJ, Schopf PS (1988) A delayed action oscillator for ENSO. J Atmos Sci 45:3283–3287

Süring R (1910) A. Berson's Bericht über die aerologische Expedition des köiglichen aeronautischen Observatoriums nach Ostafrika im Jahre 1908 (Report by A. Berson about the aerological expedition of the Royal Aeronautic Observatory to East Africa in 1908). Meteorol Z 27:536–542. (Translated and edited by Volken E, Brönnimann S, Meteorol Z 22:343–348, 2013)

Sutton RT, Dong B (2012) Atlantic Ocean influence on a shift in European climate in the 1990s. Nat Geosci 5:788–792

Sutton RT, Hodson DLR (2005) Atlantic ocean forcing of North American and European summer climate. Science 309:115–118

Svensmark H, Friis-Christensen E (1997) Variation of cosmic ray flux and global cloud coverage–a missing link in solar-climate relationships. J Atmos Sol-Terr Phys 59:1225–1232

Symons GJ (ed) (1888) The eruption of Krakatoa, and subsequent phenomena. Report of the Krakatoa Committee of the Royal Society, Trübner & Company 499 pp

Tacconi L (2003) Fires in Indonesia: causes, costs and policy implications. CIFOR Occasional Paper No. 38, p 22

Tang Q, Zhang X, Francis JA (2013) Extreme summer weather in Northern mid-latitudes linked to a vanishing cryosphere. Nat Clim Change 4:45–50

Thompson DWJ, Solomon S (2002) Interpretation of recent Southern Hemisphere climate change. Science 296:895–899

Thompson DWJ, Wallace JM (1998) The Arctic oscillation signature in the wintertime geopotential height and temperature fields. Geophys Res Lett 25:1297–1300

Thompson DWJ, Wallace JM (2000) Annular modes in the extratropical circulation. Part I: Month-to-month variability. J Clim 13:1000–1016

Thompson DWJ, Kennedy JJ, Wallace JM, Jones PD (2008) A large discontinuity in the mid-twentieth century in observed global-mean surface temperature. Nature 453:646–649

Thompson DWJ, Wallace JM, Kennedy JJ, Jones PD (2010) An abrupt drop in Northern Hemisphere sea surface temperature around 1970. Nature 467:444–447

Thompson DW, Solomon S, Kushner PJ et al (2011) Signatures of the Antarctic ozone hole in Southern Hemisphere surface climate change. Nat Geosci 4:741–749

Thompson DWJ, Seidel DJ, Randel WJ et al (2012) The mystery of recent stratospheric temperature trends. Nature 491:692–697

Thompson DM, Cole JE, Shen GT, Tudhope AW, Meehl GA (2015) Early 20th century warming linked to tropical Pacific wind strength. Nat Geosci. 8:117–121

Thompson LG, Mosley-Thompson E, Davis ME et al (2002) Kilimanjaro ice core records: Evidence of Holocene climate change in tropical Africa. Science 298:589–593

Thorne PW, Vose RS (2010) Reanalyses suitable for characterizing long-term trends. Bull Am Meteorol Soc 91:353–361

Tierney JE, Abram NJ, Anchukaitis KJ, Evans MN, Giry C, Kilbourne KH, Saenger CP, Wu HC, Zinke J (2015) Tropical sea surface temperatures for the past four centuries reconstructed from coral archives. Paleoceanography 30:226–252

Timmreck C (2012) Modeling the climatic effects of large explosive volcanic eruptions. WIREs Clim Change 3:545–564

Timmreck C, Lorenz SJ, Crowley TJ et al (2009) Limited temperature response to the very large ad 1258 volcanic eruption. Geophys Res Lett 36:L21708

Tolwinski-Ward SE, Evans MN, Hughes MK, Anchukaitis KJ (2011) An efficient forward model of the climate controls on interannual variation in tree-ring width. Clim Dyn 36:2419–2439

Toreti A, Kuglitsch FG, Xoplaki E, Luterbacher J (2012) A novel approach for the detection of inhomogeneities affecting climate time series. J Appl Meteorol Clim 51:317–326

Trachsel M, Eggenberger U, Grosjean M, Blass A, Sturm M (2008) Mineralogy-based quantitative precipitation and temperature reconstructions from annually laminated lake sediments (Swiss Alps) since AD 1580. Geophys Res Lett 35:L13707

Trachsel M, Kamenik C, Grosjean M et al (2012) Multi-archive summer temperature reconstruction for the European Alps, AD 1053–1996. Quat Sci Rev 46:66–79

Trenberth KE (1984) Signal versus noise in the Southern Oscillation. Mon Weather Rev 112:326–332

Trenberth KE (1997) The definition of El Niño. Bull Am Meteorol Soc 78:2771–2777

Trenberth KE, Dai A (2007) Effects of Mount Pinatubo volcanic eruption on the hydrological cycle as an analog of geoengineering. Geophys Res Lett 34:L15702

Trenberth KE, Hurrell JW (1994) Decadal atmosphere-ocean variations in the Pacific. Clim Dyn 9:303–319

Trenberth KE, Shea DJ (2006) Atlantic hurricanes and natural variability in 2005. Geophys Res Lett 33:L12704

Trenberth KE, Smith L (2006) The vertical structure of temperature in the tropics: Different flavors of El Niño. J Clim 19:4956–4973

Trenberth KE, Branstator GW, Karoly D et al (1998) Progress during TOGA in understanding and modeling global teleconnections associated with tropical sea surface temperatures. J Geophys Res 103:14291–14324

Trigo RM, Vaquero JM, Alcoforado MJ et al (2009) Iberia in 1816, the year without a summer. Int J Climatol 29:99–115

Trouet V, Taylor AH (2010) Multi-century variability in the Pacific North American circulation pattern reconstructed from tree rings. Clim Dyn 35:953–963

Troup AJ (1965) The 'Southern Oscillation'. Q J R Meteorol Soc 91:490–506

Turco RP, Toon OB, Ackerman TP, Pollack JB, Sagan C (1983) Nuclear winter: Global consequences of multple nuclear explosions. Science 222:1283–1292

Tyrlis E, Lelieveld J, Steil B (2013) The summer circulation over the Eastern Mediterranean and the middle east: Influence of the South Asian monsoon. Clim Dyn 40:1103–1123

Uppala SM, Kållberg PW, Simmons AJ et al (2005) The ERA-40 re-analysis. Q J R Meteorol Soc 131:2961–3012

Usteri P (1817) Eröffnungsrede. In: Verhandlungen der Schweizerischen Naturforschenden Gesellschaft, vol 3

Vautard R, Cattiaux J, Yiou P, Thépaut JN, Ciais P (2010) Northern Hemisphere atmospheric stilling partly attributed to an increase in surface roughness. Nat Geosci 3:756–761

Vecchi GA, Knutson TR (2008) On estimates of historical north Atlantic tropical cyclone activity. J Clim 21:3580–3600

Vecchi GA, Knutson TR (2011) Estimating annual numbers of Atlantic hurricanes missing from the HURDAT database (1878–1965) using ship track density. J Clim 24:1736–1746

Vecchi GA, Soden BJ, Wittenberg AT et al (2006) Weakening of tropical Pacific atmospheric circulation due to anthropogenic forcing. Nature 441:73–76

Venema VKC, Mestre O, Aguilar E et al (2012) Benchmarking homogenization algorithms for monthly data. Clim Past 8:89–115

Venetz I (1830) Sur l'ancienne extension des glaciers et sur leur retraite dans leur limites actuelles. In: Actes de la Société Helvétique des Sciences Naturelles. Quinzieme Réunion Annuelle á l'Hospice du Grand-Saint-Bernard les 21, 22 et 23 juillet 1829, vol 15, p 31

Verdon-Kidd DC, Kiem AS (2009) Nature and causes of protracted droughts in Southeast Australia: comparison between the federation, WWII, and big dry droughts. Geophys Res Lett 36:L22707

Veryard RG, Ebdon RA (1961) Fluctuations in tropical stratospheric winds. Meteorol Mag 90:125–143

Vicente-Serrano SM, López-Moreno JI, Lorenzo-Lacruz J et al (2011) The NAO impact on droughts in the Mediterranean region. In: Hydrological, socioeconomic and ecological impacts of the North Atlantic Oscillation in the Mediterranean region. Springer, Dordrecht/New York, pp 23–40

Villalba R, Lara A, Boninsegna JA et al (2003) Large-scale temperature changes across the Southern Andes: 20th-century variations in the context of the past 400 years. Climatic Change 59:177–232

Villalba R, Lara A, Masiokas MH et al (2012) Unusual Southern Hemisphere tree growth patterns induced by changes in the Southern annular mode. Nat Geosci 5:793–798

Vimont DJ, Kossin JP (2007) The Atlantic meridional mode and hurricane activity. Geophys Res Lett 34:L07709

Volney CF (1803) Tableau du climat et du sol des Etats-Unis d'Amérique, vol 1. Courcier-Dentu, Paris

Vuille M, Francou B, Wagnon P et al (2008) Climate change and tropical Andean glaciers: Past, present and future. Earth-Sci Rev 89:79–96

Wagenbach D (1989) Environmental records in alpine glaciers. In: Oeschger H, Langway CC (eds) The environmental record in glaciers and ice sheets. Wiley, New York, pp 69–73

Wagner A (1940) Klimaänderungen und Klimaschwankungen, Die Wissenschaft, vol 92. Vieweg+Teubner Verlag

Wahl ER, Smerdon JE (2012) Comparative performance of paleoclimate field and index reconstructions derived from climate proxies and noise-only predictors. Geophys Res Lett 39:L06703

Walker GT (1909) Correlation in seasonal variation of climate, II. Mem India Met Dept 20:22–45

Walker GT (1923) Correlation in seasonal variation of weather VIII: A preliminary study of world weather. Mem India Met Dept 24:75–131

Walker GT (1924) Correlation in seasonal variation of weather IX: A further study of world weather. Mem India Met Dept 25:275–332

Walker GT, Bliss EW (1932) World weather V. Mem R Meteorol Soc 4:53–84

Wallace JM, Blackmon ML (1983) Observations of low-frequency atmospheric variability. In: Hoskins B, Pearce RP (eds) Large-scale dynamical processes in the atmosphere. Academic, London, pp 55–94

Wallace JM, Gutzler DS (1981) Teleconnections in the geopotential height field during the Northern Hemisphere winter. Mon Weather Rev 109:784–812

Wallace JM, Hobbs PV (2006) Atmospheric science: An introductory survey, 2nd edn. International Geophysics. Academic Press, Amsterdam/Boston

Wallace JM, Held IM, Thompson DW, Trenberth KE, Walsh JE (2014) Global warming and winter weather. Science 343:729–730

Wallis J (1665) A relation concerning the late earthquake Neer Oxford; Together with some observations of the sealed weatherglass, and the barometer both upon that phaenomenon, and in general. Phil Trans 1:166–171

Walshaw CD (1989) GMB Dobson—the man and his work. Planet Space Sci 37:1485–1507

Wang B, Wu R, Lau KM (2001) Interannual variability of the Asian summer monsoon: Contrasts between the Indian and the Western North Pacific–East Asian monsoons. J Clim 14:4073–4090

Wang C (2004) ENSO, atlantic climate variability, and the Walker and Hadley circulations. Adv Glob Change Res 21:173–202

Wang C, Lee Sk, Enfield DB (2007a) Impact of the Atlantic warm pool on the summer climate of the Western Hemisphere. J Clim 20:5021–5040

Wang C, Kim D, Ekman AM, Barth MC, Rasch PJ (2009a) Impact of anthropogenic aerosols on Indian summer monsoon. Geophys Res Lett 36:L21704

Wang G, Eltahir EAB (2000) Role of vegetation dynamics in enhancing the low-frequency variability of the Sahel rainfall. Water Resour Res 36:1013–1021

Wang XL, Wen QH, Wu Y (2007b) Penalized maximal t test for detecting undocumented mean change in climate data series. J Appl Meteorol Clim 46:916–931

Wang X, Zwiers FW, Swail VR, Feng Y (2009b) Trends and variability of storminess in the Northeast Atlantic region, 1874–2007. Clim Dyn 33:1179–1195

Wang XL, Wan H, Zwiers FW et al (2011) Trends and low-frequency variability of storminess over western Europe, 1878–2007. Clim Dyn 37:2355–2371

Wang Y, Lee KH, Lin Y, Levy M, Zhang R (2014) Distinct effects of anthropogenic aerosols on tropical cyclones. Nat Clim Change 4:368–373

Wang YM, Lean J, Sheeley N Jr (2005) Modeling the Sun's magnetic field and irradiance since 1713. Astrophys J 625:522

Wanner H, Brönnimann S, Casty C et al (2001) North Atlantic oscillation – concepts and studies. Surv Geophys 22:321–381

Watson AJ, Lovelock JE (1983) Biological homeostasis of the global environment: the parable of Daisyworld. Tellus B 35:284–289

Weart SR (2008) The discovery of global warming: Revised and expanded edition. Cambridge, Harvard University Press

Webster PJ (2004) The elementary Hadley circulation. Adv Glob Change Res 21:9–60

Wegmann M, Brönnimann S, Bhend J et al (2014) Volcanic influence on European summer precipitation through monsoons: possible cause for "years without summer". J Clim 27:3683–3691

Wegmann M, Orsolini Y, Vázquez M, Gimeno L, Nieto R, Bulygina O, Jaiser R, Handorf D, Rinke A, Dethloff K, Sterin A, Brönnimann S (2015) Arctic moisture source for Eurasian snow cover variations in autumn. Environ Res Lett 10:054015

Werner M, Mikolajewicz U, Heimann M, Hoffmann G (2000) Borehole versus isotope temperatures on Greenland: Seasonality does matter. Geophys Res Lett 27:723–726

Wernli H, Dirren S, Liniger MA, Zillig M (2002) Dynamical aspects of the life cycle of the winter storm 'Lothar'(24–26 December 1999). Q J R Meteorol Soc 128:405–429

Wetter O, Pfister C, Weingartner R et al (2011) The largest floods in the high Rhine basin since 1268 assessed from documentary and instrumental evidence. Hydrol Sci J 56:733–758

Weusthoff T (2011) Weather type classification at MeteoSwiss: Introduction of new automatic classification schemes. Arbeitsbericht MeteoSchweiz Nr. 235, Bundesamt für Meteorologie und Klimatologie, MeteoSchweiz, Zürich

Wexler H (1958) Modifying weather on a large scale: Science 128:1059–1063

Wick L, Tinner W (1997) Vegetation changes and timberline fluctuations in the central Alps as indicators of Holocene climatic oscillations. Arct Alp Res 29:445–458

Widmann M, Goosse H, van der Schrier G, Schnur R, Barkmeijer J (2010) Using data assimilation to study extratropical Northern Hemisphere climate over the last millennium. Clim Past 6:627–644

Wigley TML, Ammann CM, Santer BD, Raper SCB (2005) Effect of climate sensitivity on the response to volcanic forcing. J Geophys Res 110:D09107

Wild M (2009) Global dimming and brightening: A review. J Geophys Res 114:D00D16

Wild M, Gilgen H, Roesch A et al (2005) From dimming to brightening: Decadal changes in solar radiation at earth's surface. Science 308:847–850

Wild M, Folini D, Schär C et al (2013) The global energy balance from a surface perspective. Clim Dyn 40:3107–3134

Willett K, Williams C, Jolliffe IT, Lund R, Alexander LV, Brönnimann S, Vincent LA, Easterbrook S, Venema VKC, Berry D, Warren RE, Lopardo G, Auchmann R, Aguilar E, Menne MJ, Gallagher C, Hausfather Z, Thorarinsdottir T, Thorne PW (2014) A framework for benchmarking of homogenisation algorithm performance on the global scale. Geosci Instrum Method Data Syst 3:187–200

Williamson H (1771) An attempt to account for the change of climate, which has been observed in the Middle Colonies in North-America. Trans Am Philos Soc 1, (Volume reprinted 1789, 336–345)

Wilson R, Tudhope A, Brohan P et al (2006) Two-hundred-fifty years of reconstructed and modeled tropical temperatures. J Geophys Res 111:C10007

Wilson R, D'Arrigo R, Buckley B et al (2007) A matter of divergence: Tracking recent warming at hemispheric scales using tree ring data. J Geophys Res 112:D17103

WMO (1992) International meteorological vocabulary. WMO, Geneva

WMO (1998) Final report, commission for basic systems. Working group on data processing, task group on WMO/CTBTO matters. WMO, Geneva

WMO (2008) Guide to meteorological instruments and methods of observation, 7th edn. WMO, Geneva

WMO (2011) Scientific Assessment of Ozone Depletion: 2010. WMO Global Ozone Research and Monitoring Project. Report No. 52. WMO, Geneva

WMO (2014) Scientific Assessment of Ozone Depletion: 2014. WMO Global Ozone Research and Monitoring Project. Report No. 55. WMO, Geneva

Woeikof A (1895) Die Schneedecke in "paaren" und "unpaaren" Wintern. Meteorol Z 12:77–78

Wood KR, Overland JE (2010) Early 20th century Arctic warming in retrospect. Int J Climatol 30:1269–1279

Woodhouse CA, Overpeck JT (1998) 2000 years of drought variability in the central United States. Bull Am Meteorol Soc 79:2693–2714

Woodruff SD, Worley SJ, Lubker SJ et al (2011) ICOADS release 2.5: extensions and enhancements to the surface marine meteorological archive. Int J Climatol 31:951–967

Woollings T, Lockwood M, Masato G, Bell C, Gray L (2010) Enhanced signature of solar variability in Eurasian winter climate. Geophys Res Lett 37:L20805

Worster D (1979) Dust Bowl: the Southern High Plains in the 1930s. Oxford University Press, Oxford

Wu B, Wang J, Walsh JE (2006) Dipole anomaly in the winter Arctic atmosphere and its association with sea ice motion. J Clim 19:210–225

Wunsch C (2002) What is the thermohaline circulation. Science 298:1179–1181

Xie SP, Carton JA (2004) Tropical atlantic variability: Patterns, mechanisms, and impacts. In: Wang C, Xie SP, Carton JA (eds) Earth's climate. Geophysical Monograph Series 147, AGU, Washington, DC, pp 121–142

Xie SP, Hu K, Hafner J et al (2009) Indian Ocean capacitor effect on Indo–Western Pacific climate during the summer following El Niño. J Clim 22:730–747

Xoplaki E, Luterbacher J, Paeth H et al (2005) European spring and autumn temperature variability and change of extremes over the last half millennium. Geophys Res Lett 32:L15713

Yoshimori M, Stocker TF, Raible CC, Renold M (2005) Externally forced and internal variability in ensemble climate simulations of the maunder minimum. J Clim 18:4253–4270

Zanchettin D, Timmreck C, Bothe O et al (2013) Delayed winter warming: a robust decadal response to strong tropical volcanic eruptions? Geophys Res Lett 40:204–209

Zebiak SE, Cane MA (1987) A model El Niño–Southern Oscillation. Mon Weather Rev 115:2262–2278

Zhang D, Liang Y (2010) A long lasting and extensive drought event over China in 1876–1878. Adv Clim Change Res 1:91–99

Zhang R, Delworth TL (2006) Impact of Atlantic multidecadal oscillations on India/Sahel rainfall and Atlantic hurricanes. Geophys Res Lett 33:L17712

Zhang R, Delworth TL, Held IM (2007) Can the Atlantic Ocean drive the observed multidecadal variability in Northern Hemisphere mean temperature? Geophys Res Lett 34:L02709

Zhang X, He J, Zhang J et al (2013) Enhanced poleward moisture transport and amplified Northern high-latitude wetting trend. Nat Clim Change 3:47–51

Zhou T, Brönnimann S, Griesser T, Fischer AM, Zou L (2010) A reconstructed dynamic Indian monsoon index extended back to 1880. Clim Dyn 34:573–585

Zorita E, von Storch H, Gonzalez-Rouco FJ et al (2004) Climate evolution in the last five centuries simulated by an atmosphere-ocean model: global temperatures, the North Atlantic Oscillation and the Late Maunder Minimum. Meteorol Z 13:271–289

Zubiaurre I, Calvo N (2012) The El Niño–Southern Oscillation (ENSO) Modoki signal in the stratosphere. J Geophys Res 117:D04104

Zumbühl HJ (1980) Die Schwankungen der Grindelwaldgletscher in den historischen Bild- und Schriftquellen des 12. bis 19. Jahrhunderts: ein Beitrag zur Gletschergeschichte und Erforschung des Alpenraumes. Birkhäuser

Zumbühl HJ (2009) Der Berge wachsend Eis... Die Entdeckung der Alpen und ihrer Gletscher durch Albrecht von Haller und Caspar Wolf. Mitteilungen der Naturforschenden Gesellschaft in Bern 66:105–132

Zumbühl HJ, Messerli B, Pfister C (1983) Die kleine Eiszeit: Gletschergeschichte im Spiegel der Kunst. Katalog zur Sonderausstellung des Schweizerischen Alpinen Museums Bern und des Gletschergarten-Museums Luzern vom 9.6.-14.8.1983 (Luzern), 24.8.-16.10.1983 (Bern)

Zumbühl HJ, Steiner D, Nussbaumer SU (2008) 19th century glacier representations and fluctuations in the central and Western European Alps: An interdisciplinary approach. Glob Plan Change 60:42–57

The manufacturer's authorised representative in the EU is Springer Nature Customer Service Centre GmbH, Europaplatz 3, 69115 Heidelberg, Germany. If you have any concerns regarding our products, please contact ProductSafety@springernature.com

Printed and bound by CPI Group (UK) Ltd, Croydon, CR0 4YY
30/09/2026
02225230-0012